T0145337

Studies in Systems, Decision and Control

Volume 243

Series Editor

Janusz Kacprzyk, Systems Research Institute, Polish Academy of Sciences, Warsaw, Poland

The series "Studies in Systems, Decision and Control" (SSDC) covers both new developments and advances, as well as the state of the art, in the various areas of broadly perceived systems, decision making and control–quickly, up to date and with a high quality. The intent is to cover the theory, applications, and perspectives on the state of the art and future developments relevant to systems, decision making, control, complex processes and related areas, as embedded in the fields of engineering, computer science, physics, economics, social and life sciences, as well as the paradigms and methodologies behind them. The series contains monographs, textbooks, lecture notes and edited volumes in systems, decision making and control spanning the areas of Cyber-Physical Systems, Autonomous Systems, Sensor Networks, Control Systems, Energy Systems, Automotive Systems, Biological Systems, Vehicular Networking and Connected Vehicles, Aerospace Systems, Automation, Manufacturing, Smart Grids, Nonlinear Systems, Power Systems, Robotics, Social Systems, Economic Systems and other. Of particular value to both the contributors and the readership are the short publication timeframe and the world-wide distribution and exposure which enable both a wide and rapid dissemination of research output.

** Indexing: The books of this series are submitted to ISI, SCOPUS, DBLP, Ulrichs, MathSciNet, Current Mathematical Publications, Mathematical Reviews, Zentralblatt Math: MetaPress and Springerlink.

More information about this series at http://www.springer.com/series/13304

El Hassan Zerrik · Said Melliani ·
Oscar Castillo

Editors

Recent Advances in Modeling, Analysis and Systems Control: Theoretical Aspects and Applications

Editors
El Hassan Zerrik
Department of Mathematics
Moulay Ismail University
Zitoune Meknès, Morocco

Said Melliani
Department of Mathematics
Sultan Moulay Slimane University
Beni Mellal, Morocco

Oscar Castillo
Division of Graduate Studies and Research
Tijuana Institute of Technology
Tijuana, Mexico

ISSN 2198-4182   ISSN 2198-4190   (electronic)
Studies in Systems, Decision and Control
ISBN 978-3-030-26151-1   ISBN 978-3-030-26149-8   (eBook)
https://doi.org/10.1007/978-3-030-26149-8

This Springer imprint is published by the registered company Springer Nature Switzerland AG
The registered company address is: Gewerbestrasse 11, 6330 Cham, Switzerland

Preface

This book contains the written versions of most of the contributions presented during the 8th Edition of the Workshop "Modelling, Analysis and Systems Control" organized by the Moroccan Network on Systems Theory; it took place at Moulay Ismail University Meknes, Morocco, from October 26 to 27, 2018.

The Workshop provided a setting for discussing recent developments in a wide variety of topics including modeling, analysis, and control of dynamical systems, and to explore its current and future developments and applications. The workshop provides a forum where researchers can share ideas, results on theory, and experiments in application problems. The literature devoted to dynamical systems is quite large, and our choice for the considered topics was motivated by the following considerations. First, the mathematical jargon for systems theory remains quite complex and we feel strongly that we have to maintain connections between the people of this research field. Second, dynamical systems cover a wider range of applications, including engineering, life sciences, and environment.

Until today, there are more than ten research laboratories in system theory all around Morocco. From 1990, more than 1000 papers were published in international journals, more than 200 Ph.D. theses were produced, and hundreds of talks in national, regional, or international conferences were given. The activity of these research laboratories is mainly focused on modelization, analysis, and control of dynamical systems. Some innovative ideas on regional analysis were developed and investigated by Moroccan researchers. Other studies on nonlinear systems, control theory, PDEs and applications to fishery, biomathematics, etc. were carried out. These laboratories are interconnected within a research network (System Theory Network) and strongly supported by the Academy Hassan II of Sciences and Technology.

The published chapters address various aspects of academic and applied research in systems theory or automatic control covering localized, lumped, or distributed parameters systems using continuous or discrete approaches.

We hope that the various contributions of this book will be useful for researchers and graduate students in applied mathematics, engineering, physics, and even for others from applications fields, assuming that the reader already knows basic concepts of analysis and control in systems theory and dynamical systems.

Zitoune Meknès, Morocco
Beni Mellal, Morocco
Tijuana, Mexico
May 2019

El Hassan Zerrik
Said Melliani
Oscar Castillo

Acknowledgements We would like to thank all the authors for their excellent contributions and for the patience in the reviewing process. We would like to thank all the reviewers for their professional work. The workshop would not have happened without the financial support of many organizations, particularly Hassan II Academy of Sciences and Technology, Moulay Ismail University, Faculty of sciences Meknes and Moroccan Society of Applied Mathematics (SM2A).

About This Book

We describe, in this book, recent developments in a wide variety of topics including modeling, analysis, and control of dynamical systems, and explore its current and future developments and applications. The book provided a forum where researchers have shared their ideas, results on theory, and experiments in application problems. The current literature devoted to dynamical systems is quite large, and our choice for the considered topics was motivated by the following considerations. First, the mathematical jargon for systems theory remains quite complex and we feel strongly that we have to maintain connections between the people of this research field. Second, dynamical systems cover a wider range of applications, including engineering, life sciences, and environment. We consider that the book will be an important contribution to the state of the art in the fuzzy and dynamical systems' areas.

Contents

Nonlinear Elliptic Boundary Value Problems by Topological Degree

Mustapha Ait Hammou and E. Azroul

Abstract Based on the recent Berkovits degree, and by way of an abstract Hammerstein equation, we study the Dirichlet boundary value problem involving nonlinear operators of the form

$$-diva(x, \nabla u) = \lambda u + f(x, u, \nabla u),$$

where a and f are Carathéodory functions satisfying some nonstandard growth conditions and $\lambda \in I\!R$. The function a satisfy also a condition of strict monotony and a condition of coercivity. We prove the existence of weak solutions of this problem in the weighted Sobolev spaces $W_0^{1,p(x)}(\Omega, \rho)$ where ρ is a weight function, satisfying some integrability conditions.

1 Introduction

Spaces with variable exponent are relevant in the study of non-Newtonian fluids. The underlying integral energy appearing in the modeling of the so called electrorheological fluids (see for instance [13, 14]) is $\int_\Omega |\nabla u|^{p(x)} dx$ or $\int_\Omega \rho(x)|\nabla u|^{p(x)} dx$. Such energies occur also in elasticity [17].

Accordingly, this naturally leads to study these fluids in the weighted variable exponent Sobolev space $W_0^{1,p(x)}(\Omega, \rho)$.

Let Ω is a bounded open subset of $I\!R^N$, $N \geq 2$. In this paper we study the problem of existence of solutions of the following nonlinear degenerated $p(x)$ elliptic problem

M. Ait Hammou (✉) · E. Azroul
Laboratory LAMA, Department of Mathematics, Faculty of Sciences Dhar El Mahraz,
University of Fez, B.P. 1796, Atlas Fez, Morocco
e-mail: mus.aithammou@gmail.com

E. Azroul
e-mail: elhoussine.azroul@usmba.ac.ma

© Springer Nature Switzerland AG 2020

1

E. H. Zerrik et al. (eds.), *Recent Advances in Modeling, Analysis and Systems Control: Theoretical Aspects and Applications*, Studies in Systems, Decision and Control 243, https://doi.org/10.1007/978-3-030-26149-8_1

$$\begin{cases} -diva(x, \nabla u) = \lambda u + f(x, u, \nabla u) & in \ \ \Omega \\ u = 0 & on \ \ \partial\Omega. \end{cases} \tag{1}$$

where $\lambda \in R$ and $p : \overline{\Omega} \to I\!R^+$ is a continuous function satisfying

$$1 \le p^- \le p(x) \le p^+ < +\infty, \tag{2}$$

and the log-Hölder continuity condition, i.e. there is a constant $C > 0$ such that for every $x, y \in \overline{\Omega}, x \ne y$ with $|x - y| \le \frac{1}{2}$ one has

$$|p(x) - p(y)| \le \frac{C}{-\log|x - y|}, \tag{3}$$

The operator $Au \equiv -diva(x, \nabla u)$ is a Leray–Lions operator defined on $W_0^{1,p(x)}$ (Ω, ρ), where ρ is a weight function, satisfying some integrability conditions and f is Carathéodory functions satisfying some nonstandard growth condition.

Our goal in this article is to prove, by using the Berkovits topological degree theory as an effective tool in the study of nonlinear equations [3–6], the existence of a least weak solution of problem (1). We then extend both a class of problems involving Leray–Lions type operators with variable exponents (see [2]) and a class of some degenerate problems involving special weights [2, 7, 9].

This paper is divided into three sections, organized as follows: in Sect. 2, we introduce some basic properties of the space $W^{1,p(x)}(\Omega, \rho)$, some class of operators and topological degree theory. In Sect. 3, we prove the existence of solutions of the problem (1).

2   Mathematical Preliminaries

2.1   Weighted Lebesgue and Sobolev Spaces with Variable Exponents

In what follows, we recall some definitions and basic properties of weighted Lebesgue and Sobolev spaces with variable exponents (more detailed description can be found in [1]).

Let Ω an open bounded subset of $I\!R^N$ ($N \ge 1$) and $p(\cdot)$ satisfying (2) and (3).

Definition 1 Let ρ be a function defined on Ω, ρ is called a weight function if it is a measurable and strictly positive a.e. in Ω.

Let introduce the integrability conditions used on the framework of weighted variable Lebesgue and Sobolev spaces

$$\rho \in L_{loc}^1(\Omega), \tag{4}$$

$$\rho^{\frac{-1}{p(x)-1}} \in L^1_{loc}(\Omega). \tag{5}$$

We define the weighted variable exponent Lebesgue space by

$$L^{p(x)}(\Omega, \rho) = \left\{ u : \Omega \to R \text{ measurable}, \int_\Omega |u(x)|^{p(x)} \rho(x) dx < \infty \right\}.$$

The space $L^{p(x)}(\Omega, \rho)$ endowed withe the norm:

$$\|u\|_{p(x),\rho} = \inf \left\{ \lambda > 0, \int_\Omega |\frac{u(x)}{\lambda}|^{p(x)} \rho(x) dx \leq 1 \right\}$$

is a uniformly convex Banach space, thus reflexive. We denote by $L^{p'(x)}(\Omega, \rho^*)$ the conjugate space of $L^{p(x)}(\Omega, \rho)$ where $\frac{1}{p(x)} + \frac{1}{p'(x)} = 1$, and

$$\rho^*(x) = \rho(x)^{1-p'(x)}.$$

As in [10] we can prove the following proposition

Proposition 1 *1. For any $u \in L^{p(x)}(\Omega, \rho)$ and $v \in L^{p'(x)}(\Omega, \rho^*)$, we have the Hölder inequality*

$$|\int_\Omega uv dx| \leq \left(\frac{1}{p_-} + \frac{1}{p'_-} \right) \|u\|_{p(x),\rho} \|v\|_{p'(x),\rho^*}.$$

2. For all p_1, p_2 continuous on $\overline{\Omega}$ such that $p_1(x) \leq p_2(x)$ a.e $x \in \overline{\Omega}$, we have $L^{p_2(x)}(\Omega, \rho) \hookrightarrow L^{p_1(x)}(\Omega, \rho)$ and the embedding is continuous.

Let us denote

$$I_{p,\rho}(u) = \int_\Omega |u|^{p(x)} \rho(x) dx, \quad \forall u \in L^{p(x)}(\Omega, \rho).$$

By taking $I_{p,\rho}(u) = I(\rho^{\frac{1}{p(x)}} u)$, where $I(u) = \int_\Omega |u|^{p(x)} dx$ and $\|\rho^{\frac{1}{p(x)}} u\|_{p(x)} = \|u\|_{p(x),\rho}$, we can prove the following result as a consequence of the corresponding one in [10].

Proposition 2 *For each $u \in L^{p(x)}(\Omega, \rho)$,*

1. $\|u\|_{p(x),\rho} < 1$ (resp. $= 1$ or > 1) $\Leftrightarrow I_{p,\rho}(u) < 1$ (resp. $= 1$ or > 1),
2. $\|u\|_{p(x),\rho} > 1 \Rightarrow \|u\|_{p(x),\rho}^{p_-} \leq I_{p,\rho}(u) \leq \|u\|_{p(x),\rho}^{p_+}$,
   $\|u\|_{p(x),\rho} < 1 \Rightarrow \|u\|_{p(x),\rho}^{p_+} \leq I_{p,\rho}(u) \leq \|u\|_{p(x),\rho}^{p_-}$,
   $I_{p,\rho}(u) \leq \|u\|_{p(x),\rho}^{p_+} + \|u\|_{p(x),\rho}^{p_-}$,
3. $\|u\|_{p(x),\rho} \to 0 \Leftrightarrow I_{p,\rho}(u) \to 0$ and $\|u\|_{p(x),\rho} \to \infty \Leftrightarrow I_{p,\rho}(u) \to \infty$.

We define the weighted variable exponent Sobolev space by

$$W^{1,p(x)}(\Omega, \rho) = \{u \in L^{p(x)}(\Omega) \text{ and } |\nabla u| \in L^{p(x)}(\Omega, \rho)\}.$$

with the norm

$$\|u\|_{1,p(x),\rho} = \|u\|_{p(x)} + \|\nabla u\|_{p(x),\rho} \quad \forall u \in W^{1,p(x)}(\Omega).$$

We denote by $W_0^{1,p(x)}(\Omega, \rho)$ the closure of $C_0^\infty(\Omega)$ in $W^{1,p(x)}(\Omega, \rho)$.

We will use the following result of compact imbedding which can be proved in a similar manner to that of Theorem 4.8.2. in [8] (see also [11])

Proposition 3 $W_0^{1,p(x)}(\Omega, \rho) \hookrightarrow\hookrightarrow L^{p(x)}(\Omega)$.

The dual space of $W_0^{1,p(\cdot)}(\Omega, \rho)$, denoted $W^{-1,p'(\cdot)}(\Omega, \rho^*)$, is equipped with the norm

$$\|v\|_{-1,p'(x),\rho^*} = \inf \left\{ \|v_0\|_{p'(x),\rho^*} + \sum_{i=1}^N \|v_i\|_{p'(x),\rho^*} \right\}$$

where the infimum is taken on all possible decompositions $v = v_0 - div F$.

Remark 1 We can see following [15, Theorem 3] that the the Poincaré inequality holds for the weighted Sobolev spaces $W_0^{1,p(\cdot)}(\Omega, \rho)$. In particular, this space has a norm $\| \cdot \|_{1,p(x)}$ given by

$$\|u\|_{1,p(x)} = \|\nabla u\|_{p(x),\rho} \text{ for all } u \in W_0^{1,p(\cdot)}(\Omega, \rho),$$

which equivalent to $\| \cdot \|_{1,p(x),\rho}$.

2.2   Some Class of Operators and Topological Degree

Let X be a real separable reflexive Banach space with dual X^* and with continuous pairing $\langle . , . \rangle$ and let Ω be a nonempty subset of X. The symbol $\rightarrow$ ($\rightharpoonup$) stands for strong (weak) convergence.

Let Y be a real Banach space. We recall that a mapping $F : \Omega \subset X \rightarrow Y$ is *bounded*, if it takes any bounded set into a bounded set. F is said to be *demicontinuous*, if for any $(u_n) \subset \Omega, u_n \rightarrow u$ implies $F(u_n) \rightharpoonup F(u)$. F is said to be *compact* if it is continuous and the image of any bounded set is relatively compact.

A mapping $F : \Omega \subset X \rightarrow X^*$ is said to be *of class* (S_+), if for any $(u_n) \subset \Omega$ with $u_n \rightharpoonup u$ and $limsup\langle Fu_n, u_n - u \rangle \leq 0$, it follows that $u_n \rightarrow u$. F is said to be *quasimonotone* , if for any $(u_n) \subset \Omega$ with $u_n \rightharpoonup u$, it follows that $limsup\langle Fu_n, u_n - u \rangle \geq 0$.

For any operator $F : \Omega \subset X \rightarrow X$ and any bounded operator $T : \Omega_1 \subset X \rightarrow X^*$ such that $\Omega \subset \Omega_1$, we say that F satisfies condition $(S_+)_T$, if for any $(u_n)\Omega$ with $u_n \rightharpoonup u, y_n := Tu_n \rightharpoonup y$ and $limsup\langle Fu_n, y_n - y \rangle \leq 0$, we have $u_n \rightarrow u$.

Let $\mathcal{O}$ be the collection of all bounded open set in X. For any $\Omega \subset X$, we consider the following classes of operators:

$$\mathcal{F}_1(\Omega) := \{F : \Omega \to X^* \mid F \text{ is bounded, demicontinuous and satisfies condition } (S_+)\},$$

$$\mathcal{F}_{T,B}(\Omega) := \{F : \Omega \to X \mid F \text{ is bounded, demicontinuous and satifies condition } (S_+)_T\},$$

$$\mathcal{F}_T(\Omega) := \{F : \Omega \to X \mid F \text{ is demicontinuous and satifies condition } (S_+)_T\},$$

$$\mathcal{F}_B(X) := \{F \in \mathcal{F}_{T,B}(\bar{G}) \mid G \in \mathcal{O}, T \in \mathcal{F}_1(\bar{G})\}.$$

Here, $T \in \mathcal{F}_1(\bar{G})$ is called an *essential inner map* to F.

Definition 2 Let G be a bounded open subset of a real reflexive Banach space X, $T \in \mathcal{F}_1(\bar{G})$ be continuous and let $F, S \in \mathcal{F}_T(\bar{G})$.
The affine homotopy $H : [0, 1] \times \bar{G} \to X$ defined by

$$H(t, u) := (1 - t)Fu + tSu \text{ for } (t, u) \in [0, 1] \times \bar{G}$$

is called an admissible affine homotopy with the common continuous essential inner map T.

Remark 2 ([4]) The above affine homotopy satisfies condition $(S_+)_T$.

We introduce the topological degree for the class $\mathcal{F}_B(X)$ due to Berkovits [4].

Theorem 1 *There exists a unique degree function*

$$d : \{(F, G, h) \mid G \in O, T \in F_1(\bar{G}), F \in F_{T,B}(\bar{G}), h \notin F(\partial G)\} \to \mathbf{Z}$$

that satisfies the following properties

1. *(Existence) if $d(F, G, h) \neq 0$, then the equation $Fu = h$ has a solution in G.*
2. *(Additivity) Let $F \in \mathcal{F}_{T,B}(\bar{G})$. If G_1 and G_2 are two disjoint open subset of G such that $h \notin F(\bar{G} \setminus (G_1 \cup G_2))$, then we have*

$$d(F, G, h) = d(F, G_1, h) + d(F, G_2, h).$$

3. *(Homotopy invariance) If $H : [0, 1] \times \bar{G} \to X$ is a bounded admissible affine homotopy with a common continuous essential inner map and $h : [0, 1] \to X$ is a continuous path in X such that $h(t) \notin H(t, \partial G)$ for all $t \in [0, 1]$, then the value of $d(H(t, .), G, h(t))$ is constant for all $t \in [0, 1]$.*
4. *(Normalization) For any $h \in G$, we have $d(I, G, h) = 1$.*

3   Main Result

3.1   *Basic Assumptions and Technical Lemmas*

Throughout the paper, we assume that the following assumptions hold:

Let Ω be a bounded open set of $I\!R^N$ ($N \geq 1$), $p \in C^+(\overline{\Omega})$ and

$$1/p(x) + 1/p'(x) = 1.$$

And let ρ a weight function in Ω which satisfies (4) and (5).

The function $a : \Omega \times I\!R^N \to I\!R^N$ is a Carathéodory function satisfying the following conditions: For all $\xi \in I\!R^N$ and for almost every $x \in \Omega$,

$$|a(x,\xi)| \leq \beta \rho^{\frac{1}{p(x)}}(k(x) + \rho^{\frac{1}{p'(x)}}|\xi|^{p(x)-1}), \tag{6}$$

$$[a(x,\xi) - a(x,\eta)](\xi - \eta) > 0 \quad \forall \xi \neq \eta, \tag{7}$$

$$a(x,\xi)\xi \geq \alpha\rho(x)|\xi|^{p(x)}, \tag{8}$$

where $k(x)$ is a positive function in $L^{p'(x)}(\Omega)$ and α and β are a positive constants.

Let $f(x,s,\xi) : \Omega \times I\!R \times I\!R^N \to I\!R$ be a Carathéodory function such that for a.e. $x \in \Omega$ and for all $s \in I\!R$, $\xi \in I\!R^N$, the growth condition

$$|f(x,s,\xi)| \leq \beta \rho^{\frac{1}{p(x)}}(h(x) + \rho^{\frac{1}{p'(x)}}|s|^{q(x)-1} + \rho^{\frac{1}{p'(x)}}|\xi|^{q(x)-1}) \tag{9}$$

for a.e. $x \in \Omega$ and all $(s,\xi) \in I\!R \times I\!R^N$, where β is a positive constant, h is a positive function in $L^{p'(x)}(\Omega)$ and $1 < q^- \leq q(x) \leq q^+ < p^-$.

Lemma 1 ([12]) *Let* $g \in L^{r(x)}(\Omega, \rho)$ *and* $g_n \in L^{r(x)}(\Omega, \rho)$ *with* $\|g_n\|_{r(x),\rho} \leq C$ *for* $1 < r(x) < \infty$. *If* $g_n(x) \to g(x)$ *a.e on* Ω, *then* $g_n \rightharpoonup g$ *in* $L^{r(x)}(\Omega, \rho)$.

Let define the operator $A : W_0^{1,p(x)}(\Omega, \rho) \to W^{-1,p'(x)}(\Omega, \rho^\star)$ by

$$Au = -div a(x, \nabla u) \tag{10}$$

then,

$$\langle Au, v \rangle = \int_\Omega a(x, \nabla u)v dx,$$

for all v in $W_0^{1,p(x)}(\Omega, \rho)$.

Lemma 2 *Under assumptions (6)–(8), A is bounded, continuous, coercive and satisfies condition* (S_+).

Proof **Step 1**:
Let's show that the operator A is bounded. Firstly, by using (4), (6) and Proposition 2 we can easily prove that $\|a(x, \nabla u)\rho(x)^{\frac{-1}{p(x)}}\|_{p'(x)}$ is bounded for all $u \in W_0^{1,p(x)}(\Omega, \rho)$. Therefore, thanks to Hölder's inequality, we have for all $u, v \in W_0^{1,p(x)}(\Omega, \rho)$,

$$\langle Au, v\rangle = \Big| \int_{\Omega} a(x, \nabla u)\nabla v dx\Big|$$
$$\leq C\|a(x, \nabla u)\|_{p'(x),\rho^*}.\|\nabla v\|_{p(x),\rho}$$
$$\leq C\|a(x, \nabla u)\rho(x)^{\frac{-1}{p(x)}}\|_{p'(x)}$$
$$\leq C\|v\|_{1,p(x),\rho},$$

which implies that the operator A is bounded.

Step 2:

To show that A is continuous, let $u_n \to u$ in $W_0^{1,p(x)}(\Omega, \rho)$. Then $\nabla u_n \to \nabla u$ in $L^{p(x)}(\Omega, \rho)$. Hence there exist a subsequence (u_k) of (u_n) and measurable function g in $(L^{p(x)}(\Omega, \rho))^N$ such that

$$\nabla u_k(x) \to \nabla u(x) \text{ and } |\nabla u_k(x)| \leq g(x)$$

for a.e. $x \in \Omega$ and all $k \in N$. Since a satisfies the Carathodory condition, we obtain that

$$a(x, \nabla u_k(x)) \to a(x, \nabla u(x)) \text{ a.e. } x \in \Omega.$$

it follows from (6) that

$$|a(x, \nabla u_k(x))| \leq \beta \rho^{\frac{1}{p(x)}}(k(x) + \rho^{\frac{1}{p'(x)}}|g(x)|^{p(x)-1})$$

for a.e. $x \in \Omega$ and for all $k \in N$. We have

$$\int_{\Omega}[\rho^{\frac{1}{p(x)}}(k(x) + \rho^{\frac{1}{p'(x)}}|g(x)|^{p(x)-1})]^{p'(x)}\rho^*(x)dx \leq C\int_{\Omega}(|k(x)|^{p'(x)} + \rho|g(x)|^{p(x)}) < \infty,$$

because $k \in L^{p'(x)}(\Omega)$ and $g \in (L^{p(x)}(\Omega, \rho))^N$.
    Then

$$x \longmapsto \beta\rho^{\frac{1}{p(x)}}(k(x) + \rho^{\frac{1}{p'(x)}}|g(x)|^{p(x)-1}) \in L^{p'(x)}(\Omega, \rho^*),$$

and taking into account the equality

$$I_{p',\rho^*}(a(x, \nabla u_k) - a(x, \nabla u)) = \int_{\Omega}|a(x, \nabla u_k(x)) - a(x, \nabla u(x))|^{p'(x)}\rho^*(x)dx,$$

the dominated convergence theorem and the equivalence 3 of the proposition 2 implies that

$$a(x, \nabla u_k) \to a(x, \nabla u) \text{ in } L^{p'(x)}(\Omega, \rho^*).$$

Thus the entire sequence $a(x, \nabla u_n)$ converges to $a(x, \nabla u)$ in $L^{p'(x)}(\Omega, \rho^*)$.

Then, $\forall v \in W_0^{1,p(x)}(\Omega, \rho); \langle Au_n, v \rangle \rightarrow \langle Au, v \rangle$, which implies that the operator A is continuous.

Step 3:

In this step, we show that A is coercive. For that Let $v \in W_0^{1,p(x)}(\Omega, \rho)$, from (8), we have by using the Proposition 2 and Remark 1

$$\frac{< Av, v >}{||v||_{1,p(x),\rho}} \geq \frac{C}{||v||_{1,p(x)}} \alpha I_{p,\rho}(\nabla v) \geq C'||v||_{1,p(x)}^r$$

for some $r > 1$. By consequent

$$\frac{< Av, v >}{||v||_{1,p(x),\rho}} \longrightarrow \infty \quad \text{as } ||v||_{1,p(x),\rho} \rightarrow \infty.$$

Step 4:

It remains to show that the operator A satisfies condition (S_+). Let $(u_n)_n$ be a sequence in $W_0^{1,p(x)}(\Omega, \rho)$ such that

$$\begin{cases} u_n \rightharpoonup u & \text{in } W_0^{1,p(x)}(\Omega, \rho) \\ \limsup_{k \to \infty} \langle Au_n, u_n - u \rangle \leq 0. \end{cases}$$

We will prove that

$$u_n \longrightarrow u \text{ in } W_0^{1,p(x)}(\Omega, \rho).$$

Since $u_n \rightharpoonup u$ in $W_0^{1,p(x)}(\Omega, \rho)$, then $(u_n)_n$ is a bounded sequence in $W_0^{1,p(x)}(\Omega, \rho)$. By using the Proposition 2 there is a subsequence still denoted by $(u_n)_n$ such that

$$u_k \rightharpoonup u \text{ in } W_0^{1,p(x)}(\Omega, \rho),$$

$$u_k \rightarrow u \text{ in } L^{p(x)}(\Omega) \text{ and a.e in } \Omega.$$

From $u_n \rightharpoonup u$, $\limsup_{k \to \infty} \langle Au_n, u_n - u \rangle \leq 0$ and (7), we have

$$\lim_{n \to \infty} \langle Au_n, u_n - u \rangle = \lim_{n \to \infty} \langle Au_n - Au, u_n - u \rangle = 0. \tag{11}$$

Let $D_n = a(x, \nabla u_n).\nabla(u_n - u)$. By (11), $D_n \rightarrow 0$ in $L^1(\Omega)$ and for a subsequence $D_n \rightarrow 0$ a.e. in Ω.

Since $u_n \rightharpoonup u$ in $W_0^{1,p(x)}(\Omega, \rho)$ and a.e in Ω, there exists a subset B of Ω, of zero measure, such that for $x \in \Omega \setminus B$, $|u(x)| < \infty, |\nabla u(x)| < \infty, k(x) < \infty, u_n(x) \rightarrow u(x)$ and $D_n(x) \rightarrow 0$.

Defining $\xi_n = \nabla u_n(x)$, $\xi = \nabla u(x)$, we have

$$
\begin{aligned}
D_n(x) &= a(x, \xi_n).(\xi_n - \xi) \\
&= a(x, \xi_n)\xi_n - a(x, \xi_n)\xi \\
&\geq \alpha\rho(x)|\xi_n|^{p(x)} - \beta\rho(x)^{\frac{1}{p(x)}}(k(x) + \rho(x)^{\frac{1}{p'(x)}}|\xi_n|^{p(x)-1})|\xi| \\
&\geq \alpha|\xi_n|^{p(x)} - C_x\big[1 + |\xi_n|^{p(x)-1}\big],
\end{aligned}
$$

where C_x is a constant which depends on x, but does not depend on n. Since $u_n(x) \to u(x)$ we have $|u_n(x)| \leq M_x$, where M_x is some positive constant. Then by a standard argument $|\xi_n|$ is bounded uniformly with respect to n, we deduce that

$$
D_n(x) \geq |\xi_n|^{p(x)}\left(\alpha - \frac{C_x}{|\xi_n|^{p(x)}} - \frac{C_x}{|\xi_n|}\right).
$$

If $|\xi_n| \to \infty$ (for a subsequence), then $D_n(x) \to \infty$ which gives a contradiction. Let now ξ^* be a cluster point of ξ_n. We have $|\xi^*| < \infty$ and by the continuity of a we obtain

$$
a(x, \xi^*).(\xi^* - \xi) = 0.
$$

In view of (7), we have $\xi^* = \xi$, which implies that

$$
\nabla u_n(x) \to \nabla u(x) \quad \text{a.e. in } \Omega.
$$

Since the sequence $a(x, \nabla u_n)$ is bounded in $(L^{p'(x)}(\Omega, \rho^*))^N$, and $a(x, \nabla u_n) \to a(x, \nabla u)$ a.e. in Ω, then by Lemma 1 we get

$$
a(x, \nabla u_n) \rightharpoonup a(x, \nabla u) \quad \text{in } (L^{p'(x)}(\Omega, \rho^*))^N \text{ a.e. in } \Omega.
$$

We set $\bar{y}_n = a(x, \nabla u_n)\nabla u_n$ and $\bar{y} = a(x, \nabla u)\nabla u$. We can write

$$
\bar{y}_n \to \bar{y} \quad \text{in } L^1(\Omega).
$$

We have

$$
a(x, \nabla u_n).\nabla u_n \geq \alpha\rho(x)|\nabla u_n|^{p(x)}.
$$

Let $z_n = \rho|\nabla u_n|^{p(.)}$, $z = \rho|\nabla u|^{p(.)}$, $y_n = \frac{\bar{y}_n}{\alpha}$, and $y = \frac{\bar{y}}{\alpha}$. By Fatou's lemma,

$$
\int_\Omega 2y\, dx \leq \liminf_{n\to\infty} \int_\Omega y + y_n - |z_n - z|\, dx;
$$

i.e., $0 \leq -\limsup\limits_{n\to\infty} \int_\Omega |z_n - z|dx$. Then

$$
0 \leq \liminf_{n\to\infty} \int_\Omega |z_n - z|dx \leq \limsup_{n\to\infty} \int_\Omega |z_n - z|dx \leq 0,
$$

this implies

$$\nabla u_n \to \nabla u \quad \text{in } (L^{p(x)}(\Omega, \rho))^N.$$

Hence $u_n \to u$ in $W_0^{1,p(x)}(\Omega, \rho)$, which completes the proof.

Lemma 3 *Under assumption* (9)*, the operator* $S : W_0^{1,p(x)}(\Omega, \rho) \to W^{-1,p'(x)}(\Omega, \rho^*)$ *setting by*

$$\langle Su, v \rangle = -\int_\Omega (\lambda u + f(x, u, \nabla u))v dx, \quad \forall u, v \in W_0^{1,p(x)}(\Omega, \rho)$$

is compact.

Proof Let $\phi : W_0^{1,p(x)}(\Omega, \rho) \to L^{p'(x)}(\Omega, \rho)$ be an operator defined by

$$\phi u(x) := -f(x, u, \nabla u) \text{ for } u \in W_0^{1,p(x)}(\Omega, \rho) \text{ and } x \in \Omega.$$

By a similar proof to that of Lemma 2, we can show that ϕ is bounded and continuous.

Since the embedding $I : W_0^{1,p(x)}(\Omega, \rho) \hookrightarrow\hookrightarrow L^{p(x)}(\Omega, \rho)$ is compact, it is known that the adjoint operator $I^* : L^{p'(x)}(\Omega, \rho^*) \to W^{-1,p'(x)}(\Omega, \rho^*)$ is also compact. Therefore, the composition $I^* o \phi : W_0^{1,p(x)}(\Omega, \rho) \to W^{-1,p'(x)}(\Omega, \rho^*)$ is compact.

Moreover, considering the operator $K : W_0^{1,p(x)}(\Omega) \to W^{-1,p'(x)}(\Omega, \rho^*)$ given by

$$\langle Ku, v \rangle = -\int_\Omega \lambda uv \, dx \text{ for } u, v \in W_0^{1,p(x)}(\Omega, \rho),$$

it can be seen that K is compact, by nothing that the embedding
$i : L^{p(x)}(\Omega, \rho) (\hookrightarrow L^{p'(x)}(\Omega, \rho^*)$ is continuous and $K = -\lambda I^* o i o I$. We conclude that $S = K + I^* o \phi$ is compact. This completes the proof.

Lemma 4 ([4, Lemma 2.2 and 2.4]) *Suppose that* $T \in \mathcal{F}_1(\bar{G})$ *is continuous and* $S : D_S \subset X^* \to X$ *is demicontinuous such that* $T(\bar{G}) \subset D_s$*, where G is a bounded open set in a real reflexive Banach space X. Then the following statement are true:*

- *If S is quasimonotone, then* $I + SoT \in \mathcal{F}_T(\bar{G})$*, where I denotes the identity operator.*
- *If S is of class* (S_+)*, then* $SoT \in \mathcal{F}_T(\bar{G})$*.*

3.2 Existence of Weak Solution

Let us first define the weak solution of problem (1).

Definition 3 A weak solution of the problem (1) is a measurable function $u \in W_0^{1,p(x)}(\Omega, \rho)$ such that

$$\int_\Omega a(x, \nabla u)\nabla v dx = \int_\Omega (\lambda u + f(x, u, \nabla u))v dx$$

for all $v \in W_0^{1,p(x)}(\Omega, \rho)$.

The main result of this paper is the following theorem.

Theorem 2 *Under assumptions (6– 9), there exists at least a weak solution of the problem (1).*

Proof Let A and $S : W_0^{1,p(x)}(\Omega, \rho) \to W^{-1,p'(x)}(\Omega, \rho^*)$ be as in (10) and Lemma 3. Then $u \in W_0^{1,p(x)}(\Omega, \rho)$ is a weak solution of (1) if and only if

$$Au = -Su \qquad (12)$$

Since the operator A is strictly monotone (by assumbtion (7), thanks to the properties seen in Lemma 2 and in view of Minty-Browder Theorem (see [16], Theorem 26A), the inverse operator $T := A^{-1} : W^{-1,p'(x)}(\Omega, \rho^*) \to W_0^{1,p(x)}(\Omega, \rho)$ is bounded, continuous and satisfies condition (S_+). Moreover, note by Lemma 3 that the operator S is bounded, continuous and quasimonotone.

Consequently, Eq. (12) is equivalent to

$$u = Tv \text{ and } v + SoTv = 0. \qquad (13)$$

To solve Eq. (13), we will apply the degree theory introducing in Sect. 2. To do this, we first claim that the set

$$B := \{v \in W^{-1,p'(x)}(\Omega, \rho^*) | v + tSoTv = 0 \text{ for some } t \in [0, 1]\}$$

is bounded. Indeed, let $v \in B$. Set $u := Tv$, then $\|Tv\|_{1,p(x),\rho} = \|\nabla u\|_{p(x),\rho}$.

If $\|\nabla u\|_{p(x),\rho} \le 1$, then $\|Tv\|_{1,p(x),\rho}$ is bounded.

If $\|\nabla u\|_{p(x)} > 1$, then we get by the Proposition 2, the assumption (8) and the growth condition (9) the estimate

$$
\begin{aligned}
\|Tv\|_{1,p(x),\rho}^{p^-} &= \|\nabla u\|_{p(x),\rho}^{p^-} \\
&\le I_{p,\rho}(\nabla u) \\
&= \frac{1}{\alpha}\langle Au, u\rangle \\
&= \frac{1}{\alpha}\langle v, Tv\rangle \\
&= \frac{-t}{\alpha}\langle SoTv, Tv\rangle \\
&= \frac{t}{\alpha}\int_\Omega f(x, u, \nabla u)u dx \\
&\le const\left(\int_\Omega |\rho^{\frac{1}{p(x)}} h(x)u(x)| dx + I_{q,\rho}(u) + \int_\Omega |\nabla u|^{q(x)-1}|u|\rho(x) dx\right).
\end{aligned}
$$

Since $h \in L^{p'(x)}(\Omega)$, then $\rho^{\frac{1}{p(\cdot)}}h \in L^{p'(x)}(\Omega, \rho^*)$. By the Hölder inequality, we have

$$\int_{\Omega} |\rho^{\frac{1}{p(x)}} h(x) u(x)| dx \leq 2\|\rho^{\frac{1}{p(\cdot)}}h\|_{p'(x),\rho^*} \|u\|_{p(x),\rho} = 2\|h\|_{p'(x)} \|u\|_{p(x),\rho}.$$

By the Young inequality, we have

$$\begin{aligned}
\int_{\Omega} |\nabla u|^{q(x)-1} |u| \rho(x) dx &= \int_{\Omega} |\nabla u|^{q(x)-1} \rho^{\frac{1}{q'(x)}} \cdot |u| \rho^{\frac{1}{q(x)}} dx \\
&\leq \int_{\Omega} \frac{1}{q'(x)} |\nabla u|^{q(x)} \rho(x) dx + \int_{\Omega} \frac{1}{q(x)} |u|^{q(x)} \rho(x) dx \\
&= \frac{1}{q'^-} I_{q,\rho}(\nabla u) + \frac{1}{q^-} I_{q,\rho}(u) \\
&\leq const (\|\nabla u\|_{q(x),\rho}^{q^+} + \|u\|_{q(x),\rho}^{q^+} + \|u\|_{q(x),\rho}^{q^-}).
\end{aligned}$$

We can then deduce

$$\|Tv\|_{1,p(x),\rho}^{p^-} \leq const (\|u\|_{p(x),\rho} + \|u\|_{q(x),\rho}^{q^+} + \|u\|_{q(x),\rho}^{q^-} + \|\nabla u\|_{q(x),\rho}^{q^+}).$$

From the Poincaré inequality and the continuous embedding $L^{p(x)} \hookrightarrow L^{q(x)}$, we can deduct the estimate

$$\|Tv\|_{1,p(x),\rho}^{p^-} \leq const (\|Tv\|_{1,p(x),\rho} + \|Tv\|_{1,p(x),\rho}^{q^+}).$$

It follows that $\{Tv | v \in B\}$ is bounded.

Since the operator S is bounded, it is obvious from (13) that the set B is bounded in $W^{-1,p'(x)}(\Omega, \rho^*)$. Consequently, there exists $R > 0$ such that

$$\|v\|_{-1,p'(x),\rho^*} < R \text{ for all } v \in B.$$

This says that

$$v + tSoTv \neq 0 \text{ for all } v \in \partial B_R(0) \text{ and all } t \in [0, 1].$$

From Lemma 4 it follows that

$$I + SoT \in \mathcal{F}_T(\overline{B_R(0)}) \text{ and } I = AoT \in \mathcal{F}_T(\overline{B_R(0)}).$$

Consider a homotopy $H : [0, 1] \times \overline{B_R(0)} \to W^{-1,p'(x)}(\Omega, \rho^*)$ given by

$$H(t, v) := v + tSoTv \text{ for } (t, v) \in [0, 1] \times \overline{B_R(0)}.$$

Applying the homotopy invariance and normalization property of the degree d stated in Theorem 1, we get

$$d(I + SoT, B_R(0), 0) = d(I, B_R(0), 0) = 1,$$

and hence there exists a point $v \in B_R(0)$ such that

$$v + SoTv = 0.$$

We conclude that $u = Tv$ is a weak solution of (1). This completes the proof.

References

1. Aydin, I.: Weighted variable sobolev spaces and capacity. J. Funct. Spaces Appl. **2012**, 17 pages (2012). Article ID 132690
2. Azroul, E., Benboubker, M.B., Barbara, A.: Quasilinear elliptic problems with nonstandard growth. EJDE **2011**(62), 1–16 (2011)
3. Berkovits, J.: On the degree theory for nonlinear mappings of monotone type. Ann. Acad. Sci. Fenn. Ser. A I Math. Dissertationes **58**, 2 (1986)
4. Berkovits, J.: Extension of the Leray-Schauder degree for abstract Hammerstein type mappings. J. Differ. Equ. **234**, 289–310 (2007)
5. Berkovits, J., Mustonen, V.: On topological degree for mappings of monotone type. Nonlinear Anal. **10**, 1373–1383 (1986)
6. Berkovits, J., Mustonen, V.: Nonlinear mappings of monotone type I. Classification and degree theory, Preprint No 2/88, Mathematics, University of Oulu
7. Boureanu, M.M., Matei, A., Sofonea, M.: Nonlinear problems with $p(x)$-growth conditions and applications to antiplane contact models. Adv. Nonlinear Stud. **14**, 295–313 (2014)
8. Diening, L., et al.: Lebesgue and Sobolev Spaces with Variable Exponents. Lecture Notes in Mathematics. Springer, Berlin (2011). https://doi.org/10.1007/978-3-642-18363-8
9. Fan, X.L., Zhang, Q.H.: Existence for $p(x)$-Laplacien Dirichlet problem. Nonlinear Anal. **52**, 1843–1852 (2003)
10. Fan, X., Zhao, D.: On the spaces $L^{p(x)}(\Omega)$ and $W^{k,p(x)}(\Omega)$. J. Math. Anal. Appl. **263**, 424–446 (2001)
11. Kovàčik, O., Rákosnik, J.: On spaces $L^{p(x)}$ and $W^{1,p(x)}$. Czechoslovak Math. J. **41**, 592–618 (1991)
12. Lahmi, B., Azroul, E., El Haiti, K.: Nonlinear degenerated elliptic problems with dual data and nonstandard growth. Math. Rep. **20**(70), 81–91 (2018)
13. Rajagopal, K.R., Růžička, M.: Mathematical modeling of electrorheological materials. Contin. Mech. Thermodyn. **13**, 59–78 (2001)
14. Růžička, M.: Electrorheological Fluids: Modeling and Mathematical Theory. Lecture Notes in Mathematics, vol. 1748. Springer, Berlin (2000)
15. Samko, S.G.: Density of $\mathcal{C}_0^\infty(I\!R^N)$ in the generalized Sobolev spaces $W^{m,p(x)}(I\!R^N)$. Dokl. Akad. Nauk. **369**(4), 451–454 (1999)
16. Zeidler, E.: Nonlinear Functional Analysis and Its Applications, II/B: Nonlinear Monotone Operators. Springer, New York (1990)
17. Zhikov, V.V.: Averaging of functionals of the calculus of variations and elasticity theory, [in Russian]. Izv. Akad. Nauk SSSR, Ser. Math. **50**(4), 675–710 (1986). English translation: Math. USSR, Izv. **29**(1), 33–66 (1987)

On Some Class of Quasilinear Elliptic Systems with Weak Monotonicity

Elhoussine Azroul and Farah Balaadich

Abstract We study the existence of solutions for a class of quasilinear elliptic systems whose right hand side is of general form. Using the concept of Young measure, we obtain the needed results under some conditions on the source term f.

1 Introduction

We would like to explore a variational approach for weak solutions of the following quasilinear ellitpic system

$$-\operatorname{div} \sigma(x, u, Du) = f(x, u, Du) \quad \text{in } \Omega, \tag{1}$$

$$u = 0 \qquad \text{on } \partial\Omega, \tag{2}$$

where $\sigma(x, u, Du) := a(u)|Du|^{p-2}Du$. The unknown $u : \Omega \to \mathbb{R}^m$ is a vector-valued function, while Ω is an open bounded domain of $\mathbb{R}^n$ ($n \geq 2$) and $a : \mathbb{R}^m \to \mathbb{R}$ is a continuous function, which is bounded from above and below by positive constants, i.e., there exists $\alpha_1, \alpha_2 > 0$ such that

$$\alpha_1 \leq a(u) \leq \alpha_2. \tag{3}$$

We note by $\mathbb{M}^{m \times n}$ the space of $m \times n$ matrices equipped with the inner product $A : B = \sum_{i,j} A_{ij} B_{ij}$ and the tensor product $a \otimes b$ of two vectors $a, b \subset \mathbb{R}^m$ is defined to be the $m \times m$ matrix of entries $(a_i b_j)_{i,j}$, with $i, j = 1, \ldots, m$.

In [7], Hungerbühler have been established the existence of a weak solution for the following system

E. Azroul · F. Balaadich (✉)
Faculty of Sciences Dhar El Mehraz-Fez, Atlas-Fez, Morocco
e-mail: balaadich.edp@gmail.com

E. Azroul
e-mail: elhoussine.azroul@gmail.com

© Springer Nature Switzerland AG 2020   15
E. H. Zerrik et al. (eds.), *Recent Advances in Modeling, Analysis and Systems Control: Theoretical Aspects and Applications*, Studies in Systems, Decision and Control 243,
https://doi.org/10.1007/978-3-030-26149-8_2

$$\begin{cases} -\mathrm{div}\,\sigma(x, u, Du) \ = f \ \ \text{in } \Omega, \\ \qquad\qquad\qquad u = 0 \ \ \text{on } \partial\Omega, \end{cases} \tag{4}$$

where f belongs to $W^{-1,p'}(\Omega; \mathbb{R}^m)$ the dual space of $W_0^{1,p}(\Omega; \mathbb{R}^m)$, and $\sigma : \Omega \times \mathbb{R}^m \times \mathbb{M}^{m\times n} \to \mathbb{M}^{m\times n}$ satisfies some weak monotonicity assumptions.

In (4), the source term f does not depend on u and Du. Hence, if we consider $\sigma(x, u, Du) := a(u)|Du|^{p-2}Du$ in (4), we find that this last term satisfy the assumptions (H0)-(H2) listed in [7]. Indeed, σ is easily seen to be a Carathédory function. The definition of the function a implies that

$$|\sigma(x, u, Du)| := \big|a(u)|Du|^{p-2}Du\big|$$
$$\leq \alpha_2|Du|^{p-1},$$

and

$$\sigma(x, u, Du) : Du := a(u)\big(|Du|^{p-2}Du\big) : Du$$
$$= a(u)|Du|^p \geq \alpha_1|Du|^p.$$

By direct calculations, it is clear that σ is monotone. It remains then to verify the assumption (H2)(b) in [7]. For this, we take the potential $W(x, u, F) = \frac{1}{p}a(u)|F|^p$, because we have

$$\frac{\partial W}{\partial F}(x, u, F) = a(u)|F|^{p-2}F =: \sigma(x, u, F), \ \forall F \in \mathbb{M}^{m\times n}.$$

Hence, when $f := f(x)$ and $\sigma(x, u, Du) := a(u)|Du|^{p-2}Du$, we deduce from the results of [7] that (4) has a weak solution u in $W_0^{1,p}(\Omega; \mathbb{R}^m)$. Our purpose in this paper is to prove the existence result when $f := f(x, u, Du)$.

Following the method used in [7] to establish the existence result for our model (1)–(2) when the data f is allowed to be of more general form and satisfying the following conditions:

(F0) $f : \Omega \times \mathbb{R}^m \times \mathbb{M}^{m\times n} \to \mathbb{R}^m$ is a Carathéodory function.
(F1) $F \mapsto f(x, u, F)$ is linear and there exist $d_1 \in L^{p'}(\Omega), 0 < \gamma < p - 1$ and $0 \leq \mu < \frac{p-1}{p'}$ such that

$$|f(x, u, F)| \leq d_1(x) + |u|^\gamma + |F|^\mu.$$

The condition (F1) together with the Galerkin approximating sequences, permits the extraction of a subsequence that generates a Young measure.

We will prove the existence of a weak solution for the problem (1)–(2).

Theorem 1 *If f satisfies the conditions (F0)–(F1), then the Dirichlet problem (1)–(2) has a weak solution $u \in W_0^{1,p}(\Omega; \mathbb{R}^m)$.*

2 Galerkin Approximation

Let $V_1 \subset V_2 \subset \cdots \subset W_0^{1,p}(\Omega; \mathbb{R}^m)$ be a sequence of finite dimensional subspaces with the property that $\underset{i \in \mathbb{N}}{\cup} V_i$ is dense in $W_0^{1,p}(\Omega; \mathbb{R}^m)$. Such a sequence (V_i) exists since $W_0^{1,p}(\Omega; \mathbb{R}^m)$ is separable. We consider the operator

$$F : W_0^{1,p}(\Omega; \mathbb{R}^m) \to W^{-1,p'}(\Omega; \mathbb{R}^m)$$

$$u \mapsto \left(w \mapsto \int_\Omega a(u)|Du|^{p-2}Du : Dwdx - \int_\Omega f(x, u, Du).wdx\right).$$

We observe that for arbitrary $u \in W_0^{1,p}(\Omega; \mathbb{R}^m)$, the functional $F(u)$ is well defined and linear. Let show that F is bounded. By the growth condition (F1), we have

$$\left| \int_\Omega f(x, u, Du).wdx \right| \leq \int_\Omega \left(d_3(x) + |u|^\gamma + |Du|^\mu\right)|w|dx$$

$$\leq \|d_1\|_{p'}\|w\|_p + \|u\|_{\gamma p'}^\gamma \|w\|_p + \|Du\|_{\mu p'}^\mu \|w\|_p$$

$$\leq \|Dw\|_p \left[\alpha\|d_1\|_{p'} + \alpha^{\gamma+1}\|Du\|_p^\gamma + \alpha\|Du\|_p^\mu\right],$$

for each $w \in W_0^{1,p}(\Omega; \mathbb{R}^m)$, where we have used the Poincaré's inequality: there exists $\alpha \geq 0$ such that $\|v\|_p \leq \alpha\|Dv\|_p$ for all $v \in W_0^{1,p}(\Omega; \mathbb{R}^m)$, and the facts that $\gamma p' < p'(p-1) = p$ and $\mu p' < p - 1 < p$. By virtue of (3), we conclude the boundedness of F.

Let V be a finite subspace of $W_0^{1,p}(\Omega; \mathbb{R}^m)$ such that $\dim V = r$ and let (φ_i) be a basis of V. Let $(u_k = a_k^i \varphi_i)$ be a sequence in V which converges to $u = a^i \varphi_i$ in V (with standard summation convention). Hence the sequence (a_k) converges to $a \in \mathbb{R}^r$. This implies $u_k \to u$ and $Du_k \to Du$ almost everywhere. On the other hand, $\|u_k\|_p$ and $\|Du_k\|_k$ are bounded by a constant C. Thus, the continuity of the function a and that of the condition (F0) allows to deduce that $a(u_k)|Du_k|^{p-2}Du_k : Dw \to a(u)|Du|^{p-2}Du : Dw$ and $f(x, u_k, Du_k).w \to f(x, u, Du).w$ almost everywhere.

From (3) and the growth condition (F1) we deduce the equiintegrability of the sequences $\left(a(u_k)|Du_k|^{p-2}Du_k : Dw\right)$ and $\left(f(x, u_k, Du_k).w\right)$. Applying the Vitali theorem, we have

$$\lim_{k \to \infty} \langle F(u_k), w \rangle = \langle F(u), w \rangle, \quad \forall w \in W_0^{1,p}(\Omega; \mathbb{R}^m).$$

Thus the restriction of F to a finite subspace of $W_0^{1,p}(\Omega; \mathbb{R}^m)$ is continuous.

Let consider V_k be a finite subspace of $W_0^{1,p}(\Omega; \mathbb{R}^m)$ such that $\dim V_k = r$ for some fixed k, and $\varphi_1, \ldots, \varphi_r$ be a basis of V_k. We define the map

$$G : \mathbb{R}^r \to \mathbb{R}^r$$

$$(a^1, \ldots, a^r) \mapsto \left(\langle F(a^i \varphi_i), \varphi_j \rangle\right)_{j=1,\ldots,r}$$

which is continuous by that of F restricted to V_k since $G(a).a = \langle F(u), u \rangle$ for $a = (a^1, \ldots, a^r)^t$ and $u = a^i \varphi_i \in V_k$.

By the growth condition (F1), we have

$$\int_{\Omega} f(x, u, Du).u dx \leq \|d_1\|_{p'} \|u\|_p + \|u\|_{\gamma p'}^{\gamma} \|u\|_p + \|Du\|_{\mu p'}^{\mu} \|u\|_p$$

$$\leq \alpha \|d_1\|_{p'} \|Du\|_p + \alpha^{\gamma+1} \|Du\|_p^{\gamma+1} + \alpha^{\mu+1} \|Du\|_p^{\mu+1},$$

where we have used the Poincaré's inequality, $\gamma p' < p'(p-1) = p$ and $\mu p' < p - 1 < p$. According to (3), we deduce that

$$\langle F(u), u \rangle \geq \alpha_1 \|Du\|_p^p - \alpha \|d_1\|_{p'} \|Du\|_p - \alpha^{\gamma+1} \|Du\|_p^{\gamma+1} - \alpha^{\mu+1} \|Du\|_p^{\mu+1}.$$

The right hand side of the above inequality tends to infinity as $\|u\|_{1,p} \to \infty$ since $p > \max\{1, \gamma + 1, \mu + 1\}$ and $\alpha_1 > 0$.

As a consequence, there exists $R > 0$ such that for all $a \in \partial B_R(0) \subset \mathbb{R}^r$ we have $G(a).a > 0$ (because $\|a\|_{\mathbb{R}^m}$ is equivalent to $\|u\|_{1,p}$). By the topological arguments (see [8]), we have then $G(x) = 0$ has a solution in $B_R(0)$. Hence, for all k there exists $u_k \in V_k$ such that

$$\langle F(u_k), v \rangle = 0 \quad \forall v \in V_k.$$

Now, we introduce the concept of Young measure which permits to explore the weak limits of sequences.

Lemma 1 *(i) If the sequence $\{Du_k\}$ is bounded in $L^p(\Omega; \mathbb{M}^{m \times n})$, then there is a Young measure v_x generated by Du_k satisfying $\|v_x\| = 1$ and the weak L^1-limit of Du_k is $\langle v_x, id \rangle = \int_{\mathbb{M}^{m \times n}} \lambda dv_x(\lambda)$.*
(ii) For almost every $x \in \Omega$, v_x satisfies $\langle v_x, id \rangle = Du(x)$.

Proof (i) We have $\langle F(u), u \rangle \to \infty$ as $\|u\|_{1,p} \to \infty$, then there exists $R > 0$ such that $\langle F(u), u \rangle > 1$ whenever $\|u\|_{1,p} > R$. Hence, for the sequence of the Galerkin approximations $u_k \in V_k$ which satisfy $\langle F(u_k), u_k \rangle = 0$, we have the uniform bound

$$\|u_k\|_{1,p} \leq R \quad \forall k \in \mathbb{N}. \tag{5}$$

Thus, there is $c \geq 0$, for any $R > 0$

$$c \geq \int_{\Omega} |Du_k|^p dx \geq \int_{\{x \in \Omega \cap B_R(0):\, |Du_k| \geq L\}} |Du_k|^p dx$$

$$\geq L^p \left| \{x \in \Omega \cap B_R(0) : |Du_k| \geq L\} \right|.$$

Evidently

$$\sup_{k \in \mathbb{N}} \left| \{x \in \Omega \cap B_R(0) : |Du_k| \geq L\} \right| \leq \frac{c}{L^p} \to 0 \text{ as } L \to \infty.$$

By vertue of [4] Theorem 5, we deduce that Du_k generates v_x such that $\|v_x\|_{\mathcal{M}} = 1$. The reflexivity of $L^p(\Omega; \mathbb{M}^{m \times n})$ allow to deduce that there exists a subsequence (still denoted by $\{Du_k\}$) weakly convergent in $L^p(\Omega; \mathbb{M}^{m \times n}) \subset L^1(\Omega; \mathbb{M}^{m \times n})$. By taking $\varphi := id$ in [4, Theorem 5(iii)], we conclude that

$$Du_k \rightharpoonup \langle v_x, id \rangle = \int_{\mathbb{M}^{m \times n}} \lambda dv_x(\lambda) \quad \text{weakly in } L^1(\Omega; \mathbb{M}^{m \times n}).$$

(ii) Since $u_k \rightharpoonup u$ in $W_0^{1,p}(\Omega; \mathbb{R}^m)$ and $u_k \to u$ in $L^p(\Omega; \mathbb{R}^m)$ (by (5)), we have $Du_k \rightharpoonup Du$ in $L^p(\Omega; \mathbb{M}^{m \times n})$. Moreover $Du_k \rightharpoonup Du$ in $L^1(\Omega; \mathbb{M}^{m \times n})$, then by the property (i), we can infer that $\langle v_x, id \rangle = Du(x)$.

3   Proof of Theorem 1

We have in one hand by the contuinuity of the function $a : \mathbb{R}^m \to \mathbb{R}$, that $a(u_k) \to a(u)$. On the other hand

$$|Du_k|^{p-2} Du_k \rightharpoonup \int_{\mathbb{M}^{m \times n}} |\lambda|^{p-2} \lambda dv_x(\lambda).$$

Hence

$$a(u_k)|Du_k|^{p-2} Du_k \rightharpoonup \int_{\mathbb{M}^{m \times n}} a(u)|\lambda|^{p-2} \lambda dv_x(\lambda) = a(u)|Du|^{p-2} Du$$

in $L^1(\Omega)$ and thus weakly in $L^{p'}(\Omega)$ (because $\||a(u_k)|Du_k|^{p-2} Du_k\|_{p'} \leq c$).

By the growth condition (F1), we may extract a subsequence still denoted by u_k verifying

$$f(x, u_k, Du_k) \rightharpoonup \chi \quad \text{in } L^{p'}(\Omega), \tag{6}$$

for some $\chi \in L^{p'}(\Omega)$.

To finish the proof, it is sufficient to identify χ with $f(x, u, Du)$. By using the property (6) together with $u_k \to u$ in $L^p(\Omega; \mathbb{R}^m)$, we then obtain that

$$\lim_{k \to \infty} \int_{\Omega} f(x, u_k, Du_k).u_k dx = \int_{\Omega} \chi.u dx.$$

By the growth condition (F1), we have $f(x, u_k, Du_k)$ is equiintegrable. Therefore, from the fact that $u_k \to u$ and $Du_k \to Du$ in measure on Ω, and the linearity of f, it follows together with Lemma 1 that

$$f(x, u_k, Du_k) \rightharpoonup \int_{\mathbb{M}^{m \times n}} f(x, u, \lambda) dv_x(\lambda) = f(x, u, .)o \int_{\mathbb{M}^{m \times n}} \lambda dv_x(\lambda)$$
$$= f(x, u, .)oDu$$
$$= f(x, u, Du).$$

Consequently

$$f(x, u_k, Du_k) \rightharpoonup \chi = f(x, u, Du) \quad \text{in} \quad L^{p'}(\Omega).$$

This is sufficient to pass to the limit in the Galerkin equations and to conclude the Proof of Theorem 1.

References

1. Attouch, H., Buttazzo, G., Michaille, G.: Variational Analysis in Sobolev and BV Spaces: Applications to PDEs and Optimization, vol. 17. SIAM (2006)
2. Augsburger, F., Hungerbühler, N.: Quasilinear elliptic systems in divergence form with weak monotonicity and nonlinear physical data. Electron. J. Differ. Equ. **2004**(144), 118 (2004)
3. Ball, J.M.: A version of the fundamental theorem for Young measures. PDEs and Continuum Models of Phase Transitions (Nice 1988). Lecture Notes in Physics, vol. 344, pp. 207–215 (1989)
4. Dolzmann, G., Hungerühler, N., Muller, S.: Nonlinear elliptic systems with measure-valued right hand side. Math. Z. **226**, 545–574 (1997)
5. Fonseca, I., Leoni, G.: Modern methods in the Calculus of Variations: LP Spaces. Springer Monographs in Mathematics. Springer, Berlin (2007)
6. Hungerühler, N.: A refinement of Balls theorem on Young measures. N.Y. J. Math. **3**, 48–53(1997)
7. Hungerbühler, N.: Quasilinear elliptic systems in divergence form with weak monotonicity. New York J. Math. **5**, 83–90 (1999)
8. Nirenberg, L.: Topics in Nonlinear Functional Analysis. Lecture Notes. Courant Institute, New York (1974)
9. Valadier, M.: A course on Young measures, workshop on measure theory and real analysis (Grado, 1993). Rend. Istit. Mat. Univ. Trieste **26**(suppl), 349–394 (1994)

A Sub-supersolutions Method for a Class of Weighted $(p(.), q(.))$-Laplacian Systems

Elhoussine Azroul and Athmane Boumazourh

Abstract In this paper we study the existence of a positive weak solutions for a quasilinear elliptic system involving weighted $(p(.), q(.))$—Laplacian operators. The approach is based on sub-supersolutions method and on fixed point theorem.

1 Introduction

Let Ω be an open bounded subset of $\mathbb{R}^N$ ($N \geqslant 2$) with smooth boundary $\partial \Omega$. We consider the quasilinear elliptic system

$$\begin{cases} -div(|x|^{-ap(x)}|\nabla u|^{p(x)-2}\nabla u) = f(u)|v|^{\alpha_1(x)} + g(u)|v|^{\beta_1(x)} \ in \ \ \Omega, \\ -div(|x|^{-aq(x)}|\nabla v|^{q(x)-2}\nabla v) = f(v)|u|^{\alpha_2(x)} + g(v)|u|^{\beta_2(x)} \ in \ \ \Omega, \\ u, v > 0 \qquad\qquad\qquad\qquad\qquad\qquad\qquad\qquad\quad in \ \ \Omega, \\ u = v = 0, \qquad\qquad\qquad\qquad\qquad\qquad\qquad\qquad\ \ on \ \partial\Omega, \end{cases} \quad (1)$$

where $\alpha_1(.), \alpha_2(.), \beta_1(.) \ and \ \beta_2(.)$ are positive variable exponents and $p, q \in C^1(\bar\Omega)$ with,

$$1 < p^- \leqslant p^+ < N \quad and \quad 1 < q^- \leqslant q^+ < N. \qquad (2)$$

We denote by

$$r^- = \inf_{x \in \Omega} r(x) \quad and \quad r^+ = \sup_{x \in \Omega} r(x) \quad for \ all \ r \in C(\bar\Omega).$$

and $a < \min\{\frac{N-p^-}{p^-}, \frac{N-q^-}{q^-}\}$.

E. Azroul · A. Boumazourh (✉)
Department of Mathematics, FEZ Laboratory LAMA, Sidi Mohamed Ben Abdellah University, P.O. Box 1796, Atlas Fez 30000, Morocco
e-mail: athmane.boumazourh@gmail.com

E. Azroul
e-mail: elhoussine.azroul@gmail.com

© Springer Nature Switzerland AG 2020   21
E. H. Zerrik et al. (eds.), *Recent Advances in Modeling, Analysis and Systems Control: Theoretical Aspects and Applications*, Studies in Systems, Decision and Control 243, https://doi.org/10.1007/978-3-030-26149-8_3

A weak solution of (1) is a pair $(u, v) \in W_0^{1,p(x)}(\Omega, |x|^{-ap(x)}) \times W_0^{1,q(x)}(\Omega, |x|^{-aq(x)})$, $u, v > 0$ a.e in Ω such that,

$$\int_\Omega |x|^{-ap(x)} |\nabla u|^{p(x)-2} \nabla u \nabla \varphi dx + \int_\Omega |x|^{-aq(x)} |\nabla v|^{q(x)-2} \nabla v \nabla \psi dx$$

$$= \int_\Omega \left[f(u)|v|^{\alpha_1(x)} + g(u)|v|^{\beta_1(x)} \right] \varphi dx + \int_\Omega \left[f(v)|u|^{\alpha_2(x)} + g(v)|u|^{\beta_2(x)} \right] \psi dx,$$

for all $(\varphi, \psi) \in W_0^{1,p(x)}(\Omega, |x|^{-ap(x)}) \times W_0^{1,q(x)}(\Omega, |x|^{-aq(x)})$.

In the case where the variable exponents $p(.)$ and $q(.)$ are reduced to be constants, the existence of a positive weak solutions was studied in [14] for the weighted (p, q)–Laplacian system

$$\begin{cases} -div(|x|^{-ap}|\nabla u|^{p-2}\nabla u) = \lambda |x|^{-(a+1)p+c_1} u^\alpha v^\gamma & in \ \ \Omega, \\ -div(|x|^{-bq}|\nabla v|^{q-2}\nabla v) = \lambda |x|^{-(b+1)q+c_2} u^\delta v^\beta & in \ \ \Omega, \\ u, v > 0 & in \ \ \Omega, \\ u = v = 0 & on \ \partial\Omega, \end{cases}$$

where $0 < a < \frac{N-p}{p}$, $0 < b < \frac{N-q}{q}$, $0 < \alpha < p - 1$ and $0 < \beta < q - 1$. Authors proved the existence of solution by applying the lower and upper-solution method, they are obtained the subsolution from the eigenfunction associated to the first eigenvalue $\lambda_{1,p}$ of the problem

$$\begin{cases} -div(|x|^{-ap(x)}|\nabla u|^{p(x)-2}\nabla u) = \lambda |x|^{-(a+1)p+c_1} u^{p(x)-2} u & in \ \ \Omega, \\ v = 0 & on \ \partial\Omega. \end{cases}$$

and the eigenfunction associated to the first eigenvalue $\lambda_{1,q}$ of the problem

$$\begin{cases} -div(|x|^{-bq(x)}|\nabla v|^{q(x)-2}\nabla v) = \lambda |x|^{-(b+1)q+c_2} v^{q(x)-2} v & in \ \ \Omega \\ v = 0 & on \ \partial\Omega. \end{cases}$$

Variable exponent Lebesgue spaces appeared in the literature for the first time in 1931 by Orlicz [15]. Quasilinear problems involving weighted $p(x)$–Laplacian appear in some physical phenomena like in electrorheological fluids which is an important class of non-Newtonian fluids, in image restoration and in elasticity see [1, 6, 16, 17].

$p(x)$–Laplacian problems are more complicated than those of p–Laplacian ones, because of the nonhomogeneity of $p(x)$–Laplacian, for example if $\Omega \subset \mathbb{R}^N$ is a bounded domain, the Rayleigh quotient

$$\lambda_{p(x)} = \inf_{u \in W_0^{1,p(x)}(\Omega)\setminus\{0\}} \frac{\int_\Omega \frac{1}{p(x)} |\nabla u|^{p(x)} dx}{\int_\Omega \frac{1}{p(x)} |u|^{p(x)} dx},$$

is zero in generally, and only under some special conditions $\lambda_{p(x)} > 0$. Many results and methods for p-Laplacian problems are invalid for $p(x)-$Laplacian problems (see [8]). To give an example of this kind of systems with variable exponents, we refer to [2], when $a = 0$ authors have shown, using sub-supersolutions method, the existence of positive solutions for the problem

$$\begin{cases} -div(|\nabla u|^{p(x)-2}\nabla u) = \lambda[a(x)g_1(v) + f_1(u)] \ in \ \ \Omega, \\ -div(|\nabla v|^{q(x)-2}\nabla v) = \lambda[a(x)f_2(u) + g_2(v)] \ in \ \ \Omega, \\ \qquad\qquad u, v > 0 \qquad\qquad\qquad\qquad in \ \ \Omega, \\ \qquad\qquad u = v = 0 \qquad\qquad\qquad\quad on \ \partial\Omega. \end{cases}$$

In [10], authors have considered the quasilinear elliptic system of the form

$$\begin{cases} -\Delta_{\phi_1} u = f_1(v)|v|^{\alpha_1} + g_1(v)|v|^{\beta_1} \ in \ \ \Omega, \\ -\Delta_{\phi_2} v = f_2(u)|u|^{\alpha_2} + g_2(u)|u|^{\beta_2} \ in \ \ \Omega, \\ \qquad\quad u, v > 0 \qquad\qquad\qquad\quad in \ \ \Omega, \\ \qquad\quad u, v = 0 \qquad\qquad\qquad\quad on \ \partial\Omega, \end{cases}$$

where α_1, α_2, β_1 and β_2 are are positive constants. Here, Δ_{ϕ_i} stands for the ϕ_i-Laplacian operator, that is, $\Delta_{\phi_i} w = div(\phi_i(|\nabla w|)\phi_w)$, for $i = 1, 2$, where $\phi_i : \mathbb{R} \longrightarrow \mathbb{R}$ are N-functions.

Our goal in this paper is to prove the existence of positive solutions for the problem (1). We assume that $f, g : (0, +\infty) \longrightarrow (0, +\infty)$ are continuous functions satisfying the conditions:

(H.f):

$$f(s) \leqslant s^{\eta_1(x)} \quad \text{for all} \quad s \in (0, +\infty).$$

(H.g):

$$g(s) \leqslant s^{\eta_2(x)} \quad \text{for all} \quad s \in (0, +\infty).$$

Our result is formulated as follows.

Theorem 1 *For $i = \{1, 2\}$, let us assume that*

$$\eta_1^- + \alpha_i^- > \min\{p^+ - 1, q^+ - 1\}, \tag{3}$$

and

$$\eta_2^- + \beta_i^- > \min\{p^+ - 1, q^+ - 1\}. \tag{4}$$

Then problem (1) has at least one positive weak solution $(u, v) \in C(\bar{\Omega}) \times C(\bar{\Omega})$.

To prove Theorem 1 we will give some comparison properties in order to show a version of sub-supersolutions method for quasilinear elliptic systems, and by applying fixed point theorem [4] we obtain our main result.

This paper is organized as follows: in Sect. 2 we recall some notations and properties of weighted Lebesgue and Sobolev spaces with variable exponents. Section 3, contains some lemmas, comparison properties that are usable to apply the sub-supersolutions method, and the last subsection is reserved to the proof of Theorem 1.

Our paper can be seen as a generalization of the work [14] in variable exponents case, and of the work [2] in the degenerated case.

2 Preliminaries and Basic Assumptions

In this section, we will give some definitions and assumptions as well as a framework of this work. Let $w : \Omega \longrightarrow \mathbb{R}$ be a weight function (that is a function which is measurable strictly positive a.e in Ω) and let $p \in C_+(\bar{\Omega})$

where $C_+(\bar{\Omega}) = \{h \in C(\bar{\Omega}) \ : \ h(x) > 1 \ for \ any \ x \ in \ \bar{\Omega}\}$.

The generalized weighted Lebesgue space $L^{p(x)}(\Omega, w)$ is defined by

$$L^{p(x)}(\Omega, w) = \{u : \Omega \longrightarrow \mathbb{R} \, measurable \ : \ \rho_{p(x),w}(u) = \int_{\Omega} w(x) \, |u(x)|^{p(x)} dx < +\infty\}.$$

It's equipped with the Luxemburg norm,

$$||u||_{L^{p(x)}(\Omega, w)} = inf\{\lambda > 0 \ : \ \rho_{p(x),w}(\frac{u}{\lambda}) \leqslant 1\}.$$

The space $\left(L^{p(x)}(\Omega, w), \ ||.||_{L^{p(x)}(\Omega,w)}\right)$ is a Banach space (see [3, 7]).

Lemma 1 (cf.[9, 11]) *Let $u \in L^{p(x)}(\Omega, w)$. We have the following assertions:*

1. *$\rho_{p(x),w}(u) > 1 \ (= 1; < 1) \ \Leftrightarrow \ ||u||_{L^{p(x)}(\Omega,w)} > 1 \ (= 1; < 1)$ respectively.*
2. *If $||u||_{L^{p(x)}(\Omega,w)} > 1$, then $||u||_{L^{p(x)}(\Omega,w)}^{p^-} \leqslant \rho_{p(x),w}(u) \leqslant ||u||_{L^{p(x)}(\Omega,w)}^{p^+}$.*
3. *If $||u||_{L^{p(x)}(\Omega,w)} < 1$, then $||u||_{L^{p(x)}(\Omega,w)}^{p^+} \leqslant \rho_{p(x),w}(u) \leqslant ||u||_{L^{p(x)}(\Omega,w)}^{p^-}$.*

Suppose that w satisfies the following integrability condition:

$$(H_1) \quad w \in L^1_{loc}(\Omega) \, ; \ w^{\frac{-1}{p(x)-1}} \in L^1_{loc}(\Omega).$$

The generalized weighted Sobolev space $W^{1,p(x)}(\Omega, w)$ is defined by

$$W^{1,p(x)}(\Omega, w) = \{u \in L^{p(x)}(\Omega) \ : \ \frac{\partial u}{\partial x_i} \in L^{p(x)}(\Omega, w), \ i = 1, ..., N\},$$

which is equipped with the norm,

$$||u||_{W^{1,p(x)}(\Omega,w)} = ||u||_{L^{p(x)}(\Omega)} + \sum_{i=1}^{N} ||\frac{\partial u}{\partial x_i}||_{L^{p(x)}(\Omega,w)}.$$

Note that this norm is equivalent to the so called Luxemburg norm

$$|||u||| = inf\{\lambda > 0 : \int_\Omega \left(\left|\frac{u}{\lambda}\right|^{p(x)} + w(x) \sum_{i=1}^{N} \left|\frac{\frac{\partial u}{\partial x_i}}{\lambda}\right|^{p(x)}\right) dx \leqslant 1\}. \quad (5)$$

To deal with the Dirichlet problems, we use the space $W_0^{1,p(x)}(\Omega, w)$ defined as the closure of $C_0^\infty(\Omega)$ with respect to the norm (5). Note that $C_0^\infty(\Omega)$ is dense in $W_0^{1,p(x)}(\Omega, w)$ and $\left(W_0^{1,p(x)}(\Omega, w), |||.|||\right)$ is a Banach reflexive space.

Proposition 1 (cf. [3])
Let w be a weight function in Ω satisfying the condition (H_1). Then, $\left(W^{1,p(x)}(\Omega, w), ||.||_{W^{1,p(x)}(\Omega, w)}\right)$ is a Banach space.

Proposition 2 (cf. [9])
Let $p \in C_+(\Omega)$ and w a weight function satisfying (H_1). Then, for all $G \in \left(W_0^{1,p(x)}(\Omega, w)\right)'$, there exist unique system of functions $(g_0, g_1, ..., g_N) \in L^{p'}(\Omega) \times \left(L^{p'(x)}(\Omega, w^{1-p'(x)})\right)^N$ such that

$$G(f) = \int_\Omega f(x)g_0(x)dx + \sum_{i=1}^{N} \int_\Omega \frac{\partial f}{\partial x_i} g_i(x)dx$$

for all $f \in W_0^{1,p(x)}(\Omega, w)$.
    Thus the dual space of $W_0^{1,p(x)}(\Omega, w)$ can be identified to $W^{-1,p'(x)}(\Omega, w^*)$ with $w^* = w^{1-p(x)}$.

Proposition 3 (cf. [9]) (Young inequality)
Let $p : \Omega \longrightarrow]1, +\infty[$ be a measurable function, and p' it's conjugate, i.e, $\frac{1}{p} + \frac{1}{p'} = 1$. Then, for all $a, b > 0$, we have,

$$ab \leqslant \frac{a^{p(x)}}{p(x)} + \frac{b^{p'(x)}}{p'(x)}$$

Proposition 4 (cf. [9]) (Generalized Hölder inequality)
Let $p : \Omega \longrightarrow]1, +\infty[$ be a measurable function, and let p' it's conjugate. Then, we have

$$\int_\Omega |u(x)v(x)| dx \leqslant r_p ||u||_{L^{p(x)}(\Omega)} ||v||_{L^{p'(x)}(\Omega)}$$

$\forall u \in L^{p(x)}(\Omega), \forall v \in L^{p'(x)}(\Omega)$, where $r_p = \frac{1}{p^-} + \frac{1}{p'^-}$

Theorem 2 (cf. [19]) (comparison principle)
Let $B : W_0^{1,r(x)}(\Omega, w) \longrightarrow W^{-1,r'(x)}(\Omega, w^*)$ be an operator such that

$$Bu - Bv \geqslant 0 \quad \forall u, v \in W_0^{1,r(x)}(\Omega, w) \cap C(\Omega)$$

If $u \geqslant v$ on $\partial\Omega$, then $u \geqslant v$ a.e. in Ω.

Theorem 3 (cf. [4]) (Schauder fixed point theorem)
Let X be a topological space, $K \subset X$ non empty, compact and convex subset. Then, for all $f : K \longrightarrow K$ continuous, there exists $x \in K$ such that $f(x) = x$.

A Caffarelli–Kohn–Nirenburg-type inequality: [13]
    Let Ω be an open bounded subset of $\mathbb{R}^N$ ($N > 2$), with smooth boundary $\partial\Omega$. For each $x = (x_1, x_2, ..., x_N) \in \Omega$ we denote

$$m = \min_{i \in \{1,2,...,N\}} x_i \qquad M = \max_{i \in \{1,2,...,N\}} x_i$$

For each $i = 1, ..., N$, let $a_i : [m, M] \longrightarrow \mathbb{R}$ be function of class C^1. consider $\overrightarrow{a} :$ $\Omega \longrightarrow \mathbb{R}^N$ defined by

$$\overrightarrow{a}(x) = (a_1(x_1), ..., a_N(x_N))$$

We assume that there exists $a_0 > 0$ such that

$$div(\overrightarrow{a}(x)) \geqslant a_0 > 0 \quad \forall x \in \Omega.$$

Let $p \in C_+(\bar{\Omega})$, $p \leqslant N$ such that

$$\overrightarrow{a}(x).\nabla p(x) \quad \forall x \in \Omega.$$

Then there exists a positive constant C such that

$$\int_\Omega |u|^{p(x)} dx \leqslant C \int_\Omega |\overrightarrow{a(x)}|^{p(x)} |\nabla u|^{p(x)} dx, \tag{6}$$

in particular, we have

$$W^{1,p(x)}(\Omega, |x|^{-ap(x)}) \hookrightarrow L^{p(x)}(\Omega). \tag{7}$$

Remark 1 In [13], authors have also shown the compact embedding of $W^{1,p(x)}(\Omega, |x|^{-ap(x)})$ in $L^{q(x)}(\Omega)$ for all $q \in (1, \frac{2Np^-}{2N+p^-})$.

3   Main Results

In this section we are concerned to prove Theorem 1, but we still need more properties and lemmas that are usable to show our result.

Lemma 2 *Let $\lambda > 0$ and let u_λ be the unique solution of the problem*

$$
\begin{cases}
-div(|x|^{-ap(x)}|\nabla u|^{p(x)-2}\nabla u) = \lambda & in \ \ \Omega, \\
\\
\qquad\qquad u = 0 & on \ \partial\Omega.
\end{cases}
\tag{8}
$$

Define $\varepsilon_0 = \dfrac{p^-}{2|\Omega|^{\frac{1}{N}}C_0}$. If $\lambda > \varepsilon_0$ then $u_\lambda \leqslant C^*\lambda^{\frac{1}{p^--1}}$ and if $\lambda < \varepsilon_0$ then $u_\lambda \leqslant C_*\lambda^{\frac{1}{p^+-1}}$.

Proof Let u_λ be the solution of (8). For $k > 0$ set $A_k = \{x \in \Omega : u_\lambda(x) > k\}$.
  Taking $(u_\lambda - k)^+$ as a test function for (8). By using Young inequality we have

$$
\begin{aligned}
\int_{A_k} |x|^{-ap(x)}|\nabla u_\lambda|^{p(x)} &= \lambda \int_{A_k}(u_\lambda - k)dx \\
&\leqslant \lambda ||(u_\lambda - k)^+||_{L^{\frac{N}{N-1}}(\Omega,|x|^{-a})} ||\chi_{A_k}||_{L^N(\Omega,|x|^{a(N-1)})} \\
&\leqslant \lambda |A_k|^{\frac{1}{N}} \int_{A_k} |x|^{-a}|\nabla u_\lambda|dx \\
&\leqslant \lambda |A_k|^{\frac{1}{N}} \left(\int_{A_k} \varepsilon^{p(x)} \frac{|x|^{-ap(x)}}{p(x)}|\nabla u_\lambda|^{p(x)}dx + \int_{A_k} \frac{\varepsilon^{-p'(x)}}{p'(x)}dx \right) \\
&\leqslant \lambda |A_k|^{\frac{1}{N}} \left(\frac{1}{p^-} \int_{A_k} \varepsilon^{p(x)}|x|^{-ap(x)}|\nabla u_\lambda|^{p(x)}dx + \frac{1}{p'^-} \int_{A_k} \varepsilon^{-p'(x)}dx \right).
\end{aligned}
\tag{9}
$$

For $\lambda > \varepsilon_0$, taking

$$
\varepsilon = \left(\frac{p^-}{2\lambda|\Omega|^{\frac{1}{N}}C} \right)^{\frac{1}{p^-}} = \left(\frac{\varepsilon_0}{\lambda} \right)^{\frac{1}{p^-}}.
$$

From (9) we get

$$
\int_{A_k} (u_\lambda - k)dx \leqslant \frac{2C\varepsilon^{-(p')^-}}{p^-} A^{1+\frac{1}{N}}.
\tag{10}
$$

The previous inequality implies that

$$
||u_\lambda||_\infty \leqslant C^*\lambda^{\frac{1}{p^--1}},
$$

where $C^* = \dfrac{(N+1)(2C)^{(p')^-}}{(p')^-(p^-)^{\frac{(p')^-}{p^-}}}$.
  For $\lambda < \varepsilon_0$ we obtain

$$
||u_\lambda||_\infty \leqslant C_*\lambda^{\frac{1}{p^+-1}},
$$

where $C_* = \dfrac{(N+1)(2C)^{(p')^+}}{(p')^+(p^-)^{\frac{(p')^+}{p^+}}}$. Similarly, the unique solution v_λ of the problem

$$\begin{cases} -div(|x|^{-aq(x)}|\nabla v|^{q(x)-2}\nabla u) = \lambda \ in \ \ \Omega, \\ \\ \qquad\qquad v = 0 \qquad\qquad on \ \partial\Omega, \end{cases} \tag{11}$$

satisfies the following.

For $\varepsilon_0 = \dfrac{q^-}{2|\Omega|^{\frac{1}{N}}C_0}$. If $\lambda > \varepsilon_0$ then $v_\lambda \leqslant C^*\lambda^{\frac{1}{q^--1}}$ and if $\lambda < \varepsilon_0$ then $v_\lambda \leqslant C_*$ $\lambda^{\frac{1}{q^+-1}}$.   □

In the following, we will give the definition of sub-supersolution.

Definition 1

(i) We say that $(\underline{u}, \underline{v}) \in W_0^{1,p(x)}(\Omega, |x|^{-ap(x)}) \times W_0^{1,q(x)}(\Omega, |x|^{-aq(x)})$ is a subsolution of (1) if

$$\begin{cases} \int_\Omega |x|^{-ap(x)}\,|\nabla \underline{u}|^{p(x)-2}\nabla \underline{u}\ \nabla\varphi dx \leqslant \int_\Omega \left[f(\underline{u})|\underline{v}|^{\alpha_1(x)} + g(\underline{u})|\underline{v}|^{\beta_1(x)} \right] \varphi dx, \\ \\ \int_\Omega |x|^{-aq(x)}\,|\nabla \underline{v}|^{q(x)-2}\nabla \underline{v}\ \nabla\psi dx \leqslant \int_\Omega \left[f(\underline{v})|\underline{u}|^{\alpha_2(x)} + g(\underline{v})|\underline{u}|^{\beta_2(x)} \right] \psi dx. \end{cases}$$

for all $(\varphi, \psi) \in W_0^{1,p(x)}(\Omega, |x|^{-ap(x)}) \times W_0^{1,q(x)}(\Omega, |x|^{-aq(x)})$.

(ii) We say that $(\overline{u}, \overline{v}) \in W_0^{1,p(x)}(\Omega, w) \times W_0^{1,q(x)}(\Omega, w)$ is a supersolution of (P_1) if

$$\begin{cases} \int_\Omega |x|^{-ap(x)}\,|\nabla \overline{u}|^{p(x)-2}\nabla \overline{u}\ \nabla\varphi dx \geqslant \int_\Omega \left[f(\overline{u})|\overline{v}|^{\alpha_1(x)} + g(\overline{u})|\overline{v}|^{\beta_1(x)} \right] \varphi dx, \\ \\ \int_\Omega |x|^{-aq(x)}\,|\nabla \overline{v}|^{q(x)-2}\nabla \overline{v}\ \nabla\psi dx \geqslant \int_\Omega \left[f(\overline{v})|\overline{u}|^{\alpha_2(x)} + g(\overline{v})|\overline{u}|^{\beta_2(x)} \right] \psi dx. \end{cases}$$

for all $(\varphi, \psi) \in W_0^{1,p(x)}(\Omega, |x|^{-ap(x)}) \times W_0^{1,q(x)}(\Omega, |x|^{-aq(x)})$.

Theorem 4 *Suppose that (H.f) and (H.g) hold true. If the assumptions of Theorem 1 are satisfied and $(\underline{u}, \underline{v})$ and $(\overline{u}, \overline{v})$ are sub-supersolutions for (1) with $\underline{u}, \underline{v} > 0$ a.e. in Ω. Then (1) has a weak positive solution (u, v), with $\underline{u} \leqslant u \leqslant \overline{u}$ and $\underline{v} \leqslant v \leqslant \overline{v}$.*

Proof Let consider the set

$$B = \{(w_1, w_2) \in L^\infty(\Omega) \times L^\infty(\Omega) : \ \underline{u} \leqslant w_1 \leqslant \overline{u} \ and \ \underline{v} \leqslant w_2 \leqslant \overline{v}\}$$

B is non empty, bounded, closed and coercive. Let define the operators $T_1 : L^{p(x)}(\Omega) \to L^\infty(\Omega)$ and $T_2 : L^{q(x)}(\Omega) \to L^\infty(\Omega)$ defined by

$$T_1 z(x) = \begin{cases} \underline{u}(x) & if \quad z(x) \leqslant \underline{u}(x), \\ z(x) & if \ \underline{u}(x) \leqslant z(x) \leqslant \overline{u}(x), \\ \overline{u} & if \quad z(x) \leqslant \overline{u}, \end{cases}$$

and

$$T_2 z(x) = \begin{cases} \underline{v}(x) & if \quad z(x) \leqslant \underline{v}(x), \\ z(x) & if \ \underline{v}(x) \leqslant z(x) \leqslant \overline{v}(x), \\ \overline{v} & if \quad z(x) \leqslant \overline{v}. \end{cases}$$

Since $T_1 z \in [\underline{u}, \overline{u}]$, $T_2 z \in [\underline{v}, \overline{v}]$ and $\underline{u}, \overline{u}, \underline{v}, \overline{v} \in L^\infty(\Omega)$, then the operators $T_1 z$ and $T_2 z$ are well defined.

Now, we consider the operators $F : [\underline{u}, \overline{u}] \times [\underline{v}, \overline{v}] \longrightarrow L^{p'(x)}(\Omega)$ and $G : [\underline{u}, \overline{u}] \times [\underline{v}, \overline{v}] \longrightarrow L^{q'(x)}(\Omega)$ defined by

$$F(u, v)(x) = f(u)|v|^{\alpha_1(x)} + g(u)|v|^{\beta_1(x)},$$

and

$$G(u, v)(x) = f(v)|u|^{\alpha_2(x)} + g(v)|u|^{\beta_2(x)}.$$

Since f, g are continuous functions, $T_1 z_1 \in [\underline{u}, \overline{u}]$, $T_2 z_2 \in [\underline{v}, \overline{v}]$, there exist constants $K_1, K_2 > 0$ such that

$$F(T_1 z_1, T_2 z_2) \leqslant K_1,$$

and

$$G(T_1 z_1, T_2 z_2) \leqslant K_2,$$

for all $(z_1, z_2) \in L^{p(x)}(\Omega, |x|^{-ap(x)}) \times L^{q(x)}(\Omega, |x|^{-aq(x)})$.

Thus, by Dominate Convergence Theorem, it follows that the mapping $(z_1, z_2) \longmapsto F(T_1 z_1, T_2 z_2)$ is continuous from $L^{p(x)}(\Omega, |x|^{-ap(x)}) \times L^{q(x)}(\Omega, |x|^{-aq(x)})$ in $L^{p'(x)}(\Omega, |x|^{ap'(x)})$ and the mapping $(z_1, z_2) \longmapsto G(T_1 z_1, T_2 z_2)$ is continuous from $L^{p(x)}(\Omega, |x|^{-ap(x)}) \times L^{q(x)}(\Omega, |x|^{-aq(x)})$ in $L^{p'(x)}(\Omega, |x|^{aq'(x)})$.

Let consider the problem

$$\begin{cases} -div(|x|^{ap(x)}|\nabla u|^{p(x)-2}\nabla u) = F(T_1 z_1, T_2 z_2) \ in \ \ \Omega, \\ -div(|x|^{aq(x)}|\nabla v|^{q(x)-2}\nabla v) = G(T_1 z_1, T_2 z_2) \ in \ \ \Omega, \\ \qquad\qquad u, v > 0 \qquad\qquad\qquad\quad in \ \ \Omega, \\ \qquad\qquad u = v = 0, \qquad\qquad\qquad on \ \partial\Omega. \end{cases} \tag{12}$$

Minty–Browder theorem [5] implies that problem (12) has a unique solution $(u_1, v_1) \in W_0^{1,p(x)}(\Omega, |x|^{-ap(x)}) \times W_0^{1,q(x)}(\Omega, |x|^{-aq(x)})$. Thus, the operator

$$\mathscr{T} : L^{p(x)}(\Omega, |x|^{-ap(x)}) \times L^{q(x)}(\Omega, |x|^{-aq(x)}) \longrightarrow L^{p(x)}(\Omega, |x|^{-ap(x)}) \times L^{q(x)}(\Omega, |x|^{-aq(x)})$$
$$(z_1, z_2) \qquad\qquad\qquad \longmapsto \mathscr{T}(z_1, z_2) = (u, v)$$

is well-defined.

In order to apply fixed point theorem, we will show that $\mathscr{T}$ is compact and continuous.

Claim 1: $\mathscr{T}$ is compact. Let $\{(z_{1,n}, z_{2,n})\} \subset L^{p(x)}(\Omega, |x|^{-ap(x)}) \times L^{q(x)}(\Omega, |x|^{-aq(x)})$ be a bounded sequence and consider $(u_n, v_n) = \mathscr{T}(z_{1,n}, z_{2,n})$. We have

$$\int_\Omega |x|^{-ap(x)}|\nabla u_n|^{p(x)-2}\nabla u_n \nabla \varphi dx = \int_\Omega F(T_1 z_{1,n}, T_2 z_{2,n})\varphi dx,$$

and

$$\int_\Omega |x|^{-aq(x)}|\nabla v_n|^{q(x)-2}\nabla v_n \nabla \psi dx = \int_\Omega G(T_1 z_{1,n}, T_2 z_{2,n})\psi dx,$$

for all $(\varphi, \psi) \in L^{p(x)}(\Omega, |x|^{-ap(x)}) \times L^{q(x)}(\Omega, |x|^{-aq(x)})$.

For $\varphi = u_n$ and $\psi = v_n$, since F, G are bounded, it follows that

$$\int_\Omega |x|^{-ap(x)}|\nabla u_n|^{p(x)}dx \leqslant K_1 \int_\Omega u_n dx,$$

and

$$\int_\Omega |x|^{-aq(x)}|\nabla v_n|^{q(x)}dx \leqslant K_2 \int_\Omega v_n dx.$$

From the embedding of $L^{p(x)}(\Omega, |x|^{-ap(x)})$ in $L^1(\Omega)$ and of $L^{q(x)}(\Omega, |x|^{-aq(x)})$ in $inL^1(\Omega)$, we conclude that u_n and v_n are bounded in $W^{1,p(x)}(\Omega, |x|^{-ap(x)})$ and in $W^{1,q(x)}(\Omega, |x|^{-aq(x)})$ respectively.

Thus, the compact embedding $W^{1,p(x)}(\Omega, |x|^{-ap(x)})$ in $L^{p(x)}(\Omega)$ and $W^{1,q(x)}(\Omega, |x|^{-aq(x)})$ in $L^{q(x)}(\Omega)$ provide the result.

Claim 2: $\mathscr{T}$ is continuous. Let $\{(z_{1,n}, z_{2,n})\} \subset L^{p(x)}(\Omega, |x|^{-ap(x)}) \times L^{q(x)}(\Omega, |x|^{-aq(x)})$ be a sequence that converge to (z_1, z_2). Define $(u_n, v_n) = \mathscr{T}(z_{1,n}, z_{2,n})$ and $(u, v) = \mathscr{T}(z_1, z_2)$, we have

$$\int_\Omega |x|^{-ap(x)}|\nabla u_n|^{p(x)-2}\nabla u_n \nabla \varphi dx = \int_\Omega F(T_1 z_{1,n}, T_2 z_{2,n})\varphi dx,$$

$$\int_\Omega |x|^{-aq(x)}|\nabla v_n|^{q(x)-2}\nabla v_n \nabla \psi dx = \int_\Omega G(T_1 z_{1,n}, T_2 z_{2,n})\psi dx,$$

$$\int_\Omega |x|^{-ap(x)}|\nabla u|^{p(x)-2}\nabla u \nabla \varphi dx = \int_\Omega F(T_1 z_1, T_2 z_2)\varphi dx,$$

and

$$\int_\Omega |x|^{-aq(x)}|\nabla v|^{q(x)-2}\nabla v \nabla \psi dx = \int_\Omega G(T_1 z_1, T_2 z_2)\psi dx,$$

for all $(\varphi, \psi) \in L^{p(x)}(\Omega, |x|^{-ap(x)}) \times L^{q(x)}(\Omega, |x|^{-aq(x)})$.

Considering $\varphi = u_n - u$ and $\psi = v_n - v$, we get

$$\int_\Omega < |x|^{-ap(x)}|\nabla u_n|^{p(x)-2}\nabla u_n - |x|^{-ap(x)}|\nabla u|^{p(x)-2}\nabla u, \nabla(u_n - u) > dx$$
$$= \int_\Omega F(T_1 z_{1,n}, T_2 z_{2,n})(u_n - u)dx$$
$$- \int_\Omega F(T_1 z_1, T_2 z_2)(u_n - u)dx,$$

and

$$\int_\Omega < |x|^{-aq(x)}|\nabla v_n|^{q(x)-2}\nabla v_n - |x|^{-aq(x)}|\nabla v|^{q(x)-2}\nabla v, \nabla(v_n - v) > dx$$
$$= \int_\Omega G(T_1 z_{1,n}, T_2 z_{2,n})(v_n - v)dx$$
$$- \int_\Omega G(T_1 z_1, T_2 z_2)(v_n - v)dx.$$

Using Hölder inequality, it follows that

$$\int_\Omega < |x|^{-ap(x)}|\nabla u_n|^{p(x)-2}\nabla u_n - |x|^{-ap(x)}|\nabla u|^{p(x)-2}\nabla u, \nabla(u_n - u) > dx$$
$$\leqslant \|u_n - u\|_{L^{p(x)}(\Omega, |x|^{-ap(x)})} \|F(T_1 z_{1,n}, T_2 z_{2,n}) - F(T_1 z_1, T_2 z_2)\|_{L^{p'(x)}(\Omega, |x|^{ap'(x)})},$$

and

$$\int_\Omega < |x|^{-aq(x)}|\nabla v_n|^{q(x)-2}\nabla v_n - |x|^{-aq(x)}|\nabla v|^{q(x)-2}\nabla v, \nabla(v_n - v) > dx$$
$$\leqslant \|v_n - v\|_{L^{q(x)}(\Omega, |x|^{-aq(x)})} \|G(T_1 z_{1,n}, T_2 z_{2,n}) - G(T_1 z_1, T_2 z_2)\|_{L^{q'(x)}(\Omega, |x|^{aq'(x)})}.$$

So, $\{u_n\}$ and $\{v_n\}$ are bounded in $W^{1,p(x)}(\Omega, |x|^{-ap(x)})$ and in $W^{1,q(x)}(\Omega, |x|^{-aq(x)})$ respectively.

Since

$$F(T_1 z_{1,n}, T_2 z_{2,n}) \to F(T_1 z_1, T_2 z_2) \quad in \quad L^{p'(x)}(\Omega, |x|^{ap'(x)}),$$

and

$$G(T_1 z_{1,n}, T_2 z_{2,n}) \to G(T_1 z_1, T_2 z_2) \quad in \quad L^{q'(x)}(\Omega, |x|^{aq'(x)}),$$

we have

$$\left| \int_\Omega < |x|^{-ap(x)}|\nabla u_n|^{p(x)-2}\nabla u_n - |x|^{-ap(x)}|\nabla u|^{p(x)-2}\nabla u, \nabla(u_n - u) > dx \right| \longrightarrow 0,$$

and

$$\left| \int_\Omega < |x|^{-aq(x)}|\nabla v_n|^{q(x)-2}\nabla v_n - |x|^{-aq(x)}|\nabla v|^{q(x)-2}\nabla v, \nabla(v_n - v) > dx \right| \longrightarrow 0.$$

Therefore,

$$u_n \longrightarrow u \quad in \quad L^{p(x)}(\Omega, |x|^{-ap(x)}),$$

and

$$v_n \longrightarrow v \quad in \ \ L^{q(x)}(\Omega, |x|^{-aq(x)}).$$

Hence $\mathscr{T}$ is continuous.

We are in position to apply the Schauder's fixed point theorem, which allows to deduce the existence of $(u, v) \in \mathscr{B}$ such that $(u, v) = \mathscr{T}(u, v)$. Thus the Proof of theorem 4 is achieve.   □

3.1 Proof of Theorem 1

In this part, we show the existence result for system (1). We will construct a super-solution and a subsolution in order to apply the result of the previous Theorem to conclude the proof.

From Lemma 2, for $\lambda > 0$ small enough, there exists a constant $K > 1$ such that

$$0 < u_\lambda(x) \leqslant K\lambda^{\frac{1}{p^+-1}}, \tag{13}$$

and

$$0 < v_\lambda(x) \leqslant K\lambda^{\frac{1}{q^+-1}}. \tag{14}$$

We will construct a supersolution $(\bar{u}, \bar{v})$ of (1). Let u_λ and v_λ be the unique solutions of (13) and (14) respectively. Then, there exists $\lambda_0 > 0$ small enough such that for $\lambda \in (0, \lambda_0]$ we have

$$f(u_\lambda)|v\lambda|^{\alpha_1(x)} + g(u_\lambda)|v\lambda|^{\beta_1(x)} \leqslant |u_\lambda|^{\eta_1(x)}|v_\lambda|^{\alpha_1(x)} + |u_\lambda|^{\eta_2(x)}|v_\lambda|^{\beta_1(x)}$$

$$\leqslant K^{\eta_1(x)}\lambda^{\frac{\eta_1(x)}{p^+-1}} K^{\alpha_1(x)}\lambda^{\frac{\alpha_1(x)}{q^+-1}} + K^{\eta_2(x)}\lambda^{\frac{\eta_2(x)}{p^+-1}} K^{\beta_1(x)}\lambda^{\frac{\beta_1(x)}{q^+-1}}$$

$$\leqslant \overline{K}\lambda^{\frac{\eta_1(x)}{p^+-1}+\frac{\alpha_1(x)}{q^+-1}} + \overline{K}\lambda^{\frac{\eta_2(x)}{p^+-1}+\frac{\beta_1(x)}{q^+-1}}$$

$$\leqslant \lambda,$$

where $\overline{K} = \max\{K^{\eta_1^+} K^{\alpha_1^+}, K^{\eta_1^+} K^{\alpha_1^-}, K^{\eta_2^+} K^{\beta_1^+}, K^{\eta_2^+} K^{\beta_1^-}\}$.

Then we conclude that

$$-\Delta_{p(x),|x|^{-ap(x)}} u_\lambda \geqslant f(u_\lambda)|v_\lambda|^{\alpha_1(x)} + g(u_\lambda)|v_\lambda|^{\beta_1(x)}.$$

Similarly, we get

$$-\Delta_{q(x),|x|^{-aq(x)}} v_\lambda \geqslant f(v_\lambda)|u_\lambda|^{\alpha_2(x)} + g(v_\lambda)|u_\lambda|^{\beta_2(x)}.$$

Hence $(\bar{u} = u_\lambda, \bar{v} = v_\lambda)$ is a supersolution of (1)

We will construct a subsolution $(\underline{u}, \underline{v})$ of (1). Let define the functions

$$
\varphi_1(x) = \begin{cases} \exp^{kd(x)} - 1 & if \quad d(x) < \sigma, \\[2ex] \exp^{k\sigma} - 1 + \int_\sigma^{d(x)} k \exp^{k\sigma} \left(\frac{2\delta - t}{2\delta - \sigma} \right)^{\frac{2}{p^- - 1}} dt & if \ \sigma \leqslant d(x) < 2\delta, \\[2ex] \exp^{k\sigma} - 1 + \int_\sigma^{2\delta} k \exp^{k\sigma} \left(\frac{2\delta - t}{2\delta - \sigma} \right)^{\frac{2}{p^- - 1}} dt & if \quad 2\delta \leqslant d(x), \end{cases}
$$

and

$$
\varphi_2(x) = \begin{cases} \exp^{kd(x)} - 1 & if \quad d(x) < \sigma, \\[2ex] \exp^{k\sigma} - 1 + \int_\sigma^{d(x)} k \exp^{k\sigma} \left(\frac{2\delta - t}{2\delta - \sigma} \right)^{\frac{2}{q^- - 1}} dt & if \ \sigma \leqslant d(x) < 2\delta, \\[2ex] \exp^{k\sigma} - 1 + \int_\sigma^{2\delta} k \exp^{k\sigma} \left(\frac{2\delta - t}{2\delta - \sigma} \right)^{\frac{2}{q^- - 1}} dt & if \quad 2\delta \leqslant d(x). \end{cases}
$$

Motivate by [12] we have

$$
-\Delta_{p(x), |x|^{-ap(x)}} (\mu \varphi_1(x)) = \begin{cases} -k(k\mu \exp^{kd(x)})^{p(x)-1} |x|^{-ap(x)} \Big[p(x) - 1 + (d(x) + \frac{\ln k\mu}{k}) \\ \qquad\qquad \times \nabla p(x) \nabla d(x) + \frac{\Delta d(x)}{k} \Big] & if \ d(x) < \sigma, \\[3ex] \left\{ \frac{1}{2\delta - \sigma} \frac{2(p(x)-1)}{p^- - 1} - \left(\frac{2\delta - d(x)}{2\delta - \sigma} \right) \Big[\ln k\mu \exp^{k\sigma} \left(\frac{2\delta - d(x)}{2\delta - \sigma} \right)^{\frac{2}{p^- - 1}} \right. \\ \qquad\qquad \left. \times \nabla p(x) \nabla d(x) + \Delta d(x) \Big] \right\} \\ (k\mu \exp^{k\sigma})^{p(x)-1} |x|^{-ap(x)} \left(\frac{2\delta - d(x)}{2\delta - \sigma} \right)^{\frac{2(p(x)-1)}{p^- - 1} - 1} & if \ \sigma \leqslant d(x) < 2\delta, \\[3ex] 0 & if \ 2\delta \leqslant d(x), \end{cases}
$$

and

$$
-\Delta_{q(x), |x|^{-aq(x)}} (\mu \varphi_1(x)) = \begin{cases} -k(k\mu \exp^{kd(x)})^{q(x)-1} |x|^{-aq(x)} \Big[q(x) - 1 + (d(x) + \frac{\ln k\mu}{k}) \\ \qquad\qquad \times \nabla q(x) \nabla d(x) + \frac{\Delta d(x)}{k} \Big] & if \ d(x) < \sigma, \\[3ex] \left\{ \frac{1}{2\delta - \sigma} \frac{2(q(x)-1)}{q^- - 1} - \left(\frac{2\delta - d(x)}{2\delta - \sigma} \right) \Big[\ln k\mu \exp^{k\sigma} \left(\frac{2\delta - d(x)}{2\delta - \sigma} \right)^{\frac{2}{q^- - 1}} \right. \\ \qquad\qquad \left. \times \nabla q(x) \nabla d(x) + \Delta d(x) \Big] \right\} \\ (k\mu \exp^{k\sigma})^{q(x)-1} |x|^{-aq(x)} \left(\frac{2\delta - d(x)}{2\delta - \sigma} \right)^{\frac{2(q(x)-1)}{q^- - 1} - 1} & if \ \sigma \leqslant d(x) < 2\delta, \\[3ex] 0 & if \ 2\delta \leqslant d(x). \end{cases}
$$

Let $\sigma = \frac{1}{k}\ln 2$ and $\mu = \exp^{-\alpha k}$ where

$$\alpha = \frac{\min\{p^- - 1, q^- - 1\}}{\max\{\sup(|\nabla p| + 1), \sup(|\nabla q| + 1)\}}.$$

Let $x \in \Omega$ with $d(x) < \sigma$. If k is large enough we have $|\nabla d(x)| = 1$, then if $d(x) < \sigma$ or $d(x) > 2\delta$ we get

$$-\Delta_{p(x),|x|^{-ap(x)}}(\mu\varphi_1(x)) \leqslant 0 \leqslant f(\mu\varphi_1)|\mu\varphi_2|^{\alpha_1(x)} + g(\mu\varphi_1)|\mu\varphi_2|^{\beta_1(x)},$$

and

$$-\Delta_{q(x),|x|^{-aq(x)}}(\mu\varphi_2(x)) \leqslant 0 \leqslant f(\mu\varphi_2)|\mu\varphi_1|^{\alpha_2(x)} + g(\mu\varphi_2)|\mu\varphi_1|^{\beta_2(x)}.$$

Now for $\sigma \leqslant d(x) < 2\delta$ we have

$$-\Delta_{p(x),|x|^{-ap(x)}}(\mu\varphi_1(x)) \leqslant \tilde{C}(k\mu)^{p^- - 1}\left|\ln\frac{k}{\exp^{\alpha k}}\right|.$$

Combining [18, Theorem 2] and the above inequality, we get

$$-\Delta_{p(x),|x|^{-ap(x)}}(\mu\varphi_1(x)) \leqslant 0 \leqslant f(\mu\varphi_1)|\mu\varphi_2|^{\alpha_1(x)} + g(\mu\varphi_1)|\mu\varphi_2|^{\beta_1(x)}.$$

A similar reasoning imply that there is $\mu > 0$ such that

$$-\Delta_{q(x),|x|^{-aq(x)}}(\mu\varphi_2(x)) \leqslant 0 \leqslant f(\mu\varphi_2)|\mu\varphi_1|^{\alpha_2(x)} + g(\mu\varphi_2)|\mu\varphi_1|^{\beta_2(x)}.$$

Hence, we conclude that $(\underline{u} = \mu\varphi_1, \underline{v} = \mu\varphi_2)$ is a subsolution of (1). As it was to be proven.

References

1. Acerbi, E., Mingione, G.: Regularity results for stationary electrorheological fluids. Arch. Ration. Mech. Anal. **164**, 213–259 (2002)
2. Afrouzi, G.A., Ghorbani, H.: Existence of positive solutions for p (x)-laplacian problems. Electron. J. Differ. Equ. (2007)
3. Azroul, E., Barbara, A., Abbassi, A.: Degenerate $p(x)$−elliptic equation with second member in L^1. ASTES J. **2**(5), 45–54 (2017)
4. Bonsall, F.F.: Lectures on Some Fixed Point Theorems of Functional Analysis. Tata Institute of Fundamental Research, Bombay, India (1962)
5. Brézis, H.: Analyse Fonctionnelle Théorie et Applications. Masson, Paris (1983)
6. Diening, L.: Theoretical and Numerical Results for Electrorheological Fluids. University of Freiberg, Germany (2002). PhD thesis
7. Edmunds, D.E., Lang, J., Nekvinda, A.: On $L^{p(x)}$ norms. Proc. R. Soc. Lond. Ser. A **455**, 219–225 (1999)

8. Fan, X.-L., Zhang, Q., Zhao, D.: Eigenvalues of p(x)-Laplacian Dirichlet problem. J. Math. Anal. Appl. **302**(2), 306–317 (2005)

9. Fan, X.L., Zhao, D.: On the generalized Orlicz-Sobolev space $W^{k,p(x)}(\Omega)$. J. Gansu Educ. Coll. **12**(1), 1–6 (1998)

10. Figueiredo, G.M., Moussaoui, A., Dos Santos, G.C.G. et al.: A sub-supersolution approach for some classes of nonlocal problems involving Orlicz spaces (2018). arXiv:1804.07977

11. Kováčik, O., Rákosnik, J.: On spaces $L^{p(x)}$ and $W^{k,p(x)}$. Czechoslovak Math. J. **41**(116), 592–618 (1991)

12. Liu, J., Zhang, Q., Zhao, C.: Existence of positive solutions for p(x) Laplacian equations with a singular nonlinear term. Electron. J. Differ. Equ. **2014**(155), 21 pp (2014)

13. Mihăilescu, M., Rădulescu, V., Stancu, D.: A Caffarelli-Kohn-Nirenberg type inequality with variable exponent and applications to PDEs. Complex Var. Elliptic Eqns. **56**, 659–669 (2011)

14. Miyagaki, O.H., Rodrigues, R.S.: On positive solutions for a class of singular quasilinear elliptic systems. J. Math. Anal. Appl. **334**, 818–833 (2007)

15. Orlicz, W.: Über konjugierte exponent enfolgen. Stud. Math. **3**, 200–212 (1931)

16. Ruzicka, M.: Electrorheological Fluids: Modeling and Mathematical Theory. Lecture Notes in Mathematics, vol. 1748. Springer, Berlin (2000)

17. Ruzicka, M.: Modeling, mathematical and numerical analysis of electrorheological fluids. Appl. Math. **49**(6), 565–609 (2004)

18. Santos, G.C.G., Figueiredo, G.M., Tavares, L.S.: Acta Appl. Math. (2017). https://doi.org/10. 1007/s10440-017-0126-1

19. Zhang, Q.: Existence and asymptotic behavior of positive solutions to p(x)-Laplacian equations with singular nonlinearities. J. Inequal. App. (2007)

Boundary Regional Controllability of Linear Boolean Cellular Automata Using Markov Chain

Sara Dridi, Samira El Yacoubi and Franco Bagnoli

Abstract The purpose of this paper is to study a special controllability problem in the case of actions exerted on the boundary of a target region that refers only on a sub-domain of the system. This problem has already been studied for distributed parameter systems described by partial differential equations. Our purpose is to consider this concept for more general spatially-distributed dynamical systems described by means of cellular automata models. In this paper we investigate the one-dimensional deterministic linear Cellular Automata by establishing a relation between controllability of CA and Markov Chains instead of using the Kalman condition.

Keywords Control theory · Regional controllability · Linear cellular automata · Markov chain

1 Introduction

Control theory is the area of application-oriented mathematics that is relevant to analysis and design of control of certain physical processes and systems. It deals with the problem of forcing or promoting a desired behaviour for a given dynamics. It studies the possibility of acting on a system in order to steer it from an initial state to a given final state in a finite number of steps. The objective can be also to stabilize a system in order to make it insensitive to certain disturbances or to

S. Dridi (✉) · S. E. Yacoubi
IMAGES, UMR Espace-Dev, University of Perpignan, Perpignan, France
e-mail: sara.dridi@univ-perp.fr

S. E. Yacoubi
e-mail: yacoubi@univ-perp.fr

F. Bagnoli
Department of Physics and Astronomy and CSDC, University of Florence, Florence, Italy
e-mail: franco.bagnoli@unifi.it

© Springer Nature Switzerland AG 2020
E. H. Zerrik et al. (eds.), *Recent Advances in Modeling, Analysis and Systems Control: Theoretical Aspects and Applications*, Studies in Systems, Decision and Control 243, https://doi.org/10.1007/978-3-030-26149-8_4

determine an optimal control when the exerted actions are not unique. The usual modelling tools were based on differential equations, partial or integro-differential equations. Although the control principles are always expressed in mathematical form, they are potentially applicable to any concrete situation and depend on basic scientific knowledge in the specific field of application; mechanical engineering, physics, astronomy, biology, medicine, social science, etc. [9, 14, 20].

Among the main considered issues in system analysis, controllability and observability were introduced by R. Kalman in the 1960 and very well studied afterward. The controllability is concerned with whether one can find or design a control in order to lead the system to a desired value or state. The observability deals with determining the state of a system without knowing the initial state [21]. The controllability has been studied for differential equations and partial differential equations. For linear finite dimensional systems, it is characterized by the well known Kalman rank condition which was derived by R.E. Kalman in the 60s [10].

While the general output controllability problem [12, 13, 15] consists in considering actions and objectives defined on the whole system's domain, in many practical applications it is more convenient and efficient to control the system only on a small region in order to reach a specific target. The so-called regional controllability concept was then introduced by El Jai et al. (1993, 1995) and mainly studied for linear distributed parameter systems [4, 22–24].

Due to the complexity of distributed parameter systems which are usually described by partial differential equations, the search of a new and more appropriate modelling approach for this system becomes necessary. In this context, cellular Automata are often considered as a good alternative to partial differential equations for these complex systems.

Cellular Automata are discrete dynamical systems whose behaviour is completely specified in terms of local interactions. They refer to an intrinsically parallel model of computation that consists of a regular lattice constituted by cells, where each cell communicates with its nearby cells in order to build a new state. In classical synchronous cellular automata the values of all sites evolve in discrete time steps according to deterministic or probabilistic transition rules that take into account the state of a specified neighborhood. First conceived by J. von Neumann [17] in the early 60's. Cellular Automata. They have been studied from a variety of aspects and are used in several fields including the simulation of ecology, chemistry, physics and biology.

CA are a good model for various problems ranged from physics to biology. Space is represented by a regular lattice where its elements called cells can be ordered as a chain in the one-dimensional (1-D) case or a grid in the 2-D case. The cell state can take values from a discrete state set $\{0, 1, \cdots, k\}$ which will be reduced to $\{0, 1\}$ in the Boolean case. The next state of a cell is assumed to change according to its own state as well as to the states of its neighbouring cells in the previous time. The cellular automaton evolution is governed by a set of deterministic/probabilistic local rules.

Some regional control problems have been studied using CA models, see for example [1, 3, 6, 19] and the references therein. In Ref. [5], a numerical approach based on genetic algorithms has been developed for a class of additive CA in both one and two dimensional cases. In Ref. [6], an interesting theoretical study has been carried out for 1-D additive real valued CA where the effect of control is given through an evolving neighbourhood. In [1], the problem of regional controllability of CA via boundary actions was addressed using the concept of Boolean derivatives for deterministic one-dimensional CA. In ref [2] the probabilistic CA has been studied.

The problem we want to address in this paper is that of regional controllability of one-dimensional deterministic CA via boundary actions on the target region, which is that of forcing the appearance of a given pattern inside a region ω by applying the controls to the sites in the boundary. We give an interesting characterization result of controllability, based on Markov chains approach instead of the well know Kalman condition. We restrict our study to the so-called elementary/Boolean cellular automata that can be wonderful examples of systems with simple rules and which may produce unusually complex behavior Ref. [18].

The paper is organized as follows. In Sect. 2 we introduce some preliminaries on cellular automata models and give some of its important properties. Section 3 is focused on the regional controllability of linear Boolean CA via boundary actions. We characterize the controllability of the CA using a new approach based on Markov Chain instead of the Kalman Condition. Finally, conclusions are drawn in the last section.

2 Cellular Automata Models and Its Properties

Definition 1 ([6]) A cellular automaton is defined by a quadruple $A = (\mathscr{L}, S, \mathscr{N}, f)$ where:

1. $\mathscr{L}$ is a cellular space which consists in a regular lattice of the domain Ω of $\mathbb{R}^d$, $d = 1, 2$ or 3.
2. S is the set of states which is a finite set of values representing all the cells states.
3. $\mathscr{N}$ is a function which characterizes the neighbourhood of a cell c and defined by :

$$\mathscr{N} : \mathscr{L} \to \mathscr{L}^r$$
$$c \to \mathscr{N}(c) = (c_{i_1}, c_{i_2}, \ldots, c_{i_r})$$

where c_{i_j} are the cells $j = 1, \ldots, r$, and r is size of the neighbourhood $\mathscr{N}(c)$.
4. f is a transition function which allows to calculate the state of a cell c at time $t + 1$ according to the states of its neighbouring cells at time t. It may be defined as follow:

$$f : S^r \to S$$
$$s_t(\mathscr{N}(c)) \to s_{t+1}(c) = f(s_t(\mathscr{N}(c)))$$

where $s_t(c)$ is the state of cell c at time t and

$$s_t(\mathcal{N}(c)) = \{s_t(c'), c' \in \mathcal{N}(c)\}$$

is the state of the neighbourhood.

Definition 2 A one-dimensional (Boolean or binary-state) cellular automaton is a collection of ones and zeros arranged on an one-dimensional strip. Each cell interacts with its immediate neighbors on its left and right.

Definition 3 We call the configuration of a cellular automaton at time t, the set

$$\{s_t(c), c \in \mathcal{L}\}$$

It is defined by the following mapping:

$$s_t : \mathcal{L} \to S$$
$$c \to s_t(c)$$

The global dynamics of a cellular automaton is given by the function:

$$F : S^{\mathcal{L}} \to S^{\mathcal{L}}$$
$$s_t \to s_{t+1}$$

F associates to the configuration at time t a new configuration at time $t + 1$.

2.1 Important Properties

Definition 4 ([5]) A global dynamics F is additive if for every pair of configurations $s_1, s_2 \in S^{\mathcal{L}}$,
$$F(s_1 + s_2) = F(s_1) + F(s_2)$$

This definition is equivalent to the local condition of additive CA's

$$f(s_t(\mathcal{N}(c))) = \sum_{1 \leq j \leq r} a_j s_t(c_{i_j})$$

for some scalar $a_0, a_1, \ldots, a_r$ in S called the weights or coefficients of cells in the neighbourhood $\mathcal{N}$.

Definition 5 ([8]) A global dynamics F is linear if for every pair of configuration $s_1, s_2 \in S^{\mathcal{L}}$

$$\forall \lambda \in S, F(\lambda s_1 + s_2) = \lambda F(s_1) + F(s_2)$$

For the Boolean linear CA, the previous definition holds by using the XOR $\oplus$ instead of using plus $(+)$ thus we have the following definition:

Definition 6 ([11]) If in a CA, the neighbourhood dependence is on EX-OR only, then the CA is called an additive CA. Specifically, a linear CA employs XOR rules only.

Example 1 (Wolfram's rules)
In the Boolean 1-D case with neighbourhood of radius 1, each of three consecutive cells can have two values, 0 or 1. Thus, the automaton is uniquely determined once its values on the eight basic elements $000, 001, \ldots, 111$ are known. So there can be at most $2^8 = 256$ such CAs and these are also known as the elementary cellular automata. To distinguish each automaton, Wolfram [18] used a numbering scheme based on the binary expansion of each number between 0 and 255.

In the case of rule 150, since $150 = 2^1 + 2^2 + 2^4 + 2^7$ we see that the so called rule number $150 = \sum_{i=0}^{7} d_i 2^i$ forces $d_i = 1$ only for $i = 1, 2, 4, 7$. Using the disjunctive normal form expression, [7], the local evolution f of the wolfram rule 150 is expressed:

$$s_i^{t+1} = f(s_{i-1}^t, s_i^t, s_{i+1}^t) = s_{i-1}^t \oplus s_i^t \oplus s_{i+1}^t$$

It can also be expressed as a table, mapping the next states from all possible combinations of inputs.

$s_t(c_{i-1})$	$s_t(c_i)$	$s_t(c_{i+1})$	$s_{t+1}(c_i)$
0	0	0	0
0	0	1	1
0	1	0	1
0	1	1	0
1	0	0	1
1	0	1	0
1	1	0	0
1	1	1	1

3 Boundary Regional Controllability of Linear Boolean Cellular Automata

3.1 Problem Statement

Let us consider a linear Boolean CA with $S = \{0, 1\}$ and $\mathcal{N} = (c_i) = (c_{i-1}, c_i, c_{i+1})$. We denote by $\omega = \{c_1, \ldots, c_n\}$ the region to be controlled, we want to prove the

Fig. 1 Regional control of one-dimensional CA

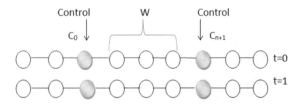

regional controllability i.e, to force the appearance of a desired state in ω by applying the control u on the boundaries of the target region i.e $\{c_0, c_{n+1}\}$. See Fig. 1

Definition 7 ([5]) The CA is said to be regionally controllable for $s_d \in S^\omega$ if there exists a control $u = (u_0, \ldots, u_{T-1})$ where $u_i = (u_i(c_0), u_i(c_{n+1}))$, $i = 0, \ldots, T - 1$ such as:

$$s_T = s_d \quad on \quad \omega$$

where s_T is the final configuration at time T and s_d is the desired configuration.

As for continuous systems, control problems that deal with a global behaviour study generally needs the use of derivatives. To approach this issue for cellular automata dynamics, one attempts to define a linear operator that acts as some sort of derivative that should be defined for Boolean functions. As a local CA rule is a function of several Boolean variables, a derivative as an instantaneous rate of change cannot work. A more appropriate notion is that of Boolean derivative that produces values 0 or 1 since a Boolean variable can only be varied by flipping its value. Before going on, let's give some definitions and related examples:

Definition 8 ([16]) The partial Boolean derivative of $f : S^r \longrightarrow S$ is defined as:

$$\frac{\partial f_i}{\partial s_j} \stackrel{not}{=} \frac{\partial f}{\partial s_j} = f(s_i, \ldots, s_j \oplus 1, \ldots, s_{i+r}) \oplus f(s_i, \ldots, s_j, \ldots, s_{i+r})$$

Where $\oplus$ is the XOR Boolean operation and $s_j = s(c_j), j = i, \cdots, i + r$ represent the values associated to the neighbourhood cells states.

For instance, the Boolean derivative with respect to s_1, for a local rule defined by $f(s_1, s_2) = s_1 \oplus s_2$ is:

$$\frac{\partial f}{\partial s_1} = s_1 \oplus s_2 \oplus (s_1 \oplus 1) \oplus s_2 = 1$$

Analogously, for $f(s_1, s_2) = s_1 s_2$, the Boolean derivative with respect to s_1 is:

$$\frac{\partial f}{\partial s_1} = s_1 s_2 \oplus (s_1 \oplus 1)s_2 = s_1 s_2 \oplus s_1 s_2 \oplus s_2 = s_2$$

Definition 9 (*Boolean Derivative*) ([16]) The Boolean derivative of F is the Jacobian matrix defined as:

$$J = \frac{\partial f_i}{\partial s_j}|_{i,j=1,...,N}$$

Where N is the lattice number of cells.

Example 1 Let $S = \{0, 1\}$ and $\mathcal{N}(c) = (c_{i-1}, c_i, c_{i+1})$. Let us consider the rule 150 defined by:

$$f_i \stackrel{not}{=} s_i^{t+1} = f(s_{i-1}^t, s_i^t, s_{i+1}^t) = s_{i-1}^t \oplus s_i^t \oplus s_{i+1}^t$$

The partial derivative of this rule with respect to s_{i-1} is

$$\frac{\partial f_i}{\partial s_{i-1}} = \frac{\partial s_i^{t+1}}{\partial s_{i-1}} = (s_{i-1} \oplus 1 \oplus s_i \oplus s_{i+1}) \oplus (s_{i-1} \oplus s_i \oplus s_{i+1})$$

and then:

$$\frac{\partial s_i^{t+1}}{\partial s_{i-1}} = 1$$

This Boolean derivative measures whether s^{t+1} changes when changing s_{i-1}.

For a CA with zero boundary conditions, the Jacobian matrix takes the form:

$$J = \begin{pmatrix} 1 & 1 & ... & ... & ... & 0 & 0 & 0 \\ 1 & 1 & 1 & ... & ... & 0 & 0 & 0 \\ 0 & 1 & 1 & 1 & ... & 0 & 0 & 0 \\ ... & ... & 1 & 1 & 1 & .. & ... & \\ ... & ... & ... & 1 & 1 & 1 & ... & ... \\ ... & ... & ... & ... & 1 & 1 & 1 & ... \\ 0 & 0 & 0 & ... & ... & 1 & 1 & 1 \\ 0 & 0 & 0 & 0 & 0 & 0 & 1 & 1 \end{pmatrix} \tag{1}$$

According to the Jacobian, the evolution of linear CA will be expressed in the form:

$$s^{t+1} = Js^t \tag{2}$$

where s^t is the CA global state or configuration at time t.

The controllability was identified by R. Kalman for finite dimensional linear systems and characterized by the Kalman condition [10]. In order to prove the controllability of linear CA, we propose an alternative method by exploring the approach based on Markov Chains that could be applicable also for nonlinear rules.

The Kalman rank condition states that if the controllability matrix defined in [10] is of full rank then the system is controllable.

This criterion has been already extended to regional controllability in [5, 19]. The Markov Chains approach allows not only to prove the regional controllability of linear CA systems but also to find the required time T for which the system is regionally controllable by using the definition of regular Markov chain. One can find also the control required on the boundaries $\{c_0, c_{n+1}\}$ in the time interval $[0, T-1]$ to reach the desired configuration s_d on ω given by $\{s_d(c_1), s_d(c_2), \ldots, s_d(c_n)\}$. Starting from an initial configuration $\{s_0(c_1), s_0(c_2), \ldots, s_0(c_n)\}$ the CA configuration will evolve to a final state at time T according to the application of a transition matrix defined in the following paragraph.

Remark 1 Note that we have taken the binary representation for the controlled region in a reverse order (the least significant bit is the first one). For instance, for a controlled region of size 3, we note: $(100)_2 = 1$, $(110)_2 = 3$ and $(001)_2 = 4$.

The evolution of CA can be seen as a Markov chain $P_{(s'|s)}$, where $s' = F(s)$ and P is the transition matrix of Markov Chain. Each element of this matrix is the probability to move from a configuration s to a configuration s'.

A controlled CA can be represented by a transition matrix P where each element $p_{(s,s')}$ represents the probability of moving from a configuration s to a configuration s' by applying the pair of control (a, b) on the target region boundaries.

For a region ω of size n, we need to define a square matrix $2^n \times 2^n$ C, where 2^n refers to all the possible configurations that can be represented in the region ω. For instance, for the CA rule 150, the controlled region is of size $|\omega| = 2$.

For the configuration $(00)_2 = 0$ $((00)_2$ is the binary conversion of $0)$, we have four possibilities of applying the control $(a, b) = \{(0, 0), (0, 1), (1, 0), (1, 1)\}$. We use the $s^{t+1} = J s^t$ to calculate the next configuration s'. Thus we have:

$$F_{150}(\mathbf{0000}) = (00)_2 = 0,$$
$$F_{150}(\mathbf{0001}) = (01)_2 = 2,$$
$$F_{150}(\mathbf{1000}) = (10)_2 = 1,$$
$$F_{150}(\mathbf{1001}) = (11)_2 = 3.$$

Starting from the configuration $(00)_2 = 0$ in the controlled region by applying the control on the boundaries, we can reach the configurations $0, 1, 2, 3$. We get then $C_{0,1} = 1, C_{0,0} = 1, C_{0,2} = 1, C_{0,3} = 1$ and fill the matrix C by the same method for the three configurations remaining $\{(01)_2 = 1, (10)_2 = 2, (11)_2 = 3\}$. The obtained matrix is:

$$C_{150} = \begin{array}{c} \\ 0 \\ 1 \\ 2 \\ 3 \end{array} \begin{array}{cccc} 0 & 1 & 2 & 3 \\ \left(\begin{array}{cccc} 1 & 1 & 1 & 1 \\ 1 & 1 & 1 & 1 \\ 1 & 1 & 1 & 1 \\ 1 & 1 & 1 & 1 \end{array} \right) \end{array}$$

to which one can associate a graph (Fig. 2).

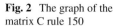 **Fig. 2** The graph of the
matrix C rule 150

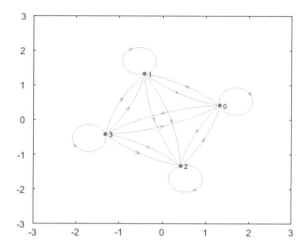

Remark 2 The matrix C is an adjacency matrix in terms of graph theory. It doesn't satisfy the property of the transition matrix of Markov chain which states that the sum of each line of the transition matrix P equals one. To obtain the matrix satisfying this propriety one can normalize the matrix C in the following way:

$$
C_{150}^{norm} =
\begin{matrix}
 & \begin{matrix} 0 & 1 & 2 & 3 \end{matrix} \\
\begin{matrix} 0 \\ 1 \\ 2 \\ 3 \end{matrix} &
\begin{pmatrix}
1/4 & 1/4 & 1/4 & 1/4 \\
1/4 & 1/4 & 1/4 & 1/4 \\
1/4 & 1/4 & 1/4 & 1/4 \\
1/4 & 1/4 & 1/4 & 1/4
\end{pmatrix}
\end{matrix}
$$

However and for more simplicity, we choose to work directly with the non normalized matrix C.

A configuration j is reachable from a configuration i, if there exists a probability strictly positive to reach a configuration j from a configuration i and we denote it $i \rightsquigarrow j$; in terms of graph theory, if there exists an arrow between i and j. Sometimes, it is impossible to reach a configuration j in one move, but in m steps. Thus we use the definition of regular Markov chains. We say that a Markov chain is regular if there exists a power $m > 0$ such as all the components of P^m are strictly positive.

Theorem 1 *A linear Cellular Automaton is regional controllable iff it exists a power C^t such as all the components are strictly positive.*

Proof The matrix C is similar to the transition matrix of Markov Chains, C^t is the matrix at time t.

For a pair of states (k_1, k_2), the state k_2 is reachable from k_1, if there exist an integer $t > 0$ such that $C_{k_1 k_2}^t > 0$. We assume that $C_{i,j}^t > 0$ for all $i, j \in \{0, \ldots, 2^{|\omega|} - 1\}$. Then all configurations in the controlled region are reachable at time t and the CA related to this transition matrix is regional controllable.

Fig. 3 Graph of the matrix
C of rule 90 at time T=1

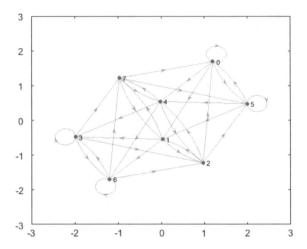

Let assume that the CA is regional controllable. Then it exists $t > 0$ such as for
each $i, j \in \{0, \ldots, 2^{|\omega|} - 1\}$ j is reachable from i in time t which implies that for
each $i, j \in \{0, \ldots, 2^{|\omega|} - 1\}$ $C_{i,j}^t > 0$. Thus the theorem holds.

Example 2 Consider a linear CA governed by Wolfram rule 90 [18]. Its local evolution rule is given by:

$$f_i = s_i^{t+1} = f(s_{i-1}^t, s_i^t, s_{i+1}^t) = s_{i-1}^t \oplus s_{i+1}^t$$

The controlled region of this CA is of $size = 3$. For $T = 1$, the CA is not regionally
controllable; there is no arrow from the vertex 4 to 1, which means that configuration
$1 = (100)_2$ is not reachable from the configuration $4 = (001)_2$. For final time $T =
2$, the matrix $C^2 > 0$ i.e, all configurations are reachable. The CA is regionally
controllable by acting on the boundaries of the target region at time $T = 2$ (Figs. 3
and 4).

One can also find the required controls on the boundaries $\{c_0, c_{n+1}\}$. For instance,
starting from the configuration $(011)_2 = 6$ one can achieve the desired configuration
$(001)_2 = 4$ at time $T = 2$. The sequence of required controls is $\{1, 1\}$ at time $T = 0$
and $\{1, 0\}$ at time $T = 1$.

Remark 3 The required time T to reach the regional controllability of linear CA can
also be obtained by investigating the biggest cost route between the vertices.

To check the controllability in minimal time, we can use this definition.

Definition 10 Markov chain is called an ergodic chain if it is possible to go from
any state to any other one (not necessarily in one step).

To complete this concern, we just give the following result without proof:

Theorem 2 *A linear CA is regionally controllable if it is ergodic.*

Fig. 4 Graph of the matrix C^2 of rule 90 at time T=2

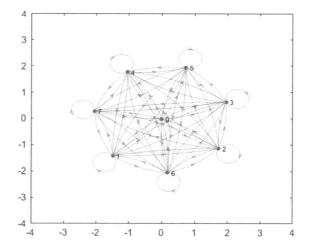

4   Conclusion

In this paper, we have studied the problem of regional controllability of Cellular automata models. We have proposed a new approach to characterize the controllability of linear Boolean CA, based on Markov chain instead of the Kalman condition. It should be stressed that linear CA rules play a central role due to the number of remaining systems described by linear CA that are still not entirely understood, not even in the elementary 1D case. Furthermore, the proposed approach can be easily applied to non linear CA and in two-dimensional case. This work is currently under study.

Acknowledgements   Sara DRIDI would like to thank the Algerian government for the funding and the international Network of Systems Theory for organizing and supporting her participation to the international conference on Modelling, Analysis and Control (Macs8), held in Meknes—Morocco, October, 26–27.

References

1. Bagnoli, F., El Yacoubi, S., Rechtman, R.: Toward a boundary regional control problem for boolean cellular automata. Nat. Comput. **99**(2–3), 235–394 (2017)
2. Bagnoli, F., Dridi, S., El Yacoubi, S., Rechtman, R.: Regional control of probabilistic cellular automata. In: International Conference on Cellular Automata for Research and Industry. Lecture Notes on Computer Sciences, vol. 11115, pp. 243–254. Springer (2018)
3. Bel Fekih, A., El Jai, A.: Regional analysis of a class of cellular automata models. In: El Yacoubi, S., Chopard, B., Bandini, S. (eds.) Cellular Automata, ACRI 2006, Proceedings of the 7th International Conference on Cellular Automata for Research and Industry. Lecture Notes in Computer Science, Springer Proceedings in Mathematics, vol. 1, pp. 623–631. Springer (2006)

4. El Jai, A., Simon, M.C., Zerrik, E., Pritchard, A.J.: Regional controllability of distributed parameter systems. Int. J. Control **62**(6), 1351–1365 (1995)

5. El Yacoubi, S., El Jai, A., Ammor, N.: Regional controllability with cellular automata models. Lect. Notes Comput. Sci. **14**(3), 227–253 (2002)

6. El Yacoubi, S., Jacewicz, P., Ammor, N.: Analysis and control of cellular automata. Ann. Univ. Craiova-Math. Comput. Sci. Ser. **231**(7), 2851–2864 (2003)

7. El Yacoubi, S., Mingarelli, A.B.: Controlling the dynamics of fca rule 90 in [0, 1]t. J. Cellular Autom. **6** (2011)

8. Garzon, M.: Models of Massive Parallelism: Analysis of Cellular Automata and Neural Networks. Springer Science & Business Media, Boca Raton, Florida (2012). ISBN 0-8493-7375-1

9. Intriligtor, M.D.: The applications of control theory to economics. Anal. Optim. Syst. **209**(1), 105–138 (1980)

10. Kalman, R.E.: On the general theory of control systems. Proc. 1st World Congr. Int. Fed. Autom. Control **188**(2), 462–471 (1960)

11. Khan, R.A., Dihidar, K. et al.: VLSI architecture of a cellular automata machine (1997)

12. Lions, J.: Exact controllability of distributed systems. CRAS, Srie **77**(11), 694–706 (1986). Fld.3996

13. Lions, J.-L.: Exact controllability for distributed systems. Some trends and some problems. Applied and Industrial Mathematics, pp 59–84. Springer (1991)

14. PALM, William, J.: An application of control theory to population dynamics. Differ. Games and Control Theory **21**(2), 526–558 (1975)

15. Russell, D.: Controllability and stabilizability theory for linear partial differential equations recent progress and open questions. SIAM Rev. **160**(1), 318–335 (1978)

16. Vichniac, G.Y.: Boolean Derivatives on Cellular Automata. MIT Press (1991)

17. Von Neumann, J., Burks, A.W. et al.: Theory of self-reproducing automata. IEEE Trans. Neural Netw. **5**(1), 3–14 (1966)

18. Wolfram, S.: Phys. D **10**, 1 (1984)

19. el Yacoubi, S.: A mathematical method for control problems on cellular automata models. Int. J. Syst. Sci. **39**(5), 529–538 (2008)

20. Yuan, J.-C.: The application of control theory in mechanical engineering. **230**(22), 7957–7963

21. Zerrik, E., Bourray, H., El Jai, A.: Regional observability for semilinear distributed parabolic systems. J. Dyn. Control Syst. (2004)

22. Zerrik, E., Boutoulout, A., El Jai, A.: Actuators and regional boundary controllability for parabolic systems. Int. J. Syst. Sci. **184**(1), 266–298 (2000)

23. Zerrik, E., Ouzahra, M., Ztot, K.: Regional stabilization for infinite bilinear systems. IEEE: Control Theory Appl. **15**, 263–268 (2004)

24. Ztot, K., Zerrik, E., Bourray, H.: Regional control problem for distributed bilinear systems: approach and simulations. Int J Appl Math Comput Sci **142**(2), 521–561 (2011)

Fuzzy Solutions for Impulsive Evolution Equations

Abdelati El Allaoui, Said Melliani and Lalla Saadia Chadli

Abstract This paper investigates the existence to the impulsive fuzzy evolution equation via fixed point theorem for absolute retract, take into consideration that E^n can be embedded isomorphically as a cone in a Banach space.

1 Introduction

Fuzzy differential equations are one of the most thoroughly studied classes of differential equations (see [1–3]). They occur in several real world problems. The proposed union of impulsive differential equations and fuzzy differential equations would be of immense value. For example, the interest rate models in bond pricing, where the interest rates are unpredictable and vague could be modeled by means of impulsive fuzzy differential equations. In [5, 14, 19, 24, 25] some results on this topic have been reported.

Melliani and al. in [18] discussed the fuzzy evolution equation with nonlocal condition. They proved the existence result using Banach fixed point theorem.

Motivated by the above-mentioned discussions, the prupose of this paper is to study the following impulsive fuzzy evolution equation:

$$
\begin{cases}
u'(t) = Au(t) + f(t, u(t)), \ t \in J = [0, a], \ t \neq t_k, \ k = 1, \cdots, p, \\
u(t_k^+) = I_k \left(u \left(t_k^- \right) \right), \ k = 1, \cdots, p, \\
u(0) = u_0 \in E^n,
\end{cases}
\tag{1}
$$

A. El Allaoui (✉) · S. Melliani · L. S. Chadli
LMACS, Laboratoire de Mathématiques Appliquées and Calcul Scientifique,
Sultan Moulay Slimane University, BP 523, 23000 Beni Mellal, Morocco
e-mail: elallaoui199@gmail.com

S. Melliani
e-mail: said.melliani@gmail.com

L. S. Chadli
e-mail: chadli@fstbm.ac.ma

© Springer Nature Switzerland AG 2020
E. H. Zerrik et al. (eds.), *Recent Advances in Modeling, Analysis and Systems Control: Theoretical Aspects and Applications*, Studies in Systems, Decision and Control 243,
https://doi.org/10.1007/978-3-030-26149-8_5

where $a > 0, 0 = t_0 < t_1 < \cdots < t_p < t_{p+1} = a$, $J_k = [t_k, t_{k+1}]$, $k = 0, \cdots, p$.

$$I_k : E^n \to E^n \text{ continuous for } k = 1, \cdots, p$$

$$f : J \times E^n \to E^n \text{ continuous in } (J \setminus \{t_1, \cdots, t_p\}) \times E^n$$

is such that there exist the limits

$$\lim_{t \to t_k} f(t, u) = f(t_k, u), \; \lim_{t \to t_k^+} f(t, u) \text{ for } k = 1, \ldots, p \text{ and } u \in E^n,$$

and A is the generator of a strongly continuous fuzzy semigroup.

The rest of the paper is organized as follows: In Sect. 1, we give associated notations and preliminaries. In Sect. 2 we talk about fuzzy semigroups, a fixed point theorem for absolute retract is used to investigate the existence of fuzzy solutions for problem (1). At last we give conclusion.

2 Preliminaries

In this section, we introduce some basic concepts. Let $\mathscr{P}_k(\mathbb{R}^n)$ denote the family of all nonempty compact, convex subsets of $\mathbb{R}^n$.

Denote by

$$E^n = \Big\{u : \mathbb{R}^n \to [0, 1] \text{ such that satisfies (i)-(iv) mentioned below}\Big\},$$

(i) u is normal: there exists $t_0 \in \mathbb{R}$ with $u(t_0) = 1$;
(ii) u is fuzzy convex: for all $t, s \in \mathbb{R}$ and $\lambda \in [0, 1]$,

$$u(\lambda t + (1 - \lambda)s) \geq \min\{u(t), u(s)\};$$

(iii) u is upper-semicontinuous;
(iv) $[u]^0 = cl(\{u \in \mathbb{R} : u(t) > 0\})$ is a compact set.

The level sets of $u \in E^n$,

$$[u]^\alpha = \{t \in \mathbb{R} : u(t) \geq \alpha\}, \; \alpha \in (0, 1],$$

and $[u]^0$ are nonempty compact convex sets in $\mathbb{R}^n$ (see [6]).

Let A and B be two nonempty bounded subsets of $\mathbb{R}^n$. The distance between A and B is defined by the Hausdorff metric

$$d_H(A, B) = \max\left\{\sup_{a \in A} \inf_{b \in B} \|a - b\|, \sup_{b \in B} \inf_{a \in A} \|a - b\|\right\},$$

where $\|\ \|$ denotes the usual Euclidean norm in $\mathbb{R}^n$. Then $(\mathscr{P}_k(\mathbb{R}^n), d_H)$ is a complete and separable metric space [21].

The space E^n is a complete metric space with the metric d_∞ given by

$$d_\infty(u, v) = \sup\{d_H([u]^\alpha, [v]^\alpha) : \alpha \in [0, 1]\}, \quad \text{for } u, v \in E^n.$$

Also for all $u, v, w \in E^n$ and $\lambda \in \mathbb{R}$ we have

$$d_\infty(u + w, v + w) = d_\infty(u, v) \text{ and } d_\infty(\lambda u, \lambda v) = |\lambda| d_\infty(u, v).$$

The addition and the multiplication by an scalar is defined on E^n by using the Zadeh's Extension Principle, in such a way that these operations are reduced level-wise to interval operations

$$[u + v]^\alpha = [u]^\alpha + [v]^\alpha, \quad [cu]^\alpha = c[u]^\alpha, \quad u, v \in E^n, \ c \in \mathbb{R} \setminus \{0\}, \text{ for each } \alpha \in [0, 1].$$

We define $\widehat{0} \in E^n$ as

$$\widehat{0}(x) = \begin{cases} 1 \ if \ x = 0 \\ 0 \ if \ x \neq 0. \end{cases}$$

it is well-known that (E^n, d_∞) can embedded isometrically as a cone in a Banach space X, i.e. there exists an embedding $j : E^n \to X$ (see [13, 22, 23]) defined by

$$j(u) = <u, \widehat{0}>, \quad \text{where } u \in E^n,$$

here $< ., . >$ is defined in [21]. And

$$\| <u, v> \|_X = d_\infty(u, v) \text{ for } u, v \in E^n,$$

so in particular

$$\|ju\|_X = d_\infty(u, \widehat{0}) \text{ for } u \in E^n.$$

We denote by $\mathscr{C}(J, E^n)$ the space of continuous fuzzy valued functions from J into E^n.

The supremum metric H on $\mathscr{C}(J, E^n)$ is defined by

$$H(u, v) = \sup_{t \in J} d_\infty(u(t), v(t)).$$

$(\mathscr{C}(J, E^n), H)$ is a complete metric space.

Now since $j : E^n \to C \subset X$ we can define a map $\tilde{j} : \mathscr{C}(J, E^n) \to \mathscr{C}(J, X)$ by

$$(\tilde{j}u)(t) = j(u(t)) = ju(t) \text{ for } t \in J.$$

By definition of j we have,

$$\|\tilde{j}u - \tilde{j}v\|_{\mathscr{C}(J,X)} = \sup_{t \in I} \|(\tilde{j}u)(t) - (\tilde{j}v)(t)\|_X$$

$$= \sup_{t \in J} \|ju(t) - jv(t)\|_X$$

$$= \sup_{t \in J} d_\infty (u(t), v(t))$$

$$= H(u, v), \text{ for } u, v \in \mathscr{C}(J, E^n).$$

Also, we can check that

$$\tilde{j} : \mathscr{C}(J, E^n) \to \tilde{j}\left(\mathscr{C}(J, E^n)\right) \subset X,$$

is a homeomorphism. Indeed
   Let $u_n, u \in \mathscr{C}(J, E^n), n \in \mathbb{N}$ with

$$u_n \to u \text{ as } n \to +\infty.$$

Then

$$\|\tilde{j}u_n, \tilde{j}u\|_{\mathscr{C}(J,X)} = \sup_{t \in J} \|ju_n(t)(t) - ju(t)\|_X$$

$$= \sup_{t \in J} d_\infty(u_n(t), u(t))$$

$$= H(u_n, u) \to 0 \text{ as } n \to +\infty.$$

So $\tilde{j}$ is continuous.
   Let $v_n, v \in \tilde{j}(\mathscr{C}(J, E^n))$ with

$$\|v_n - v\|_{\mathscr{C}(J,X)} \to 0 \text{ as } n \to +\infty.$$

Then there exists $u_n, u \in \mathscr{C}(J, E^n)$ with

$$\tilde{j}u_n = v_n \text{ and } \tilde{j}u = v.$$

We obtain

$$H\left(\tilde{j}^{-1}v_n, \tilde{j}^{-1}v\right) = \sup_{t \in J} d_\infty \left(\tilde{j}^{-1}v_n(t), \tilde{j}^{-1}v(t)\right)$$

$$= \sup_{t \in J} d_\infty (u_n(t), u(t))$$

$$= \sup_{t \in J} \|ju_n(t) - ju(t)\|_X$$

$$= \|\tilde{j} u_n - \tilde{j} u(t)\|_{\mathscr{C}(J,X)}$$

$$= \|v_n - v\|_{\mathscr{C}(J,X)}.$$

Thus $\tilde{j}^{-1}$ is continuous.

We recall some measurability, integrability properties for fuzzy set-valued mapping.

Definition 1 A map $f : J \to E^n$ is strongly measurable if, for all $\alpha \in [0, 1]$, the multi-valued map $f_\alpha : J \to \mathscr{P}_k(\mathbb{R}^n)$ defined by

$$f_\alpha(t) = [f(t)]^\alpha$$

is Lebesgue measurable, when $\mathscr{P}_k(\mathbb{R}^n)$ is endowed with the topology generated by the Hausdorff metric d_H.

Definition 2 A map $f : J \to E^n$ is called levelwise continuous at $t = t_0 \in J$ if the multi-valued map $f_\alpha(t) = [f(t)]^\alpha$ is continuous at $t = t_0$ with respect to the Hausdorff metric d_H for all $\alpha \in [0, 1]$.

A map $f : J \to E^n$ is called integrably bounded if there exists an integrable function h such that $\|x\| \le h(t)$ for all $x \in f_0(t)$.

Definition 3 Let $f : J \to E^n$. The integrale of f over J, denoted $\int_J f(t)dt$ is defined by the equation

$$\left[\int_J f(t)dt \right]^\alpha = \int_J f_\alpha(t)dt = \left\{ \int_J u(t)dt : u : J \to \mathbb{R}^n \text{ is measurable selection for } f_\alpha \right\},$$

for all $\alpha \in (0, 1]$.

A strongly measurable and integrably bounded map $f : J \to E^n$ is said to be integrable over J, if $\int_J f(t)dt \in E^n$.

If $f : J \to E^n$ is measurable and integrably bounded, then f is integrable (see [4]).

We consider differentiability of fuzzy-valued functions in the sense of Hukuhara, by using the following difference.

Definition 4 Given $u, v \in E^n$, if there exists $w \in E^n$ such that $u = v + w$, we call $w = u - v$ the Hukuhara difference of u and v.

Hence we recall the following concept.

Theorem 1 *Given $f : J \to E^n$, we say that f is Hukuhara differentiable at $t_0 \in J$ if there exists a $f'(t_0) \in E^n$ such that the limits*

$$\lim_{h \to 0^+} \frac{f(t_0 + h) - f(t_0)}{h} \quad and \quad \lim_{h \to 0^+} \frac{f(t_0) - f(t_0 - h)}{h}$$

exist and are equal to $f'(t_0)$. Here the limit is taken in the metric space (E^n, d_∞). At the end points of J, we consider only the one-side derivatives.

If $f : J \to E^n$ is differentiable at $t_0 \in J$, then we say that $f'(t_0)$ is the fuzzy derivative of $f(t)$ at the point t_0.

A detailed study of fuzzy measurability and fuzzy continuity we refer to [12].

Definition 5 A map $f : J \times E^n \to E^n$ is called levelwise continuous at point $(t_0, x_0) \in J \times E^n$ provided, for any fixed $\alpha \in [0, 1]$ and arbitrary $\epsilon > 0$, there exists a $\delta_\epsilon^\alpha > 0$ such that

$$d_H \left([f(t, x)]^\alpha, [f(t_0, x_0)]^\alpha \right) < \epsilon$$

whenever $|t - t_0| < \delta_\epsilon^\alpha$ and $d_H \left([x]^\alpha, [x_0]^\alpha \right) < \delta_\epsilon^\alpha$, for all $t \in J, x \in E^n$.

Definition 6 A space X is called an absolute retract (or simply an $\mathscr{A}\mathscr{R}$) whenever

(i) X is metrizable,
(ii) for any metrizable space U and any embedding $f : X \to U$ the set $f(X)$ is a retract of U.

The class of absolute retracts is denoted by $\mathscr{A}\mathscr{R}$.

It is evident that if X is an $\mathscr{A}\mathscr{R}$, then every space homeomorphic to X is also an $\mathscr{A}\mathscr{R}$. We call the following fixed point result.

Theorem 2 Let X an absolute retract and $\mathscr{O} : X \to X$ a continuous and completly continuous map. Then $\mathscr{O}$ has a fixed point.

For its proof, we refer to [5, 10].

3 Main Results

3.1 Fuzzy Semigroups

First of all, let as establish some notion which are used throughout afterward.

Let $T(t) : E^n \to E^n, t \geq 0$ be a family of mappings satisfies the following properties:

1. $T(0) = id_{E^n}$, the identity mapping on E^n.
2. $T(t + s) = T(t)T(s)$ for all $t, s \geq 0$.
3. The function $g : [0, +\infty[\to E^n$, defined by $g(t) = T(t)x$ is continuous at $t = 0$ for all $x \in E^n$

$$\lim_{t \to 0^+} T(t) = x.$$

4. There exist two constants $M > 0$ and $\omega \in \mathbb{R}$ such that

$$d_\infty (T(t)u, T(t)v) \le M e^{\omega t} d_\infty(u, v), \quad \text{for } t \ge 0, \ u, v \in E^n.$$

We define the family of operators $S(t)$ on the Banach space X by

$$S(t) = jT(t)j^{-1}.$$

Below we verify that $\{S(t), t \ge 0\}$ satisfies the properties of a semigroup in X. Indeed

1. For $u \in X$, we have $j^{-1}u \in E^n$ and

$$S(0)u = jT(0)j^{-1}u = jj^{-1}u = u.$$

2. For $u \in X$, $t, s \ge 0$, we have

$$S(t+s)u = jT(t+s)j^{-1}u = jT(t)j^{-1}jT(s)j^{-1}u = S(t)S(s)u.$$

3. For $u \in X$, $t \ge 0$, we have

$$\|S(t)u - u\|_X = \|jT(t)j^{-1}u - jj^{-1}u\|_X = d_\infty\left(T(t)j^{-1}u, j^{-1}u\right).$$

So $t \to S(t)u$ is continuous for $u \in X$
4. For $u, v \in X$, $t \ge 0$, we have

$$\begin{aligned}
\|S(t)u - S(t)v\|_X &= \|jT(t)j^{-1}u - jT(t)j^{-1}v\|_X \\
&= d_\infty\left(T(t)j^{-1}u, T(t)j^{-1}v\right) \\
&\le M e^{wt} d_\infty\left(j^{-1}u, j^{-1}v\right) \\
&= M e^{wt} \|u - v\|_X
\end{aligned}$$

Thus, $\{S(t), t \ge 0\}$ is a semigroup on X. Then we will call $\{T(t), t \ge 0\}$ is a fuzzy strongly continuous semigroup.
    In particular if $M = 1$ and $w = 0$, we say that $\{T(t), t \ge 0\}$ is a contraction fuzzy semigroup.

Definition 7 Let $\{T(t), t \ge 0\}$ be a fuzzy strongly continuous semigroup and $u \in E^n$. If for h sufficiently small, the Hukuhara difference $T(h)u - u$ exists, we define the operator

$$Au = \lim_{h \to 0^+} \frac{T(h)u - u}{h},$$

whenever this limit exists in the metric space (E^n, d_∞).

The operator A defined on

$$D(A) = \left\{ u \in E^n \ : \ \lim_{h \to 0^+} \frac{T(h)u - u}{h} \ \text{exists} \right\}$$

is called the infinitesimal generator of the fuzzy semigroup $\{T(t), t \geq 0\}$.

Remark 1 Let A be the generator of a fuzzy semigroup $\{T(t), t \geq 0\}$. Its easy to check that $B = jAj^{-1}$ is the infinitesimal generator of the semigroup $\{S(t) = jT(t)j^{-1}, t \geq 0\}$ on X.

Lemma 1 *Let A be the generator of a fuzzy semigroup $\{T(t), t \geq 0\}$, then for all $u \in E^n$ such that $T(t)u \in D(A)$ for all $t \geq 0$, the mapping $t \to T(t)u$ is differentiable and*

$$\frac{d}{dt}(T(t)u) = AT(t)u, \ \text{for } t \geq 0.$$

Example 1 We define on E^n the family of operator $\{T(t), t \geq 0\}$ by

$$T(t)x = e^{kt}x, \ \ k \in \mathbb{R}.$$

For $k \geq 0$, $\{T(t), t \geq 0\}$ is a fuzzy strongly continuous semigroup on E^n, and the linear operator A defined by $Ax = kx$ is the infinitesimal generator of this fuzzy semigroup.

For the concepts of fuzzy semigroups we refer to [7–9, 11, 17, 20].

3.2 Impulsive Fuzzy Evolution Equations

In this section we are concerned with the existence of fuzzy solutions for problem (2).

For $u : J \to E^n$, $u \in \mathscr{C}(J, E^n)$, the function $u_k : J_k \to E^n$, $k = 1, \cdots, p+1$, defined by

$$u_k(t) = u(t), \ \text{for } t \in (t_{k-1}, t_k] \text{ and } u_k(t_{k-1}) = u(t_{k-1}^+),$$

is such that $u_k \in \mathscr{C}(J_k, E^n)$.

To define a solution for this problem, we consider the following space:

$$\mathscr{PC} = \left\{ u : J \to E^n : u \text{ is continuous in } J \setminus \{t_1, \cdots, t_p\} \text{ and left continuous at} \right.$$
$$\left. t = t_k \text{ and the right limit } u(t_k) \text{ exists for } k = 1, \cdots, p \right\}.$$

$\mathscr{PC}^1 = \{u \in \mathscr{PC} : u'$ exists is continuous in $J \setminus \{t_1, \cdots, t_p\}$ and left continuous at $t = t_k$ and the right limit $u'(t_k)$ exists for $k = 1, \cdots, p\}$.

which are complete metric spaces considering, respectively, the distances

$$H(u, v) = \sup \{d_\infty(u(t), v(t)) : t \in J\} \text{ and } \rho(u, v) = H(u, v) + H(u', v').$$

We give the following definition motivated by the relation between a fuzzy semigroup and the semigroup definde by the embedding j on the Banach space X.

Definition 8 A solution to (2) is a function $u \in \mathscr{PC}$ which is differentiable on each (t_{k-1}, t_k), $k = 1, \cdots, p + 1$, $u(t) \in D(A)$ and satisfies the integral equation

$$u(t) = T(t - t_k)I_k\left(u(t_k^-)\right) + \int_{t_k}^t T(t - s)f(s, u(s))ds, \ t \in J_k, \ k = 0, \cdots, p$$

with $I_0 = u_0$.

By the assumptions on f, a solution to (2) belongs to $\mathscr{PC}^1$.

To establish our results, we introduce the following assumptions:
$(\mathbf{H}_1)$ There exists a continuous nondecreasing function $\psi : [0, +\infty[\rightarrow]0, +\infty[$ and a continuous function $\varphi : J \rightarrow [0, +\infty[$ such that

$$d_\infty\left(f(t, u), \widehat{0}\right) \le \varphi(t)\psi\left(d_\infty(u, \widehat{0})\right), \text{ for } t \in J \text{ and each } u \in E^n,$$

with

$$Me^{\omega a} \int_{t_k}^{t_{k+1}} \varphi(s)ds < \int_{Me^{\omega a}d_\infty(I_k(u(t_k^-)),\widehat{0})}^a \frac{d\tau}{\psi(\tau)}, \ k = 0, \cdots, p. \tag{2}$$

$(\mathbf{H}_2)$ For each $t \in J_k$, $k = 0, \cdots, p$ the set

$$\left\{T(t - t_k)I_k\left(u(t_k^-)\right) + \int_{t_k}^t T(t - s)f(s, u(s))ds : u \in \mathscr{F}_k\right\},$$

is a totally bounded subset of E^n, where

$$\mathscr{F}_k = \left\{u \in \mathscr{C}(J_k, E^n) : d_\infty(u(t), \widehat{0}) \le a_k(t), \ t \in J_k\right\},$$

with

$$a_k(t) = \Phi_k^{-1}\left(Me^{\omega a} \int_{t_k}^t \varphi(s)ds\right) \text{ and } \Phi_k(x) = \int_{Me^{\omega a}d_\infty(I_k(u(t_k)),\widehat{0})}^x \frac{d\tau}{\psi(\tau)}.$$

Now, we can establish our existence result.

Theorem 3 *Let assumptions* (H_1) *and* (H_2) *be satisfied. Suppose, in addition, that the following property is verified:*

Then the problem (2) has at least one mild solution on J.

Proof We distinguish in the proof several steps.

Step1:   For $t \in J_0$, consider the following problem

$$
\begin{cases} u'(t) = Au(t) + f(t, u(t)), \ t \in J_0, \\ u(0) = u_0. \end{cases} \tag{3}
$$

Transform the problem (3) into a fixed point problem. Consider the operator

$$
\mathcal{O} : \mathscr{C}(J_0, E^n) \to \mathscr{C}(J_0, E^n),
$$

defined by:

$$
\mathcal{O}(u)(t) = T(t)u_0 + \int_0^t T(t-s)f(s, u(s))ds.
$$

Let

$$
\mathscr{F}_0 \cong \mathscr{G}_0 \equiv \left\{ \tilde{j}u \in \mathscr{C}(J_0, X) : u \in \mathscr{F}_0 \right\} \equiv \tilde{j}(\mathscr{F}_0).
$$

$\mathscr{G}_0$ is a convex subset of the Banach space $\mathscr{C}(J_0, X)$, so in particular $\mathscr{G}_0$ is an absolute retract. As a result $\mathscr{F}_0$ is an absolute retract. we shall prove that the operator $\mathcal{O}$ maps $\mathscr{F}_0$ into itself, continuous and completly continuous.

Substep 1: $\mathcal{O} : \mathscr{F}_0 \to \mathscr{F}_0$. Indeed
   Let $u \in \mathscr{F}_0$ and $t \in J_0$. We have

$$
d_\infty \left(\mathcal{O}u(t), \widehat{0} \right) = d_\infty \left(T(t)u_0 + \int_0^t T(t-s)f(s, u(s))ds, \widehat{0} \right)
$$

$$
\leq d_\infty \left(\int_0^t T(t-s)f(s, u(s))ds, \widehat{0} \right) + d_\infty \left(T(t)u_0, \widehat{0} \right)
$$

$$
\leq \int_0^t d_\infty \left(T(t-s)f(s, u(s)), \widehat{0} \right) ds + M e^{\omega a} d_\infty \left(u_0, \widehat{0} \right)
$$

$$
\leq M e^{\omega a} \int_0^t d_\infty \left(f(s, u(s)), \widehat{0} \right) ds + M e^{\omega a} d_\infty \left(u_0, \widehat{0} \right)
$$

$$
\leq M e^{\omega a} \int_0^t \varphi(s)\psi \left(d_\infty \left(u(s), \widehat{0} \right) \right) ds + M e^{\omega a} d_\infty \left(u_0, \widehat{0} \right)
$$

$$
\leq M e^{\omega a} \int_0^t \varphi(s)\psi \left(a_0(s) \right) ds + M e^{\omega a} d_\infty \left(u_0, \widehat{0} \right).
$$

Since for $s \in J_0$, we have

$$
\begin{aligned}
a_0'(s) &= \left(\Phi_k^{-1} \left(\int_0^s Me^{\omega a} \varphi(\tau) d\tau \right) \right)' \\
&= Me^{\omega a} \varphi(s) \left(\Phi_k^{-1} \right)' \left(\int_0^s Me^{\omega a} \varphi(\tau) d\tau \right) \\
&= Me^{\omega a} \varphi(s) \frac{1}{\Phi_k' \left(\Phi_k^{-1} \left(\int_0^s Me^{\omega a} \varphi(\tau) d\tau \right) \right)} \\
&= Me^{\omega a} \varphi(s) \frac{1}{\Phi_k' (a_0(s))} \\
&= Me^{\omega a} \varphi(s) \psi(a_0(s)).
\end{aligned}
$$

Consequently,

$$
\begin{aligned}
d_\infty \left(\mathscr{O}u(t), \widehat{0} \right) &\le \int_0^t a_0'(s) ds + Me^{\omega a} d_\infty \left(u_0, \widehat{0} \right) \\
&= a_0(t) - a_0(0) + Me^{\omega a} d_\infty \left(u_0, \widehat{0} \right) \\
&= a_0(t).
\end{aligned}
$$

Hence $\mathscr{O}$ maps $\mathscr{F}_0$ into $\mathscr{F}_0$.
Let $u \in \mathscr{F}_0, t \in J_0$, we have

$$
d_\infty \left(\mathscr{O}u(t), \widehat{0} \right) \le Me^{\omega a} \int_0^t \varphi(s) \psi(a_0(s)) \, ds + Me^{\omega a} d_\infty \left(u_0, \widehat{0} \right).
$$

Denote

$$
b_0(t) = Me^{\omega a} \int_0^t \varphi(s) \psi(a_0(s)) \, ds + Me^{\omega a} d_\infty \left(u_0, \widehat{0} \right).
$$

We have $b_0'(t) = Me^{\omega a} \varphi(t) \psi(a_0(t)) = a_0'(t)$ for $t \in J_0$ and $b_0(0) = Me^{\omega a} d_\infty \left(u_0, \widehat{0} \right) = a_0(0)$.
Then, From (2)

$$
\int_{a_0(0)}^{a_0(t)} \frac{d\tau}{\psi(\tau)} = \int_0^t \frac{a_0'(s)}{\psi(a_0(s))} ds = Me^{\omega a} \int_0^t \varphi(s) ds
$$

$$
\le Me^{\omega a} \int_0^{t_1} \varphi(s) ds < \int_{Me^{\omega a} d_\infty(u_0, \widehat{0})}^a \frac{d\tau}{\psi(\tau)} = \int_{a_0(0)}^a \frac{d\tau}{\psi(\tau)}.
$$

Consequently, there exists a constant c_0 such that $a_0(t) \le c_0 < a, t \in J_0$. This shows that $\mathscr{F}_0$ and $\mathscr{O}(\mathscr{F}_0)$ are bounded.

Substep 2: $\mathscr{O}$ is continuous. Indeed

Let $\{u_n\} \in \mathscr{F}_0$ be sequence such that $u_n \to u \in \mathscr{F}_0$ in $\mathscr{C}(J_0, E^n)$, we have

$$
\begin{aligned}
d_\infty\left(\mathscr{O}u_n(t), \mathscr{O}u(t)\right) &= d_\infty\left(T(t)u_0 + \int_0^t T(t-s)f(s, u_n(s))ds,\, T(t)u_0 + \int_0^t T(t-s)f(s, u(s))ds\right) \\
&= d_\infty\left(\int_0^t T(t-s)f(s, u_n(s))ds,\, \int_0^t T(t-s)f(s, u(s))ds\right) \\
&\leq Me^{\omega a}\int_0^{t_1} d_\infty\left(f(s, u_n(s)),\, f(s, u(s))\right).
\end{aligned}
$$

Then

$$
H\left(\mathscr{O}u_n, \mathscr{O}u\right) = \sup_{t \in J_0} d_\infty\left(\mathscr{O}u_n(t), \mathscr{O}u(t)\right) \leq Me^{\omega a}\int_0^{t_1} d_\infty\left(f(s, u_n(s)),\, f(s, u(s))\right).
$$

Let us set

$$
r_n(s) = d_\infty\left(f(s, u_n(s)),\, f(s, u(s))\right).
$$

As f is continuous, then

$$
r_n(s) \to 0 \text{ as } n \to +\infty \text{ for } s \in J_0
$$

And we have

$$
\begin{aligned}
r_n(s) &= d_\infty\left(f(s, u_n(s)),\, f(s, u(s))\right) \\
&\leq d_\infty\left(f(s, u_n(s)), \widehat{0}\right) + d_\infty\left(f(s, u(s)), \widehat{0}\right) \\
&\leq \varphi(s)\psi\left(d_\infty\left(u_n(s), \widehat{0}\right)\right) + \varphi(s)\psi\left(d_\infty\left(u(s), \widehat{0}\right)\right) \\
&\leq 2\varphi(s)\psi\left(a_0(s)\right) \text{ for } s \in J_0.
\end{aligned}
$$

As a result, by the dominated convergence theorem:

$$
\lim_{n \to +\infty}\int_0^{t_1} r_n(s)ds = \int_0^{t_1} \lim_{n \to +\infty} r_n(s)ds = 0.
$$

Then $H\left(\mathscr{O}u_n, \mathscr{O}u\right) \to 0$ as $n \to +\infty$.

Hence $\mathscr{O} : \mathscr{F}_0 \to \mathscr{F}_0$ is continuous.

Substep 3: $\mathscr{O}(\mathscr{F}_0)$ is an equicontinuous set of $\mathscr{C}(J_0, E^n)$. Indeed

Let $l_1, l_2 \in J_0$. $l_1 < l_2$, and $u \in \mathscr{F}_0$. We have

$$
\begin{aligned}
d_\infty\left(\mathscr{O}u(l_2), \mathscr{O}u(l_1)\right) &= d_\infty\left(T(t)u_0 + \int_0^{l_2} T(t-s)f(s, u(s))ds,\, T(t)u_0 + \int_0^{l_1} T(t-s)f(s, u(s))ds\right) \\
&= d_\infty\left(\int_0^{l_1} T(t-s)f(s, u(s))ds + \int_{l_1}^{l_2} T(t-s)f(s, u(s))ds,\, \int_0^{l_1} T(t-s)f(s, u(s))ds\right) \\
&= d_\infty\left(\int_{l_1}^{l_2} T(t-s)f(s, u(s))ds, \widehat{0}\right)
\end{aligned}
$$

$$\leq Me^{\omega a} \int_{l_1}^{l_2} d_{\infty} \left(f(s, u(s)), \widehat{0} \right)$$

$$\leq Me^{\omega a} \int_{l_1}^{l_2} \varphi(s) \psi \left(a_0(s) \right) ds$$

$$= \int_{l_1}^{l_2} a_0'(s) ds$$

$$= a_0(l_2) - a_0(l_1).$$

Thus, $\mathscr{O}$ is equicontinuous. And we know that, A subset of a complet metric space is totally bounded if and only if it is relatively compact. So as a consequence of $\mathbf{H_2}$, $\mathscr{O}(\mathscr{F}_0)$ is relatively compact. As a result by Arzela-Ascoli theorem, $\mathscr{O}$ is completly continuous.

It follows that $\mathscr{O} : \mathscr{F}_0 \to \mathscr{F}_0$ is continuous and completly continuous and by theorem 2. $\mathscr{O}$ hase a fixed point u_1 wich is solution of the problem (3).

Step 2: For $t \in J_1$.

Let us consider the following problem

$$\begin{cases} u'(t) = Au(t) + f(t, u(t)), \ t \in J_1, \\ u(t_1^+) = I_1 \left(u(t_1^-) \right) = I_1 \left(u_1(t_1^-) \right). \end{cases} \tag{4}$$

Consider the operator $\mathscr{O}_1 : \mathscr{C}(t_1, t_2], E^n) \to \mathscr{C}(t_1, t_2], E^n)$, defined by

$$\mathscr{O}_1 u(t) = T(t - t_1) I_1 \left(u(t_1^-) \right) + \int_{t_1}^{t} T(t - s) f(s, u(s)) ds.$$

We denote

$$\mathscr{F}_1 \cong \mathscr{G}_1 \equiv \left\{ \tilde{j} u \in \mathscr{C}(J_1, X) : u \in \mathscr{F}_1 \right\} \equiv \tilde{j}(\mathscr{F}_1).$$

$\mathscr{G}_1$ is a convex subset of the Banach space $\mathscr{C}(J_1, X)$, so in particular $\mathscr{G}_1$ is an absolute retract. As a result $\mathscr{F}_1$ is an absolute retract. we shall prove that the operator $\mathscr{O}_1$ maps $\mathscr{F}_1$ into itself, continuous and completly continuous.

Substep 1: $\mathscr{O}_1 : \mathscr{F}_1 \to \mathscr{F}_1$. Indeed

Let $u \in \mathscr{F}_1$ and $t \in J_1$. We have

$$d_{\infty} \left(\mathscr{O}_1 u(t), \widehat{0} \right) = d_{\infty} \left(T(t - t_1) I_1(u(t_1^-)) + \int_{t_1}^{t} T(t - s) f(s, u(s)) ds, \widehat{0} \right)$$

$$\leq Me^{\omega a} \int_{t_1}^{t} d_{\infty} \left(f(s, u(s)), \widehat{0} \right) ds + Me^{\omega a} d_{\infty} \left(I_1(u(t_1^-)), \widehat{0} \right)$$

$$\leq Me^{\omega a} \int_{t_1}^{t} \varphi(s) \psi \left(a_1(s) \right) ds + Me^{\omega a} d_{\infty} \left(I_1(u(t_1^-)), \widehat{0} \right)$$

$$= \int_{t_1}^{t} a_1'(s) ds + Me^{\omega a} d_{\infty} \left(I_1(u(t_1^-)), \widehat{0} \right)$$

$$= a_1(t) - a_1(t_1) + Me^{\omega a} d_\infty \left(I_1(u(t_1^-)), \widehat{0} \right)$$
$$= a_1(t).$$

Hence, we conclude that $\mathcal{O}_1$ maps $\mathscr{F}_1$ into $\mathscr{F}_1$.
Let $u \in \mathscr{F}_1, t \in J_1$, we have

$$d_\infty \left(\mathcal{O}_1 u(t), \widehat{0} \right) \leq Me^{\omega a} \int_{t_1}^t \varphi(s) \psi(a_1(s)) \, ds + Me^{\omega a} d_\infty \left(I_1(u(t_1^-)), \widehat{0} \right).$$

Denote

$$b_1(t) = Me^{\omega a} \int_{t_1}^t \varphi(s) \psi(a_1(s)) \, ds + Me^{\omega a} d_\infty \left(I_1(u(t_1^-)), \widehat{0} \right).$$

We have $b_1'(t) = Me^{\omega a} \varphi(t) \psi(a_1(t)) = a_1'(t)$ for $t \in J_1$ and $b_1(t_1) = Me^{\omega a} d_\infty$ $\left(I_1(u(t_1^-)), \widehat{0} \right) = a_1(t_1)$.
Then, From (2)

$$\int_{a_1(t_1)}^{a_1(t)} \frac{d\tau}{\psi(\tau)} = \int_{t_1}^t \frac{a_1'(s)}{\psi(a_1(s))} ds = Me^{\omega a} \int_{t_1}^t \varphi(s) ds \leq Me^{\omega a} \int_{t_1}^{t_2} \varphi(s) ds$$
$$< \int_{Me^{\omega a} d_\infty \left(I_1(u(t_1^-)), \widehat{0} \right)}^a \frac{d\tau}{\psi(\tau)} = \int_{a_1(t_1)}^a \frac{d\tau}{\psi(\tau)}.$$

Consequently, there exists a constant c_1 such that $a_1(t) \leq c_1 < a, t \in J_1$. This shows that $\mathscr{F}_1$ and $\mathcal{O}(\mathscr{F}_1)$ are bounded.
Substep 2: $\mathcal{O}_1$ is continuous. Indeed
Let $\{u_n\} \in \mathscr{F}_1$ be sequence such that $u_n \to u \in \mathscr{F}_1$ in $\mathscr{C}(J_1, E^n)$, we have

$$d_\infty \left(\mathcal{O}_1 u_n(t), \mathcal{O}_1 u(t) \right) = d_\infty \left(T(t - t_1) I_1(u(t_1^-)) + \int_{t_1}^t T(t - s) f(s, u_n(s)) ds, \right.$$
$$\left. T(t - t_1) I_1(u(t_1^-)) + \int_{t_1}^t T(t - s) f(s, u(s)) ds \right)$$
$$\leq Me^{\omega a} \int_{t_1}^{t_2} d_\infty \left(f(s, u_n(s)), f(s, u(s)) \right).$$

Then

$$H \left(\mathcal{O}_1 u_n, \mathcal{O}_1 u \right) \leq Me^{\omega a} \int_{t_1}^{t_2} d_\infty \left(f(s, u_n(s)), f(s, u(s)) \right).$$

Therefore $\mathcal{O}_1 : \mathscr{F}_1 \to \mathscr{F}_1$ is continuous.

Substep 3: $\mathscr{O}_1(\mathscr{F}_1)$ is an equicontinuous set of $\mathscr{C}(J_1, E^n)$. Indeed Let $l_1, l_2 \in J_1. l_1 < l_2$, and $u \in \mathscr{F}_1$. We have

$$d_\infty\left(\mathscr{O}_1 u(l_2), \mathscr{O}_1 u(l_1)\right) = d_\infty\left(T(t - t_1)I_1(u(t_1^-)) + \int_{t_1}^{l_2} T(t - s)f(s, u(s))ds,\right.$$

$$\left. T(t - t_1)I_1(u(t_1^-)) + \int_{t_1}^{l_1} T(t - s)f(s, u(s))ds\right)$$

$$= d_\infty\left(\int_{l_1}^{l_2} T(t - s)f(s, u(s))ds, \widehat{0}\right)$$

$$\leq Me^{\omega a}\int_{l_1}^{l_2} \varphi(s)\psi(a_1(s))\,ds$$

$$= \int_{l_1}^{l_2} a_1'(s)ds$$

$$= a_1(l_2) - a_1(l_1).$$

It follows that $\mathscr{O}_1 : \mathscr{F}_1 \to \mathscr{F}_1$ is continuous and completly continuous and by theorem 2. $\mathscr{O}_1$ hase a fixed point u_2 wich is solution of the problem (4).

Step 3: We continue with the same procedure, we construct solutions $u_k \in \mathscr{C}$ $(J, E^n), k = 2, \cdots, p$ to

$$\begin{cases} u'(t) = Au(t) + f(t, u(t)), \ t \in J_k, \\ u(t_k^+) = I_k\left(u(t_k^-)\right) = I_k\left(u_{k-1}(t_k^-)\right). \end{cases}$$

The solution u of the problem (1) is then defined by

$$u(t) = \begin{cases} u_1(t), & \text{if } t \in [0, t_1], \\ u_2(t), & \text{if } t \in (t_1, t_2], \\ \vdots \\ u_p(t), & \text{if } t \in (t_p, a]. \end{cases}$$

Conclusion

In this work, we studied a class of fuzzy evolution equations with impulses. Indeed, using the approach of semigroups and the isometric embedding between E^n and the Banach space X. we achieved to show that problem (1) possesses at least one solution under some appropriate conditions. Our results will be used in further works to verify numerical solutions and why not extend other results as [15, 16] which their results will be appeared somewhere in our future studies.

Acknowledgements The authors express their sincere thanks to the anonymous referees for numerous helpful and constructive suggestions which have improved the manuscript.

References

1. Agarwal, R.P., Benchohra, M., O'Regan, D., Ouahab, A.: Fuzzy solutions for multi-point boundary value problems. Mem. Differ. Equ. Math. Phys. **35**, 1–14 (2005)
2. Agarwal, R.P., O'Regan, D., Lakshmikantham, V.: Viability theory and fuzzy differential equations. Fuzzy Sets Syst. **151**, 563–580 (2005)
3. Agarwal, R.P., O'Regan, D., Lakshmikantham, V.: A stacking theorem approach for fuzzy differential equations. Nonlinear Anal. **55**, 299–312 (2003)
4. Aumann, R.J.: Integrals of set-valued functions. J. Math. Anal. Appl. **12**, 1–12 (1965)
5. Benchohra, M., Nieto, J.J., Ouahab, A.: Fuzzy solutions for impulsive differential equations. Commun. Appl. Anal. **11**, 379–394 (2007)
6. Diamond, P., Kloeden, P.E.: Metric Spaces of Fuzzy Sets: Theory and Applications. World Scientific, Singapore (1994)
7. El Allaoui, A., Melliani, S., Chadli, L.S.: Fuzzy dynamical systems and Invariant attractor sets for fuzzy strongly continuous semigroups. J. Fuzzy Set Valued Anal. **2**, 148–155 (2016)
8. El Allaoui, A., Melliani, S., Chadli, L.S.: Fuzzy $\alpha-$semigroups of operators. Gen. Lett. Math. **2**(2), 42–49 (2017)
9. Gal, C.G., Gal, S.G.: Semigroups of Operators on Spaces of Fuzzy-Number-Valued Functions with Applications to Fuzzy Differential Equations, vol. 17 (2013). arXiv:1306.3928v1
10. Granas, A., Dugundji, J.: Fixed Point Theory. Springer, New York (2003)
11. Kaleva, Osmo: Nonlinear iteration semigroups of fuzzy Cauchy problems. Fuzzy Sets Syst. **209**, 104–110 (2012)
12. Kim, Y.K.: Measurability for fuzzy valued functions. Fuzzy Sets Syst. **129**, 105–109 (2002)
13. Klement, E., Puri, M., Ralescu, D.: Limit theorems for fuzzy random variables. Proc. Roy. Soc. Lond. Ser. A **407**, 171–182 (1986)
14. Lakshmikantham, V., Mcrae, F.A.: Basic results for impulsive fuzzy differential equations. Math. Inequal. Appl. **4**, 239–246 (2001)
15. Melliani, S., Chadli, L.S., El Allaoui, A.: Periodic boundary value problems for controlled nonlinear impulsive evolution equations on Banach spaces. Int. J. Nonlinear Anal. Appl. **8**(1), 301–314 (2017). https://doi.org/10.22075/ij-naa.2017.1460.1370
16. Melliani, S., El Allaoui, A., Chadli, L.S.: A general class of periodic boundary value problems for controlled nonlinear impulsive evolution equations on Banach spaces. Adv. Differ. Equ. **2016**, 290 (2016). https://doi.org/10.1186/s13662-016-1004-2
17. Melliani, S., El Allaoui, A., Chadli, L.S.: Relation between fuzzy semigroups and fuzzy dynamical systems. Nonlinear Dyn. Syst. Theory **17**(1), 60–69 (2017)
18. Melliani, S., Eljaoui, E.H., Chadli, L.S.: Fuzzy differential equation with nonlocal conditions and fuzzy semigroups. Advances in Difference Equations, p. 35 (2016)
19. Nieto, J.J., Rodriguez-Lopez, R., Villanueva-Pesqueira, M.: Exact solution to the periodic boundary value problem for a first-order linear fuzzy differential equation with impulses. Fuzzy Optim. Decis. Making **10**, 323–339 (2011)
20. Pazy, A.: Semigroups of Linear Operators and Applications to Partiel Differential Equations. Springer, New York (1983)
21. Puri, M., Ralescu, D.: Fuzzy random variables. J. Math. Anal. Appl. **114**, 409–422 (1986)
22. Puri, M., Ralescu, D.: The concept of normality for fuzzy random variables. Ann. Probab. **13**, 1373–1379 (1985)
23. Puri, M., Ralescu, D.: Convergence theorem for fuzzy martingales. J. Math. Anal. Appl. **160**, 107–122 (1991)

24. Rodriguez-Lopez, Rosana: Periodic boundary value problems for impulsive fuzzy differential equations. Fuzzy Sets Syst. **159**, 1384–1409 (2008)
25. Vatsala, A.S.: Impulsive hybrid fuzzy differential equations. FACTA UNI-VERSITATIS Ser.: Mech. Autom. Control Robot. **3**, 851–859 (2003)

Regional Robustness Optimal Control Via Strong Stabilization of Semilinear Systems

Abdessamad El Alami and Ali Boutoulout

Abstract The aim of this paper is to treat the problem of regional optimal stabilization of a class of nonlinear systems by using a switching feedback. Firstly, we proof that the switching control strongly stabilize the system on subregion includes in the whole domain. Secondly, under a perturbation of the control operator we show the robustness of our result. In the last part the stabilizing feedback is characterized by the minimization of a regional cost even under a small perturbation. We conclude by giving different applications to hyperbolic and parabolic equations.

1 Introduction

The problem of regional stabilization has been the object of many works, and it includes in studying the action of a distributed system, not in its whole geometrical evolution domain, but just in a subregion which may be inside or on the boundary of this domain [6, 18]. Many approaches were used to characterize unlike kinds of stabilization, and mostly characterization of control which attain the stability and minimizing a given cost standard. Afterwards the notion of optimal stabilization was developed for bilinear system by Ouzahra [14].

This paper proposes to extend the above results to the following semilinear system:

$$\begin{cases} \dfrac{dy(t)}{dt} = Ay(t) + Ny(t) + v(t)By(t) \\ y(0) = y_0 \end{cases} \tag{1}$$

The state space H endowed with the inner product $\langle ., . \rangle$, and the corresponding norm $\|.\|$, $v(t)$ is a scaler valued control. The dynamic A is an unbounded operator with

A. El Alami (✉) · A. Boutoulout
TSI Team, MACS Laboratory, Department of Mathematics and Computer Science,
Faculty of Sciences, Moulay Ismail University, Meknes, Morocco
e-mail: elalamiabdessamad@gmail.com

A. Boutoulout
e-mail: boutouloutali@yahoo.fr

© Springer Nature Switzerland AG 2020
E. H. Zerrik et al. (eds.), *Recent Advances in Modeling, Analysis and Systems Control: Theoretical Aspects and Applications*, Studies in Systems, Decision and Control 243,
https://doi.org/10.1007/978-3-030-26149-8_6

domain $D(A) \subset H$ and generates a semigroup of contractions $(S(t))_{t \geq 0}$ on H. N is a nonlinear operator from H into H which is dissipative, such that $N(0) = 0$, and B is a linear operator from H to H. Note that the assumption $N(0) = 0$ implies that 0 remains an equilibrium for (1).This problem was treated by Haraux for the wave equation [6]. The author show that if a second order semilinear conservative equation with essentially oscillatory solutions such as the wave equation is perturbed by a possibly non monotone damping term which is effective in a non negligible sub-region for at least one sign of the velocity, all solutions of the perturbed system converge weakly to 0 as time tends to infinity. In [11] Gugat et al. considers The Schrödinger governed by a nonlinear reaction-diffusion partial differential equation with a cubic nonlinearity that determines three constant equilibrium states. They give exponential stability result of the closed loop system with respect to the L^2-norm. In particular, it is shown that with the boundary feedback law the unstable constant equilibrium point can be stabilized. Optimal control, including quadratic, nonlinear and processes, has been treated in Ouzahra [13]. The problem of stabilization of switched systems includes two aspects: The first one is how to make switched systems stable under arbitrary switching laws, and the other is how to design a switching law within which switched systems are stabilized. Extensions to infinite dimensional switched systems have been considered by using rolling mode control methodology, Lyapunov functions, and Galerkins method (see for example: [7–10, 17]). The goal of this paper is to design a switching law under which a finite or infinite dimensional type of semilinear system could be stabilized.

In [14, 15], the constrained control defined by $v(t) = -\dfrac{\langle y(t), By(t) \rangle}{\|y(t)\|^2}$ if $y(t) \neq 0$ and $v(t) = 0$, it is used to obtain the exponential stability of (1) with $N = 0$ using exact observability assumption:

$$\int_0^T |\langle BS(t)y, S(t)y \rangle| dt \geq \delta \|y\|^2, \ \forall y \in H, (T, \delta > 0). \tag{2}$$

The problem of weak stabilization has been studied in Ball and Slemrod [3] with $N = 0$ and B nonlinear, provided that B satisfies the following weak observability assumption:

$$\langle BS(t)y, S(t)y \rangle = 0, \ \forall t \geq 0 \Longrightarrow y = 0, \tag{3}$$

using the quadratic control $v(t) = -\langle y(t), By(t) \rangle$.

The regional stabilisation for the semilinear systems is motivated by the existence of a class of systems that are unstable in the whole domain but they may be stable in some subregion. For example let Ω be an open bounded subset of $\mathbb{R}^n$ with smooth boundary $\partial \Omega$ and consider the following system:

$$\begin{cases} \dfrac{\partial y(x, t)}{\partial t} = \dfrac{\partial^2 y(x, t)}{\partial x^2} - \dfrac{y(x, t)}{1 + y(x, t)^2} + K(x, t), \ in \ ]0, 2[\times]0, +\infty[, \\ y(x, 0) = y_0 \in L^2(]0, 2[), \end{cases} \tag{4}$$

where $K(x, t) = e^{(x-1)t}((x-1) - t^2) + \dfrac{e^{(x-1)t}}{1 + e^{2(x-1)t}}$.

The system (4) is not stable in the whole domain $\Omega =]0, 2[$, but it is stable in a subregion $\omega \subset \Omega$, indeed, the solution of the systems (4) is $y(x, t) = e^{(x-1)t}$ which does not tend to zero as $t \longrightarrow +\infty$ in $\Omega =]0, 2[$, but for a subregion $\omega =]0, a[\subset]0, 1[$ we have

$$\|\chi_\omega y(t)\| \le e^{(a-1)t}\|y_0\|, \; \forall t > 0,$$

where ω is an open and positive Lebesgue measurable subset of ω, χ_ω is the restriction operator in ω, and set $i_\omega = \chi_\omega^* \chi_\omega$,
with the restriction operator χ_ω is defined by

$$\begin{aligned}\chi_\omega : H_\Omega &\longrightarrow H_\omega \\ y &\longrightarrow \chi_\omega y = y_{|\omega},\end{aligned} \tag{5}$$

while χ_ω^* denotes the adjoint operator given by

$$\chi_\omega^* y(x) = \begin{cases} x \;, \text{ on } \omega \\ 0 \;, \text{ on } \Omega\backslash\omega \end{cases}$$

Concerning the kind of this work we find recently, the regional exponential stabilization problem of distributed semilinear systems has been resolved (see [6]). Then it has been proved that under the assumption

$$\int_0^T |\langle i_\omega BS(t)y, S(t)y\rangle|dt \ge \delta\|\chi_\omega y\|^2, \; \forall \; y \in H, \text{ (for some } T, \delta > 0), \tag{6}$$

the feedback defined by

$$v_j(t) = \frac{-B^* i_\omega y(t)}{R_j(i_\omega y(t))}, \;\; (j = 1, 2), \text{ where } R_1(i_\omega y) = 1 + \|B^* i_\omega y\|, \tag{7}$$

and $R_2(i_\omega y) = sup(1, \|B^* i_\omega y\|)$ guarantees the regional exponential stabilization. In this paper, we study the regional optimal stabilization and its robustness of the semilinear system (1), using the switching control $v(t) = -\eta \, sign(\langle y(t), i_\omega By(t)\rangle)$, $\forall y \in H$, $\eta > 0$. This feedback profits from the following advantages. For instance, it is uniformly bounded with respect to initial states and, thus, may be applied as a constrained control.

In this work we adopt the following assumptions:

- (A_1) The operator $i_\omega A$ is dissipative.
- (A_2) The nonlinear operator N is locally Lipschitz, with K is a Lipschitz constant of N, and N is dissipative.
- (A_3) Assume that the operator B satisfies:

$$\int_0^T |\langle i_\omega B S(t) y_0, S(t) y_0 \rangle| dt \geq \alpha \|\chi_\omega y_0\|^2, \ \forall y_0 \in H, \ (T, \alpha > 0), \quad (8)$$

with $\alpha \leq \alpha(B) = \inf_{\|y_0\|=1} |\langle i_\omega B S(t) y_0, S(t) y_0 \rangle|.$

This paper deals with a robustness of regional optimal stabilization governed by a class of the system (1) using switching control.whose state belongs to an abstract Hilbert space H. The paper is structured as follows: In Sect. 2, provides the regional strong stabilization and robustness of semilinear evolution equations. Section 3, is devoted the idea of robustness of regional optimal stabilization results for the systems (1). In the last section, we give illustrations through examples, governed by a wave equation and a Schrödinger equation.

2   Regional Strong Stabilization and Robustness

In this section, we discuss the regional strong stabilization of the system (1) by using the following switching feedback

$$v(t) = -\eta \, sign(\langle y(t), i_\omega B y(t) \rangle), \forall y \in H. \quad (9)$$

2.1   Regional Strong Stabilization

The next result concerns the regional strong stabilization of (1).

Definition 1 The system (1) is weakly (resp.strongly) stabilizable if there exists a feedback control $v(t) = f(y(t))$, $t \geq 0$, $f : H \longrightarrow K := \mathbf{R}, \mathbf{C}$ such that the corresponding mild solution $y(t)$ of the system (1) satisfies the properties:

1. For each initial state y_0 of the system (1) there exists a unique mild solution defined for all $t \in \mathbf{R}^+$ of the system (1).
2. $\{0\}$ is an equilibrium state of the system (1).
3. $y(t) \to 0$, weakly (resp. strongly), as $t \to +\infty$, for all $y_0 \in H$.

Theorem 1 *Let us consider the following assumptions:*

1. *A generate a linear C_0-contraction semigroup $(S(t))_{t \geq 0}$ on H such that (37) holds.*
2. *The nonlinear operator N satisfies (A_2).*

Then the system(1) is strongly stabilize by the feedback

$$v(t) = -\eta \, sign(\langle y(t), i_\omega B y(t) \rangle), \forall y \in H. \quad (10)$$

Proof Let $y(t)$ denote the corresponding solution of (1). For $t \geq 0$ we define the function

$$\tau \longrightarrow z(\tau) := \int_t^\tau v(s)S(\tau - s)By(s) + S(\tau - s)Ny(s)ds$$

Applying the variation of constant formula with $y(t)$ as the initial state, we get:

$$y(\tau) = S(\tau - t)y(t) + z(\tau), \forall \tau \in [t, t + T]. \tag{11}$$

We obtain:
$$\|\chi_\omega y(\tau)\| \leq \|\chi_\omega S(\tau - t)y(t)\| + \|\chi_\omega z(\tau)\| \text{ Then we have:}$$

$$\|\chi_\omega y(\tau)\| \leq \|\chi_\omega y(t)\| + \|\chi_\omega z(\tau)\|.$$

Since N is sequentially continuous and B continuous, then $F = vBy + N$ it's also

$$F(y(t)) \to 0, \quad \text{as } t \to +\infty.$$

Let $\varepsilon > 0$ and $T > 0$ such that $\|F(y(t))\| < \frac{\varepsilon}{\rho T}, \forall t \geq 0$. we have the relation

$$\langle i_\omega BS(\tau - t)y(t), S(\tau - t)y(t)\rangle = \langle i_\omega B(y(\tau) - z(\tau)), S(\tau - t)y(t)\rangle$$
$$= - \langle i_\omega Bz(\tau), S(\tau - t)y(t)\rangle - \langle i_\omega By(\tau), z(\tau)\rangle$$
$$+ \langle i_\omega By(\tau), y(\tau)\rangle.$$

It follows that:

$$|\langle i_\omega BS(\tau - t)y(t), S(\tau - t)y(t)\rangle| \leq \|B\|\|\chi_\omega y(t)\|\|\chi_\omega z(\tau)\|$$
$$+ \|B\|\|\chi_\omega z(\tau)\|\|\chi_\omega y(\tau)\| + |\langle i_\omega By(\tau), y(\tau)\rangle|.$$

Therefore

$$|\langle i_\omega BS(\tau - t)y(t), S(\tau - t)y(t)\rangle| \leq \|B\|\|z(\tau)\|(\|y(t)\| + \|y(\tau)\|)$$
$$+ |\langle i_\omega By(\tau), y(\tau)\rangle|.$$

By integrating this inequality over $[t, t + T]$ we obtain the estimate

$$\int_t^{T+t} |\langle i_\omega BS(\tau - t)y(t), S(\tau - t)y(t)\rangle| d\tau \leq \varepsilon + \int_t^{t+T} |\langle i_\omega By(\tau), y(\tau)\rangle| d\tau.$$

Thus

$$\int_0^T |\langle i_\omega BS(\tau)y(t), S(\tau)y(t)\rangle| d\tau \leq \varepsilon + \int_t^{t+T} |\langle i_\omega By(\tau), y(\tau)\rangle| d\tau.$$

We have

$$\int_t^{t+T} |\langle i_\omega B y(\tau), y(\tau)\rangle| d\tau \longrightarrow 0, \text{ as } t \longrightarrow +\infty.$$

And using (37) we deduce that $\|\chi_\omega y(t)\| \longrightarrow 0, \text{ as } t \longrightarrow +\infty.$

2.2 Robustness of Regional Strong Stabilization

Consider the perturbed system

$$\frac{dy(t)}{dt} = Ay(t) + Ny(t) + v(t)(B + b)y(t), \quad y(0) = y_0, \tag{12}$$

where b is a perturbation of B. In this part, we consider the robustness of the feedback control law (10). Let us define the set of admissible operators of perturbations

$$\Lambda = \left\{ b \in \mathcal{L}(U, H_\omega); \|b^*\| \leq \frac{\alpha(B)}{2T\|B^*\|} \right\},$$

and N Lipschitz with $\tilde{B}_b y(t) = v(t)(B + b)y(t)$, which is lipschitz. Then we have the following result.

Proposition 1 *Let us consider the following assumptions:*

(i) *A generates a linear C_0-semigroup of contraction $(S(t))_{t\geq 0}$on H such that (A_3)holds.*

(ii) *$b \in \Lambda$.*

(ii) *We assume that*

$$\int_0^T \|B^*\chi_\omega S(t)y_0\|^2 dt \geq \alpha\|\chi_\omega y_0\|^2, \quad \forall y_0 \in H, \ (T, \alpha > 0). \tag{13}$$

Then the feedback (10) is strongly robust on ω.

Proof We have

$$\|(B + b)^*\chi_\omega S(t)y_0\|_U \geq \|B^*\chi_\omega S(t)y_0\| - \|b^*\chi_\omega S(t)y_0\|.$$

Then

$$\|(B + b)^*\chi_\omega S(t)y_0\|^2 \geq \|B^*\chi_\omega S(t)y_0\|_H^2 - 2\|B^*\chi_\omega S(t)y_0\|\|b^*\chi_\omega S(t)y_0\| + \|b^*\chi_\omega S(t)y_0\|^2$$
$$\geq \|B^*\chi_\omega S(t)y_0\|_H^2 - 2\|B^*\|_{\mathcal{L}(H)}^2\|b^*\|_{\mathcal{L}(H)}^2\|\chi_\omega y_0\|^2, \tag{14}$$

Integrating this inequality and using (13), we get

$$\int_0^T \|(B+b)^* \chi_\omega S(t) y_0\|_H^2 \, dt \geq (\alpha(B) - 2T \|B^*\|_{\mathcal{L}(H)} \|b^*\|_{\mathcal{L}(H)}) \|\chi_\omega y_0\|^2, \ \forall y_0 \in H, \ (T, \alpha > 0),$$

$$(15)$$

which implies that $B + b$ verifies (13), with $\alpha = \alpha(B) - 2T \|B\|_{\mathcal{L}(H)} \|b\|_{\mathcal{L}(H)}$. From Theorem (1) by replacing (A_3) with (13), we deduce that the feedback (10) is strongly robust on ω.

3 Regional Optimal Stabilization and Robustness

3.1 Regional Optimal Stabilization

In this part, we consider the following problem:

$$\begin{cases} \min_{v \in V_{ad}(\Omega)} Q(v) = \int_0^{+\infty} \frac{1}{\eta} |\langle S(t)y, BS(t)y \rangle| |v(t)|^2 dt \\ \qquad + \int_0^{+\infty} \eta |\langle S(t)y, BS(t)y \rangle| + 2|\langle S(t)y, NS(t)y \rangle| + \langle S(t)y, RS(t)y \rangle dt, \end{cases}$$

$$(16)$$

where R is a linear self-adjoint and positive operator and for some $r > 0$, we notice

$$V_{ad}(\Omega) = \{v(t) / \ |v(t)| \leq r, \forall t > 0, and \ Q(v) < \infty\}.$$

Remark 1 Stabilizing a system regionally is cheaper than stabilising it in the whole domain, and we have

$$\min_{V_{ad}(\omega)} Q(v) \leq \min_{V_{ad}(\Omega)} Q(v).$$

The goal of this section is to show that the switching control (10) is the unique solution of the following cost:

$$\begin{cases} \min_{v \in V_{ad}^r(\omega)} Q_\omega(v) = \int_0^{+\infty} \frac{1}{\eta} |\langle S(t)y, i_\omega BS(t)y \rangle| |v(t)|^2 dt \\ \qquad + \int_0^{+\infty} \eta |\langle S(t)y, i_\omega BS(t)y \rangle| + 2|\langle S(t)y, i_\omega NS(t)y \rangle| + \langle S(t)y, i_\omega RS(t)y \rangle dt \end{cases}$$

$$(17)$$

where

$$V_{ad}^r(\omega) = \{v(t) / \ |v(t)| \leq r, \forall t > 0 \ and \ Q(\omega) < \infty\},$$

with $\eta, r > 0$.

Theorem 2 *let us consider that:*

- *A generates a semigroup $(S(t))_{t \geq 0}$ of isometries, and satisfies*

$$2Re\langle i_\omega AS(t)y, S(t)y \rangle + \langle S(t)y, i_\omega RS(t)y \rangle = 0, \forall y \in \mathcal{D}(A), \qquad (18)$$

   for a linear self-adjoint and positive operator R.
- *The operator B is linear operator and (A_3) holds.*
   Then, there exists $\eta > 0$ as it is the case with (10) which is the unique solution of (17).

Proof Let us consider $r > 0$. It follows from (A_3) that (10) is an admissible control for all η with $\eta, r > 0$, so $V_{ad}^r \neq 0$. We take $v \in V_{ad}^r$ and $y(t)$ denote the corresponding solution of (1), for $t \geq 0$ we define the function

$$\tau \longrightarrow z(\tau) := \int_t^\tau v(s)S(\tau - s)By(s) + S(\tau - s)Ny(s)ds.$$

Applying the variation of constant formula with $y(t)$ as the initial state, we get:

$$y(\tau) = S(\tau - t)y(t) + z(\tau), \forall \tau \in [t, t + T]. \qquad (19)$$

we have $\|\chi_\omega y(\tau)\| \leq \|\chi_\omega S(\tau - t)y(t)\| + \|\chi_\omega z(\tau)\|$, and:

$$\|\chi_\omega y(\tau)\| \leq \|\chi_\omega y(t)\| + \|\chi_\omega z(\tau)\|.$$

Using A_1, we obtain:

$$\|\chi_\omega y(\tau)\| \leq \|\chi_\omega y(t)\| + (r\|B\| + K)\int_t^\tau \|\chi_\omega y(s)\|ds.$$

The Gronwall inequality then yields

$$\|\chi_\omega y(\tau)\| \leq \|\chi_\omega y(t)\|e^{(r\|B\|+K)T}, \forall \tau \in [t, t + T]. \qquad (20)$$

According to the relation

$$\langle i_\omega BS(\tau - t)y(t), S(\tau - t)y(t) \rangle = \langle i_\omega B(y(\tau) - z(\tau)), S(\tau - t)y(t) \rangle$$
$$= -\langle i_\omega Bz(\tau), S(\tau - t)y(t) \rangle - \langle i_\omega By(\tau), z(\tau) \rangle + \langle i_\omega By(\tau), y(\tau) \rangle.$$

It follows that:

$$|\langle i_\omega BS(\tau - t)y(t), S(\tau - t)y(t) \rangle| \leq \|B\|\|\chi_\omega y(t)\|\|\chi_\omega z(\tau)\|$$
$$+ \|B\|\|\chi_\omega z(\tau)\|\|\chi_\omega y(\tau)\| + |\langle i_\omega By(\tau), y(\tau) \rangle|.$$

Therefore,

$$|\langle i_\omega BS(\tau - t)y(t), S(\tau - t)y(t)\rangle| \leq \|B\|\|\chi_\omega z(\tau)\|(\|\chi_\omega y(t)\| + \|\chi_\omega y(\tau)\|)$$
$$+ |\langle i_\omega By(\tau), y(\tau)\rangle|,$$

using (20) we deduce that:

$$|\langle i_\omega BS(\tau - t)y(t), S(\tau - t)y(t)\rangle| \leq (r\|B\|^2 + \|B\|K)T\|\chi_\omega y(t)\|^2(1 + e^{(r\|B\|+K)T})$$
$$+ |\langle i_\omega By(\tau), y(\tau)\rangle|.$$

By integrating this inequality over [t, t + T] we obtain the estimate

$$\int_0^T \langle i_\omega BS(s)y(t), S(s)y(t)\rangle ds \leq (r\|B\|^2 + \|B\|K)T^2(1 + e^{(r\|B\|+K)T})\|\chi_\omega y(t)\|^2$$
$$+ \int_t^{t+T} |\langle i_\omega By(\tau), y(\tau)\rangle| d\tau,$$

Since $Q(v) < +\infty$, we have

$$\int_t^{t+T} |\langle i_\omega By(\tau), y(\tau)\rangle| d\tau \longrightarrow 0, \text{ as } t \longrightarrow +\infty.$$

Taking $r > k > 0$ such that:

$$(r\|B\|^2 + \|B\|K)T^2(1 + e^{(r\|B\|+K)T}) < \alpha,$$

and with (A_3), we deduce that $\|\chi_\omega y(t)\| \longrightarrow 0$, as $t \longrightarrow +\infty$, using (18), we get the following inequality

$$\frac{d\|\chi_\omega y(t)\|^2}{dt} \leq 2v(t)\langle i_\omega By(t), y(t)\rangle + 2\langle i_\omega Ny(t), y(t)\rangle - \langle i_\omega Ry(t), y(t)\rangle, \quad (21)$$

then, we have

$$\frac{d\|\chi_\omega y(t)\|^2}{dt} \leq -2\eta|\langle i_\omega By(t), y(t)\rangle| - 2|\langle i_\omega Ny(t), y(t)\rangle| - \langle i_\omega Ry(t), y(t)\rangle,$$
$$(22)$$

which gives

$$\int_0^t 2\eta|\langle i_\omega BS(s)y_0, S(s)y_0\rangle| + 2|\langle i_\omega NS(s)y_0, S(s)y_0\rangle|$$

$$+ \langle i_\omega RS(s)y_0, S(s)y_0\rangle ds \leq \|\chi_\omega y_0\|^2, \forall t, s \geq 0; s \leq t.$$
$$(23)$$

Inequality (23) is checked $\forall y_0 \in H$, since $y_0 \longrightarrow S(.)y_0$ is continuous from H to $L^2(0, t; H)$. Then $v^* \in V_{ad}^r$ and $V_{ad}^r \neq \phi$. Let $y(.)$ be the corresponding solution of (1). For $y_0 \in D(A)$ and $s \in [0, t]$, we have $y(s) \in D(A)$ and $y(s)$ is differentiable (see [1, 16], pp. 187–189). Using Derivation of the energy $E(t) = \|\chi_\omega y(t)\|^2$, we get:

$$\frac{d\|\chi_\omega y(t)\|^2}{dt} = \frac{1}{\eta} |\langle S(t)y, i_\omega BS(t)y\rangle| |v(t) + \eta \, sign(\langle S(t)y, i_\omega BS(t)y\rangle)|^2$$
$$- |\langle S(t)y, i_\omega BS(t)y\rangle|(\eta + \frac{1}{\eta} |v(t)|^2) - 2|\langle S(t)y, i_\omega NS(t)y\rangle| - \langle i_\omega RS(t)y, S(t)y\rangle,$$

which gives:

$$\int_0^t |\langle S(s)y, i_\omega BS(s)y\rangle|(\eta + \frac{1}{\eta} |v(s)|^2) + 2|\langle S(s)y, i_\omega NS(s)y\rangle| + \langle i_\omega RS(s)y, S(s)y\rangle ds + \|\chi_\omega y(t)\|^2 - \|\chi_\omega y_0\|^2$$
$$= \int_0^t \frac{1}{\eta} |\langle S(s)y, i_\omega BS(s)y\rangle| |v(s) + \eta \, sign(\langle S(s)y, i_\omega BS(s)y\rangle)|^2 ds.$$

This relation can be obtained using an approximation argument in [2]. Letting $t \longrightarrow +\infty$ in the last equation, we obtain:

$$Q_\omega(v) - \|\chi_\omega y_0\|^2 = \int_0^\infty \frac{1}{\eta} |\langle S(t)y, i_\omega BS(t)y\rangle| |v(t) + \eta \, sign(\langle S(t)y, i_\omega BS(t)y\rangle)|^2 dt,$$

which implies that (10) is the unique solution of the issue (17) and that $Q_\omega(v) \geq Q_\omega(v^*) = \|\chi_\omega y_0\|^2$.

Remark 2 In the problem (16), if $R = 0$ one can replace the condition (18) of A by the case $i_\omega A$ is skew-adjoint operator.

3.2   Robustness of Regional Optimal Stabilization

For a linear self-adjoint and positive operator R, we justify that the control (10) is the unique solution of the following problem:

$$\begin{cases} \min_{v \in V_{ad}^r(\omega)} Q(v) = \int_0^{+\infty} \frac{1}{\eta} |\langle y(t), i_\omega By(t)\rangle| |v(t)|^2 dt \\[2mm] \qquad + \int_0^{+\infty} \eta \, |\langle y(t), i_\omega By(t)\rangle| + 2|\langle y(t), i_\omega Ny(t)\rangle| + \langle i_\omega Ry(t), y(t)\rangle dt, \end{cases} \tag{24}$$

where $\eta, r > 0$ and

$$V_{ad}^r(\omega) = \{v(t), \; |v(t)| \leq r, \forall t > 0 \; and \; Q_\omega(v) < \infty\},$$

which, is associated with the perturbation of the systems (12).

So, we have the following result:

Theorem 3 *Let us consider the following assumption:*

- *A generates a semigroup $(S(t))_{t \geq 0}$ of isometries.*
- *$b \in \Lambda$.*
- *We assume that (18) holds and b satisfies*

$$sign(\langle y(t), i_\omega By(t) \rangle) = -sign(\langle y(t), i_\omega by(t) \rangle), \forall y \in H. \tag{25}$$

- *The operator B is linear and (A_3) holds.*

Then, there exists $\eta > 0$ such that (10) is the unique solution of (24).

Proof Now, let us consider the nonlinear semigroup $\Gamma(t)y_0 = y(t)$. under the assumption $i_\omega A$ is dissipatives, we obtain the following inequality

$$\frac{d\|\chi_\omega \Gamma(t)y_0\|^2}{dt} \leq 2v(t)\langle i_\omega B\Gamma(t)y_0, \Gamma(t)y_0 \rangle + 2\langle i_\omega N\Gamma(t)y_0, \Gamma(t)y_0 \rangle - \langle i_\omega R\Gamma(t)y_0, \Gamma(t)y_0 \rangle, \tag{26}$$

then we have

$$\frac{d\|\chi_\omega \Gamma(t)y_0\|^2}{dt} \leq -2\eta |\langle i_\omega B\Gamma(t)y_0, \Gamma(t)y_0 \rangle|$$
$$- 2\eta \, sign(\langle i_\omega B\Gamma(t)y_0, \Gamma(t)y_0 \rangle)\langle i_\omega b\Gamma(t)y_0, \Gamma(t)y_0 \rangle - 2|\langle i_\omega Ny(t), y(t) \rangle|, \tag{27}$$

using (18) we have

$$\frac{d\|\chi_\omega y(t)\|^2}{dt} \leq -2\eta |\langle i_\omega By(t), y(t) \rangle| - 2|\langle i_\omega Ny(t), y(t) \rangle| - \langle i_\omega R\Gamma(t)y_0, \Gamma(t)y_0 \rangle, \tag{28}$$

which gives

$$\int_0^t 2\eta |\langle i_\omega B\Gamma(s)y_0, \Gamma(s)y_0 \rangle| + 2|\langle i_\omega N\Gamma(s)y_0, \Gamma(s)y_0 \rangle| + \langle i_\omega R\Gamma(s)y_0, \Gamma(s)y_0 \rangle ds \leq \|\chi_\omega y_0\|^2, \forall t, s \geq 0; s \leq t. \tag{29}$$

This last inequality holds $\forall y_0 \in H$. Since $y_0 \longrightarrow \Gamma(.)y_0$ is continuous from H to $L^2(0, t; H)$. Then $v^* \in V_{ad}^r$ and $V_{ad}^r \neq \phi$. Let $y(.)$ be the corresponding solution of (1). For $y_0 \in D(A)$ and $s \in [0, t]$, we have $y(s) \in D(A)$ and $y(s)$ is differentiable (see [1, 16], pp. 187–189). We drive the energy $E(t) = \|\chi_\omega y(t)\|^2$, we obtain:

$$\frac{d\|\chi_\omega y(t)\|^2}{dt} = \frac{1}{\eta} |\langle y(t), i_\omega By(t) \rangle||v(t) + \eta \, sign(\langle y(t), i_\omega By(t) \rangle)|^2 - |\langle y(t), i_\omega By(t) \rangle|(\eta + \frac{1}{\eta}|v(t)|^2)$$
$$- 2\eta \, sign(\langle y(t), i_\omega By(t) \rangle)\langle i_\omega by(t), y(t) \rangle - 2|\langle y(t), i_\omega Ny(t) \rangle| - \langle i_\omega Ry(t), y(t) \rangle,$$

which gives:

$$\int_0^t |\langle y(t), i_\omega By(t)\rangle|(\eta + \frac{1}{\eta}|v(s)|^2) + 2|\langle y(t), i_\omega Ny(t)\rangle| + \langle i_\omega Ry(t), y(t)\rangle ds + \|\chi_\omega y(t)\|^2 - \|\chi_\omega y_0\|^2$$
$$= \int_0^t \frac{1}{\eta}|\langle y(t), i_\omega By(t)\rangle||v(s) + \eta\, sign(\langle y(t), i_\omega By(t)\rangle)|^2 - 2\eta\, sign(\langle y(s), i_\omega By(s)\rangle)\langle i_\omega by(t), y(s)\rangle ds.$$

Letting $t \longrightarrow +\infty$ in the last equation, then, we have:

$$\begin{cases} Q_\omega(v) - \|\chi_\omega y_0\|^2 = \int_0^\infty \frac{1}{\eta}|\langle y(t), i_\omega By(t)\rangle||v(t) + \eta\, sign(\langle y(t), i_\omega By(t)\rangle)|^2 \\ \qquad\qquad -2\,\eta\, sign(\langle y(t), i_\omega By(t)\rangle)\langle i_\omega by(t), y(t)\rangle dt, \end{cases}$$
$$(30)$$

using (25) we proved that (10) is the unique solution to the problem (24) and that:
$Q_\omega(v) \geq Q_\omega(v^*) = \|\chi_\omega y_0\|^2$.

4 Application

4.1 Example (Equation 1)

We consider the following semilinear transport equation on $\Omega =]0, +\infty[$:

$$\begin{cases} \dfrac{\partial z(x,t)}{\partial t} = -\dfrac{\partial z(x,t)}{\partial x} - \dfrac{|y(t)|^2}{(1+|y(t)|^2)}\int_0^1 z(x)sin(x)dx + v(t)z(x,t) & \Omega \times]0, +\infty[, \\ z(x,0) = z_0(x) & \Omega, \end{cases}$$
$$(31)$$

Here $H = L^2(\Omega)$, $v(.) \in L^2(0, \infty)$ and $Az = -\frac{\partial z}{\partial x}$ with domain
$\mathcal{D}(A) = \{z \in H^1(\Omega) \mid z(0) = 0,\ z(x) \to 0 \text{ as } x \to +\infty\}$.
Let $\omega =]0, a[$ be a subregion of Ω and system (31) is augmented with the output.
We have

$$\langle \chi_\omega^* \chi_\omega Az, z\rangle = -\int_0^a z'(x)z(x)dx \qquad (32)$$

$$= -\frac{z^2(a)}{2} \leq 0, \qquad (33)$$

so, the assumption (A_1) holds.
The operator A generates the following semigroup of contractions

$$(S(t)z_0)(x) = \begin{cases} z_0(x-t) & \text{if } x \geq t \\ 0 & \text{if } x < t \end{cases} \qquad (34)$$

For $T = 1$, we have

$$\int_0^1 \langle \chi_\omega^* \chi_\omega S(t)z_0, S(t)z_0 \rangle dt = \int_0^1 \int_0^{a-t} |z_0(x)|^2 dx dt \geq \|\chi_\omega z_0\|^2,$$

so (37) is verified, $Ny(t) = -\dfrac{|y(t)|^2}{(1+|y(t)|^2)} \int_0^1 z(x) \sin(x) dx$ which is locally Lipschitz. We conclude by Theorem (1) that the control (10) is strongly stabilizes the systems (31) on ω.

4.2 Example (Wave Equation)

We consider the following semilinear equation and $\Omega =]0, 1[$:

$$\begin{cases} \dfrac{\partial^2 y(t)}{\partial t^2} = \Delta y(t) - \dfrac{|y(t)|^2 y(t)}{\zeta} + v(t) \dfrac{\partial y(t)}{\partial t}, & \text{in } \Omega \times]0, +\infty[, \\ y(0) = y_0, \, y_t(0) = y_1, & \text{on } \partial\Omega \times]0, +\infty[, \end{cases} \tag{35}$$

we take the state space $H = H_0^1(\Omega) \times L^2(\Omega)$ and the operators A and B are defined by

$$A = \begin{pmatrix} 0 & I \\ \Delta & 0 \end{pmatrix}, B = \begin{pmatrix} 0 & 0 \\ 0 & I \end{pmatrix},$$

which A generate a semigroup $S(t)$ of isometries where $\Delta y(t) = \dfrac{\partial^2 y(t)}{\partial x^2}$ with

$$\mathcal{D}(A) = (H^2(\Omega) \cap H_0^1(\Omega)) \times H_0^1(\Omega).$$

In [5], we consider the operator $A_1 z = \Delta z$ in $H = L^2(\Omega)$, (Ω is a bounded open set in $\mathcal{R}^n$, $n \geq 1$) and domain

$$\mathcal{D}(A_1) = \{ z \in L^2(\Omega) | \, z \text{ is absolutely continuous, } \Delta z \in L^2(\Omega) \text{ et } z = 0 \text{ sur } \partial\Omega \}.$$

we have $\overline{\mathcal{D}(A_1)} = L^2(\Omega)$, $A_1^* = A_1$ and $\mathcal{R}e(\langle A_1 z, z \rangle) = -\|\nabla z\|^2 \leq 0$. Then A_1 is dissipative, and consequently A is also but A is not necessarily dissipative.

Note that under the so called Geometric Control Condition (GCC) (see [4]), we have for $T = 2$ with $\omega \subset \Omega$ the measure $0 < |\omega| < 1$, that

$$\int_0^T |\langle i_\omega B S(t)y_0, S(t)y_0 \rangle| dt \geq \alpha \|y_0\|^2, \, \forall y_0 \in H, \, (\alpha > 0). \tag{36}$$

Then

$$\int_0^T |\langle i_\omega BS(t)y_0, S(t)y_0\rangle| dt \geq \alpha \|\chi_\omega y_0\|^2, \ \forall y_0 \in H, \ (\alpha > 0), \qquad (37)$$

so (A_3) holds. Taking $Ny = \begin{pmatrix} 0 \\ -\dfrac{|y(t)|^2 y(t)}{\zeta} \end{pmatrix}$, which is dissipative and is Lipschitz,

with $\zeta > \dfrac{8}{\alpha}$. Then for $0 < \eta < r$ the control (10) is ω-strongly stabilizes the systems (35) and minimizes the following cost of performance:

$$\left\{ \min_{v \in V_{ad}^r(\omega)} Q(v) = \int_0^{+\infty} \|\chi_\omega y(t)\|^2 \left(\frac{1}{\eta}|v(t)|^2 + \eta + \frac{2|y(t)|^2}{\zeta} + 4 \right) dt \qquad (38) \right.$$

where $R = 4\, i_\omega$ and $V_{ad}^r = \{v(t), \ |v(t)| \leq r, \forall t > 0 \ and \ Q_\omega(v) < \infty\}$ with $0 < \eta \leq r$.

4.3 Example (Schrödinger Equation)

We consider the following semilinear Schröedinger equation and $\Omega =]0, 1[$:

$$\begin{cases} i\dfrac{\partial y(t)}{\partial t} = \Delta y(t) - \dfrac{y(t)}{\gamma(1 + y^2(t))} + v(t)\chi_\omega y(t), & in \ \Omega \times]0, +\infty[, \\ y(0) = y_0, \quad y_t(0) = y_1, & on \ \partial\Omega \times]0, +\infty[, \end{cases}$$
$$(39)$$

where ω is an open subset of Ω that indicates the subdomain where the control is supported, χ_ω indicates the restriction operator on ω ($i \in \mathbb{C}$, $i^2 = -1$). Here, the state space $H = L^2(0, 1)$ is endowed with a natural complex inner product, the dynamic operator A is defined by $Ay = -i\Delta$, with A generate a semigroup $S(t)$ of isometries.

From [12] one can deduce: for $\omega \subset]0, 1[$ the measure $0 < |\omega| < 1$, (A_3) holds. The domain of A is

$$\mathcal{D}(A) = H^2(0, 1) \cap H_0^1(0, 1).$$

Here, $B = \chi_\omega$, B is linear. Taking $Ny = -\dfrac{y(t)}{\gamma(1 + y^2(t))}$ which is dissipative and Lipschitz on $B_1 = \{y \in H/\|y\| \leq 1\}$, with $\gamma > \dfrac{2T^2}{\alpha}$.

Then for $0 < \eta < r$ the control (10) is ω-strongly stabilizes the systems (39), its also solution to the problem:

$$\left\{ \min_{v \in V_{ad}^r(\omega)} Q(v) = \int_0^{+\infty} \|\chi_\omega y(t)\|^2 (\frac{1}{\eta}|v(t)|^2 + \eta + \frac{2}{\gamma(1 + y^2(t))}) dt, \qquad (40) \right.$$

where $V_{ad}^r = \{v(t), |v(t)| \leq r, \forall t > 0 \text{ and } Q_\omega(v) < \infty\}$, and $-i\Delta$ is skew-adjoint which gives $Re(< Ay, y >) = 0$, then we have $R = 0$.

5 Conclusion

In the previous two sections, we have established of the regional optimal stabilization and its robustness for a class of semi-linear infinite dimensional systems with sliding mode control. The idea of robustness of the regional optimal stabilization is very investing. Also, the switching feedback is the unique minimizing control of an appropriate functional cost. Many questions are still open, such as the stabilization of semilinear systems without perturbation of nonlinear operator. This will be the aim of future research papers.

Acknowledgements This work has been carried out with a grant from Hassan II Academy of Sciences and Technology project N° 630/2016.

References

1. Ball, J.: Strongly continuous semi-groups, weak solutions, and the variation of constants formula. Proe. Amer. Math. Soc. **63**, 370–373 (1977)
2. Ball, J.: On the asymptotic behaviour of generalized processes, with applications to nonlinear evolution equations. J. Differ. Equ. **27**, 224–265 (1978)
3. Ball, J., Slemrod, M.: Feedback stabilization of distributed semilinear control systems. J. Appl. Math. Opt. **5** (1979)
4. Bardos, C., Lebeau, G., Rauch, J.: Sharp sufficient conditions for the observation, control and stabilization of waves from the boundary. SIAM J. Control Optim. **30**, 1024–1065, 169–179 (1992)
5. Curtain, R.F., Zwart, H.J.: An Introduction to Infinite Dimensional Linear Systems Theory. Springer, Berlin (1991)
6. El Harraki, I., El Alami, A., Boutoulout, A., Serhani, M.: Regional stabilization for semilinear parabolic systems. IMA J. Math. Control Inf. 2015–197 (2016)
7. El-Farra, N.H., Christofides, P.D.: Coordinating feedback and switching for control of spatially distributed processes. Comput. Chem. Eng. **28**, 111–128 (2004)
8. Gugat, M., Sigalotti, M.: Stars of vibrating strings: switching boundary feedback stabilization. Netw. Heterog. Media **5**, 299–314 (2010)
9. Gugat, M.: Optimal switching boundary control of a string to rest in finite time. ZAMM J. Appl. Math. Mech. **88**, 283–305 (2008)
10. Gugat, M., Tucsnak, M.: An example for the switching delay feedback stabilization of an infinite dimensional system: the boundary stabilization of a string. Syst. Control Lett. **60**, 226–233 (2011)
11. Gugat, M., Troltzsch, F.: Boundary feedback stabilization of the Schlogl system. Automatica **51**, 192–199 (2015)
12. Lebeau, G.: Contrôle de l'équation de Schrödinger. J. Math. Bures Appl **71**, 267–291 (1992)
13. Ouzahra, M.: Global stabilization of semilinear systems using switching controls. Automatica **48**, 837–843 (2012)

14. Ouzahra, M.: Exponential and weak stabilization of constrained bilinear systems. SIAM J. Control Optim **48**, 3962–3974 (2010)
15. Ouzahra, M.: Exponential stabilization of distributed semilinear systems by optimal control. J. Math. Anal. Appl **380**, 117–123 (2011)
16. Pazy, A.: Semi-groups of Linear Operators and Applications to Partial Differential Equations. Springer, New York (1983)
17. Sasane, A.: Stability of switching infinite-dimensional systems. Automatica **41**, 75–78 (2005)
18. Zerrik, E., Ouzahra, M., Ztot, K.: Regional stabilisation for infinite bilinear systems. EE Proc. Control Theory Appl. **151**, 109–116 (2004)

Solving Generalized Fractional Schrodinger's Equation by Mean Generalized Fixed Point

S. Melliani, M. Elomari and L. S. Chadli

Abstract The present paper is devoted to the existence and uniqueness results of the generalized fractional Schrodinger's equation

$$\begin{cases} \frac{1}{i}\partial_t^\alpha u(t,x) - \triangle u(t,x) + v(x)u(t,x) = 0, & x \in \mathbb{R}, \ t \geq 0 \\ v(x) = \delta(x), \quad u(0,x) = \delta(x) \end{cases}$$

by using the generalized fixed point.

1 Introduction

Fractional calculus is a generalization of ordinary differentiation and integration to arbitrary non integer order (see [6]). Moreover fractional processes have been increased many developments in the last decade. For instance, they are suitable for describing the long memory properties of many time series. A strong motivation for investigating fractional differential equations comes from physics. Fractional diffusion equations describe anomalous diffusion on fractals (physical objects of fractional dimension, like some amorphous semiconductors or strongly porous materials; see [5, 7]. Colombeau algebras (usually denoted by the letter $\mathscr{G}$) are differential (quotient) algebras with unit, and were introduced by Colombeau [2, 3]. This algebra plays a crucial role in order to give a sense of multiplication of distributions [4, 7]. As a nonlinear extension of distribution theory to deal with nonlinearities and singularities of data and coefficients in PDE theory [7]. These algebras contain the space of distributions D' as a subspace with an embedding realized through convolution

S. Melliani (✉) · M. Elomari · L. S. Chadli
Sultan Moulay Slimane University, BP 523, Beni Mellal, Morocco
e-mail: s.melliani@usms.ma; saidmelliani@gmail.com

M. Elomari
e-mail: m.elomari@usms.ma

L. S. Chadli
e-mail: sa.chadli@yahoo.fr

© Springer Nature Switzerland AG 2020
E. H. Zerrik et al. (eds.), *Recent Advances in Modeling, Analysis and Systems Control: Theoretical Aspects and Applications*, Studies in Systems, Decision and Control 243,
https://doi.org/10.1007/978-3-030-26149-8_7

with a suitable mollifier. Elements of these algebras are classes of nets of smooth functions. The reason for introducing fractional derivatives was the possibility of solving nonlinear problems with singularities and derivatives of arbitrary real order. Fixed point theory has fascinated many researchers since 1922 with the celebrated Banach fixed point theorem. There exists a vast literature on the topic field and this is very active field of research at present. Fixed point theorems are very important tools for proving the existence and uniqueness of the solutions to various mathematical models (integral and partial equations, variational inequalities, etc). It can be applied to, for example, variational inequalities, optimization, and approximation theory. The fixed point theory has been continually studied by many researchers see for example [1]. But it is rare to find a paper that presented the fixed point theory in Colombeau algebra, we are inspired from Martin in [8] in order to give a sense of this concept in the such algebra, by using the topology of locally convex spaces. The present paper is devoted to the study of existence and uniqueness result to the following problem

$$
\begin{cases}
\frac{1}{i}\partial_t^\alpha u(t,x) - \Delta u(t,x) + v(x)u(t,x) = 0, & x \in \mathbb{R}, \ t \geq 0 \\
v(x) = \delta(x), \quad u(0,x) = \delta(x)
\end{cases}
\tag{1}
$$

where δ is the Dirac function and $\alpha \in (0,1)$, by means of so-called generalized fixed point, one can establish a comprehensive and elegant solution theory for (1).

The present paper is organized as follows: After this introduction, we will recall some concept concerning the Colombeaus algebra and fractional calculus in Sect. 2. The new notion of generalized semigroup and some properties take place in Sect. 3. In Sect. 4 we provided the theorem of fixed point in Colombeau algebra. Finally, the existence-uniqueness result for a fractional Schrodinger's equation is proven in Sect. 5.

2 Preliminaries

Here we list some notations and formulas to be used later. The elements of Colombeau algebras $\mathscr{G}$ are equivalence classes of regularization, i.e., sequences of smooth functions satisfying asymptotic conditions in the regularization parameter ε. Therefore, for any set X, the family of sequences $(u_\varepsilon)_{\varepsilon \in (0,1)}$ of elements of a set X will be denoted by $X^{(0,1)}$, such sequences will also be called nets and simply written as u_ε. Let $n \in \mathbb{N}^*$, as in [4] we define the set

$$
\mathscr{E}(\mathbb{R}^n) = \left(\mathscr{C}^\infty(\mathbb{R}^n)\right)^{(0,1)}.
$$

The set of moderate functions is given as follows

$$\mathscr{E}_M(\mathbb{R}^n) = \left\{ (u_\varepsilon)_{\varepsilon>0} \subset \mathscr{E}(\mathbb{R}^n) / \forall K \subset\subset \mathbb{R}^n, \forall \alpha \in \mathbb{N}_0^n, \exists N \in \mathbb{N} \quad \text{such that} \right.$$
$$\left. \sup_{x \in K} |D^\alpha u_\varepsilon(x)| = \mathscr{O}_{\varepsilon \to 0}(\varepsilon^{-N}) \right\}.$$

The ideal of negligible functions is defined by

$$\mathscr{N}(\mathbb{R}^n) = \left\{ (u_\varepsilon)_{\varepsilon>0} \subset \mathscr{E}(\mathbb{R}^n) / \forall K \subset\subset \mathbb{R}^n, \forall \alpha \in \mathbb{N}_0^n, \forall p \in \mathbb{N} \quad \text{such that} \right.$$
$$\left. \sup_{x \in K} |D^\alpha u_\varepsilon(x)| = \mathscr{O}_{\varepsilon \to 0}(\varepsilon^p) \right\}.$$

The Colombeau algebra is defined as a factor set

$$\mathscr{G}(\mathbb{R}^n) = \mathscr{E}_M(\mathbb{R}^n) / \mathscr{N}(\mathbb{R}^n)$$

Also we define the following sets

$$\left| \mathscr{E}_M(\mathbb{R}^n) \right| = \left\{ (|u_\varepsilon|)_\varepsilon, \ u_\varepsilon \in \mathscr{E}_M(\mathbb{R}^n) \right\},$$

and

$$\left| \mathscr{N}(\mathbb{R}^n) \right| = \left\{ (|u_\varepsilon|)_\varepsilon, \ u_\varepsilon \in \mathscr{N}(\mathbb{R}^n) \right\}.$$

The set of all generalized real numbers is defined by

$$\widetilde{\mathbb{R}} = \mathscr{E}(\mathbb{R}) / N(\mathbb{R}),$$

where

$$\mathscr{E}(\mathbb{R}) := \left\{ (x_\varepsilon)_\varepsilon \in (\mathbb{R})^{(0,1)} / \exists m \in \mathbb{N}, |x_\varepsilon| = \mathscr{O}_{\varepsilon \to 0}(\varepsilon^{-m}) \right\},$$

and

$$N(\mathbb{R}) := \left\{ (x_\varepsilon)_\varepsilon \in (\mathbb{R})^{(0,1)} / \forall m \in \mathbb{N}, |x_\varepsilon| = \mathscr{O}_{\varepsilon \to 0}(\varepsilon^m) \right\}.$$

We note that $\widetilde{\mathbb{R}}$ is a ring obtained by factoring moderate families of real numbers with respect to negligible families. It is easy to prove that

Proposition 1 *The space $\mathscr{E}(\mathbb{R})$ is an algebra, and $N(\mathbb{R})$ is an ideal of $\mathscr{E}(\mathbb{R})$.*

In the same we define

$$\left| \mathscr{E}(\mathbb{R}) \right| = \left\{ (|r_\varepsilon|)_\varepsilon, \ r_\varepsilon \in \mathscr{E}(\mathbb{R}) \right\},$$

and

$$\left| N(\mathbb{R}) \right| = \left\{ (|r_\varepsilon|)_\varepsilon, \ r_\varepsilon \in N(\mathbb{R}) \right\}.$$

A fractional integral is defined by:

$$J^\alpha f(t) = \frac{1}{\Gamma(\alpha)} \int_0^t (t-\tau)^{\alpha-1} f(\tau) d\tau \qquad \alpha > 0.$$

Fractional calculus is a branch of mathematical analysis that studies the several different possibilities of defining real number powers or complex number powers of the differentiation operator D. For example, one may ask the question of meaningfully interpreting $D^{\frac{1}{2}}$. It is known that there are many types of derivatives of non-integral order, but in this time we will work with Caputo approach. The fractional derivative of order $\alpha > 0$ in the Caputo sense is defined by:

$$D^\alpha f(t) = \frac{1}{\Gamma(m-\alpha)} \int_0^t \frac{f^{(m)}(\tau) d\tau}{(t-\tau)^{\alpha+1-m}} \qquad m-1 < \alpha < m$$

Let (f_ε) is a representative of $F \in \mathscr{G}$ then

$$D^\alpha f_\varepsilon(t) = \frac{1}{\Gamma(1-\alpha)} \int_0^t \frac{f_\varepsilon'(\tau) d\tau}{(t-\tau)^\alpha} \qquad 0 < \alpha < 1.$$

We have

$$\sup_{t\in[0,T]} \left| D^\alpha f_\varepsilon(t) \right| \leq \frac{1}{\Gamma(1-\alpha)} \sup_{t\in[0,T]} \left| \int_0^t \frac{f'(\tau) d\tau}{(t-\tau)^\alpha} \right|$$

$$\leq \frac{1}{\Gamma(1-\alpha)} \|f'\|_{L^\infty([0,T])} \sup_{t\in[0,T]} \int_0^t \frac{d\tau}{(t-\tau)^\alpha} d\tau$$

$$\leq \frac{1}{\Gamma(1-\alpha)} \varepsilon^{-N} \frac{T^{1-\alpha}}{1-\alpha}$$

$$\leq C_{\alpha,T} \varepsilon^{-N}.$$

In general, for $m-1 < \alpha < m$

$$\sup_{t\in[0,T]} \left| D^\alpha f_\varepsilon(t) \right| \leq \frac{1}{\Gamma(m-\alpha)} \sup_{t\in[0,T]} \int_0^t \frac{|f^{(m)}(\tau)|}{(t-\tau)^{\alpha+1-m}} d\tau$$

$$\leq \frac{1}{\Gamma(m-\alpha)} \|f^{(m)}\|_{L^\infty([0,T])} \sup_{t\in[0,T]} \int_0^t \frac{1}{(t-\tau)^{\alpha+1-m}} d\tau$$

$$\leq \frac{1}{\Gamma(m-\alpha)} \varepsilon^{-N} \frac{T^{m-\alpha}}{m-\alpha}$$

$$\leq C_{\alpha,T} \varepsilon^{-N},$$

the constant $C_{\alpha,T}$ depends on two parameters α and T. In order to prove moderateness for higher derivatives a similar calculation is applied.

Let $G_1, G_2 \in \mathscr{G}(\mathbb{R}^n)$ and $G_{1,\varepsilon}, G_{2,\varepsilon}$ their representatives respectively.

We say that $G_1, G_2 \in \mathscr{G}(\mathbb{R}^n)$ are associated and we write $G_1 \approx G_2$, if for every $\varphi \in \mathscr{D}(\mathbb{R}^n)$

$$\lim_{\varepsilon \to 0} \int_{\mathbb{R}^n} (G_{1,\varepsilon} - G_{2,\varepsilon})\varphi(x)dx = 0.$$

We will end this preliminaries by the Grönwall's inequality

Lemma 1 *Let I denote an interval of the real line of the form $[a, b)$ or $[a, b]$ or $[a, b)$ with $a < b$. Let δ, η and u be real-valued functions defined on I. Assume that η and u are continuous and that the negative part of δ is integrable on every closed and bounded subinterval of I.*

1. *If η is non-negative and if u satisfies the integral inequality*

$$u(t) \leq \delta(t) + \int_a^t \eta(s)u(s)ds, \ \forall t \in I$$

   then

$$u(t) \leq \delta(t) + \int_a^t \delta(s)\eta(s) \exp\left(\int_s^t \eta(r)dr\right)ds, \ t \in I$$

2. *If, in addition, the function δ is non-decreasing, then*

$$u(t) \leq \delta(t) \exp\left(\int_a^t \eta(s)ds\right), \ t \in I$$

3 Generalized Semigroups

The notion of a semigroup plays a crucial role in order to study an evolutionary problem. As we have known a lot of research is devoted to the linking relationship between semigroups of an operator and its infinitesimal generator, the famous relationship is given by Hille–Yosida [9]. In this section we will benefit the classical case and the method of building the Colombeau algebra for giving a sense of the generalized semigroups. We will start by some properties of locally convex spaces.

3.1 Locally Convex Spaces

In this subsection, we recall the concept of locally convex spaces and the notion of completeness in this type of space.

Definition 1 Let X be a vector space with a seminorms familly $(p_i)_{i \in I}$. If τ_i is the topology defined by the only semi-norm p_i. If τ is the super bound of topology τ_i. The space provided with this topology τ is called a locally convex space.

A basis of 0-neighbourhood is the set of all "balls" of the seminorms $(p_i)_{i \in I}$

$$B(i, r) = \left\{ x \in X / \quad p_i(x) < r \right\}, \quad \forall i \in I \text{ and } r > 0.$$

Then, $(x_n)n \in \mathbb{N}$ is a Cauchy sequence if and only if.

$$\left(\forall \varepsilon > 0 \right)\left(\forall i \in I \right)\left(\exists n_0 \in \mathbb{N} \right)\left(\forall n, p \in \mathbb{N} \text{ if } n \geq n_0 \Rightarrow p_i(x_{n+p} - x_n) < \varepsilon \right),$$

and X is sequentially complete if any Cauchy sequence converges to an element e in X.

Definition 2 We said that $\mathscr{D}$ is dense in locally convex space X if and only if

$$\left(\forall x \in X \right) \quad \left(\exists y \in \mathscr{D} \right) \quad \left(\forall \varepsilon > 0 \right) \quad \left(\forall i \in I \right) \quad \text{we have} \quad p_i(x - y) < \varepsilon.$$

3.2 Generalized Semigroups

As we known the semigroup of operator is defined on Banach spaces, but so far we haven't this concept in algebra of Colombeau, so, in order to define this we needed to exploit the previous subsection for manipulating such notion. This subsection is devoted to defining the generalized semigroups and its properties.

Definition 3 Let X be a locally convex space with a seminorm family $(p_i)_{i \in I}$. We define

$$\mathscr{E}_M(X) := \left\{ (x_\varepsilon)_\varepsilon \in (X)^{(0,1)} / \exists m \in \mathbb{N}, \forall i \in I, \, p_i(x_\varepsilon) = \mathscr{O}_{\varepsilon \to 0}(\varepsilon^{-m}) \right\},$$

and

$$\mathscr{N}(X) := \left\{ (x_\varepsilon)_\varepsilon \in (X)^{(0,1)} / \forall m \in \mathbb{N}, \forall i \in I, \quad p_i(x_\varepsilon) = \mathscr{O}_{\varepsilon \to 0}(\varepsilon^m) \right\}.$$

We define the Colombeau algebra type by:

$$\tilde{X} = \mathscr{E}_M(X) / \mathscr{N}^s(X).$$

First, we are looking if it is possible to define a map $A : \tilde{X} \longrightarrow \tilde{X}$ by means of a given family $(A_\varepsilon)_{\varepsilon \in (0,1)}$ of maps $A_\varepsilon : X \longrightarrow X$ where A_ε is a linear and continuous operator. The general requirement is given in the following lemma

Lemma 2 *Let* $(A_\varepsilon)_{\varepsilon\in(0,1)}$ *be a given family of maps* $A_\varepsilon : X \longrightarrow X$. *For each* $(x_\varepsilon)_\varepsilon \in \mathscr{E}_M(X)$ *and* $(y_\varepsilon)_\varepsilon \in \mathscr{N}(X)$, *suppose that*

1. $\left(A_\varepsilon x_\varepsilon\right)_\varepsilon \in \mathscr{E}_M(X)$,
2. $\left(A_\varepsilon(x_\varepsilon + y_\varepsilon)\right)_\varepsilon - \left(A_\varepsilon x_\varepsilon\right)_\varepsilon \in \mathscr{N}(X)$.

Then

$$A : \begin{cases} \tilde{X} \longrightarrow \tilde{X} \\ x = \left[x_\varepsilon\right] \longmapsto Ax = \left[A_\varepsilon x_\varepsilon\right], \end{cases}$$

is well defined.

Proof From the first property we see that the class $\left[(A_\varepsilon x_\varepsilon)_\varepsilon\right] \in \tilde{X}$. Let $x_\varepsilon + y_\varepsilon$ be another representative of $x = [x_\varepsilon]$.

From the second property we have

$$\left(A_\varepsilon(x_\varepsilon + y_\varepsilon)\right)_\varepsilon - \left(A_\varepsilon x_\varepsilon\right)_\varepsilon \in \mathscr{N}(X),$$

and

$$\left[\left(A_\varepsilon(x_\varepsilon + y_\varepsilon)\right)_\varepsilon\right] = \left[\left(A_\varepsilon x_\varepsilon\right)_\varepsilon\right] \text{ in } \tilde{X}.$$

Then A is well defined. $\qquad \blacksquare$

Now we will give the definition of generalized semigroups on the Colombeau's algebra.

Definition 4 Let $\mathscr{S}\mathscr{E}_M\left(\mathbb{R}_+ : \mathscr{L}_c(X)\right)$ is the space of nets $\left(S_\varepsilon\right)_\varepsilon$ of strongly continuous mappings

$$S_\varepsilon : \mathbb{R}_+ \longrightarrow \mathscr{L}_c(X), \quad \varepsilon \in (0,1),$$

with the property that for every $T > 0$ there exists $a \in \mathbb{R}$ such that

$$\sup_{t\in[0,T)} p_i\left(S_\varepsilon(t)\right) = \mathscr{O}_{\varepsilon\to 0}(\varepsilon^a), \ \forall i \in I, \tag{2}$$

and $\mathscr{S}\mathscr{N}(\mathbb{R}_+ : \mathscr{L}_c(X))$ is the space of nets $(N_\varepsilon)_\varepsilon$ of strongly continuous mappings $N_\varepsilon : \mathbb{R}_+ \longrightarrow \mathscr{L}_c(X)$, $\varepsilon \in (0,1)$ with the properties:
For every $b \in \mathbb{R}$ and $T > 0$

$$\sup_{t\in[0,T)} p_i\left(N_\varepsilon(t)\right) = \mathscr{O}_{\varepsilon\to 0}(\varepsilon^b). \tag{3}$$

There exist $t_0 > 0$ and $a \in \mathbb{R}$ such that

$$\sup_{t<t_0} p_i\left(\frac{N_\varepsilon(t)}{t}\right) = \mathscr{O}_{\varepsilon\to 0}(\varepsilon^a), \ \forall i \in I. \tag{4}$$

There exists a net $\left(H_\varepsilon\right)_\varepsilon$ in $\mathscr{L}_c(X)$ and $\varepsilon_0 \in (0, 1)$ such that

$$\lim_{t \to 0} \frac{N_\varepsilon(t)}{t} = H_\varepsilon x, \ x \in X. \tag{5}$$

For every $b > 0$,

$$p_i\left(H_\varepsilon\right) = \mathscr{O}_{\varepsilon \to 0}(\varepsilon^b), \ \forall i \in I. \tag{6}$$

The following Proposition show that the previous notion is in type Colombeau's algebra. Namely this concept take place in our context.

Proposition 2 $\mathscr{S\!E}_M\left(\mathbb{R}_+ : \mathscr{L}_c(X)\right)$ *is algebra with respect to composition and* $\mathscr{S\!N}\left(\mathbb{R}_+ : \mathscr{L}_c(X)\right)$ *is an ideal of* $\mathscr{S\!E}_M\left(\mathbb{R}_+ : \mathscr{L}_c(X)\right)$

Proof Let $\left(S_\varepsilon(t)\right)_\varepsilon \in \mathscr{S\!E}_M\left(\mathbb{R}_+ : \mathscr{L}_c(X)\right)$ and $\left(N_\varepsilon(t)\right)_\varepsilon \in \mathscr{S\!N}_M\left(\mathbb{R}_+ : \mathscr{L}_c(X)\right)$. We will prove only the second assertion, i.e., That

$$\left(S_\varepsilon(t)N_\varepsilon(t)\right)_\varepsilon, \left(N_\varepsilon(t)S_\varepsilon(t)\right)_\varepsilon \in \mathscr{S\!N}_M\left(\mathbb{R}_+ : \mathscr{L}_c(X)\right),$$

where $S_\varepsilon(t)N_\varepsilon(t)$ denotes the composition. Let $\varepsilon \in (0, 1)$. By the properties (2) and (5) of the Definition 4, for some $a \in \mathbb{R}$ and every $b \in \mathbb{R}$, we have

$$p_i\left(S_\varepsilon(t)N_\varepsilon(t)\right) \le p_i\left(S_\varepsilon(t)\right) p_i\left(N_\varepsilon(t)\right) = \mathscr{O}_{\varepsilon \to 0}(\varepsilon^{a+b}), \ \forall i \in I.$$

The same holds for $p_i\left(N_\varepsilon(t)S_\varepsilon(t)\right)$, $\forall i \in I$. Further, the properties (2) and (5) of the definition yield

$$\sup_{t < t_0} p_i\left(\frac{S_\varepsilon(t)N_\varepsilon(t)}{t}\right) \le \sup_{t < t_0} p_i\left(S_\varepsilon(t)\right) \sup_{t < t_0} p_i\left(N_\varepsilon(t)\right)$$
$$= \mathscr{O}_{\varepsilon \to 0}(\varepsilon^a), \quad \forall i.$$

for some $t_0 > 0$ and $a \in \mathbb{R}$. Also,

$$\sup_{t < t_0} p_i\left(\frac{S_\varepsilon(t)N_\varepsilon(t)}{t}\right) = \mathscr{O}_{\varepsilon \to 0}(\varepsilon^a), \ \forall i.$$

for some $t_0 > 0$ and $a \in \mathbb{R}$.

Let now $\varepsilon \in (0, 1)$ be fixed. For all $i \in I$, we have

$$p_i\left(\frac{S_\varepsilon(t)N_\varepsilon(t)}{t}x - S_\varepsilon(0)H_\varepsilon x\right) = p_i\left(S_\varepsilon(t)\frac{N_\varepsilon(t)}{t}x - S_\varepsilon(t)H_\varepsilon x + S_\varepsilon(t)H_\varepsilon x - S_\varepsilon(0)H_\varepsilon x\right)$$
$$\le p_i\left(S_\varepsilon(t)\right)\left(\frac{N_\varepsilon(t)}{t}x - S_\varepsilon(t)H_\varepsilon x\right)$$
$$+ p_i\left(S_\varepsilon(t)H_\varepsilon x - S_\varepsilon(0)H_\varepsilon x\right).$$

By the (2) and (4) of the Definition 4 as well as by the continuity of $t \longrightarrow S_\varepsilon(t)(H_\varepsilon x)$ at zero, it follows that the last expression tend to zero as $t \mapsto 0$. Similarly, we have

$$p_i\left(\frac{N_\varepsilon(t)S_\varepsilon(t)}{t}x - H_\varepsilon S_\varepsilon(0)x\right) = \left(\frac{N_\varepsilon(t)}{t}S_\varepsilon(t)x - \frac{N_\varepsilon(t)}{t}S_\varepsilon(0)x + \frac{N_\varepsilon(t)}{t}S_\varepsilon(0)x - H_\varepsilon S_\varepsilon(0)x\right)$$

$$\leq p_i\left(\frac{N_\varepsilon(t)}{t}\right)p_i\left(S_\varepsilon(t)x - S_\varepsilon(0)x\right)$$

$$+ p_i\left(\frac{N_\varepsilon(t)}{t}(S_\varepsilon(0)x) - H_\varepsilon(S_\varepsilon(0)x)\right).$$

Assumptions (2), (4) and (5) imply that the last expression tends to zero as $t \mapsto 0$. Thus (5) is proved in both cases.

Now we define Colombeau type algebra as the factor algebra

$$\mathscr{SG}\left(\mathbb{R}_+ : \mathscr{L}(X)\right) = \mathscr{SE}_M\left(\mathbb{R}_+ : \mathscr{L}(X)\right)/\mathscr{SN}\left(\mathbb{R}_+ : \mathscr{L}(X)\right).$$

Elements of $\mathscr{SG}\left(\mathbb{R}_+ : \mathscr{L}(X)\right)$ will be denoted by $S = [S_\varepsilon]$, where $(S_\varepsilon)_\varepsilon$ is a representative of the above class.

Definition 5 $S \in \mathscr{SG}\left(\mathbb{R}_+ : \mathscr{L}(X)\right)$ is a called a Colombeau C_0-Semigroup if it has a representative $(S_\varepsilon)_\varepsilon$ such that, for some $\varepsilon_0 > 0$, S_ε is a C_0-Semigroup, for every $\varepsilon < \varepsilon_0$.

In the sequel we will use only representatives $\left(S_\varepsilon\right)_\varepsilon$ of a Colombeau C_0-semigroup S which are C_0-semigroups, for ε small enough.

Proposition 3 Let $\left(S_\varepsilon\right)_\varepsilon$ and $\left(\tilde{S}_\varepsilon\right)_\varepsilon$ be representatives of a Colombeau C_0-semigroup S, with the infinitesimal generators A_ε, $\varepsilon < \varepsilon_0$, and $\tilde{A}_\varepsilon$, $\varepsilon < \tilde{\varepsilon}_0$, respectively, where ε_0 and $\tilde{\varepsilon}_0$ correspond (in the sense of Definition 5 to $\left(S_\varepsilon\right)_\varepsilon$ and $\left(\tilde{S}_\varepsilon\right)_\varepsilon$, respectively. Then, $D(A_\varepsilon) = D\left(\tilde{A}_\varepsilon\right)$, for every $\varepsilon < \bar{\varepsilon} = \min\left\{\varepsilon_0, \tilde{\varepsilon}_0\right\}$ and $A_\varepsilon - \tilde{A}_\varepsilon$ can be extended to an element of $\mathscr{L}(X)$, denoted again by $A_\varepsilon - \tilde{A}_\varepsilon$. Moreover, for every $a \in \mathbb{R}$,

$$p_i\left(A_\varepsilon - \tilde{A}_\varepsilon\right) = \mathscr{O}_{\varepsilon\to0}(\varepsilon^a), \forall i. \tag{7}$$

Proof Denote $N_\varepsilon(S_\varepsilon - \tilde{S}_\varepsilon)_\varepsilon \in \mathscr{SN}(\mathbb{R}_+, \mathscr{L}(X))$. Let $\varepsilon < \tilde{\varepsilon}_0$ be fixed and $x \in X$. we have

$$\frac{S_\varepsilon(t)x - x}{t} - \frac{\tilde{S}_\varepsilon(t)x - x}{t} = \frac{N_\varepsilon(t)}{t}x.$$

This implies by letting $t \mapsto 0$, that $D(A_\varepsilon) = D(\tilde{A}_\varepsilon)$. Now we have

$$\left(A_\varepsilon - \tilde{A}_\varepsilon\right)x = \lim_{t \to 0} \frac{S_\varepsilon(t)x - x}{t} \tag{8}$$

$$- \lim_{t \to 0} \frac{\tilde{S}_\varepsilon(t)x - x}{t} \tag{9}$$

$$= \lim_{t \to 0} \frac{N_\varepsilon(t)}{t}x = H_\varepsilon x, \quad x \in D(A_\varepsilon), \tag{10}$$

since $D(A_\varepsilon)$ is dense in X, properties (4), (5) and (7) imply that for every $a \in \mathbb{R}$,

$$p_i\left(A_\varepsilon - \tilde{A}_\varepsilon\right) = \mathcal{O}_{\varepsilon \to 0}(\varepsilon^a).$$

Now we define the infinitesimal generator of a Colombeau C_0-semigroup S. Denote by $\mathscr{A}$ the set of pairs $((A_\varepsilon)_\varepsilon, (D(A_\varepsilon))_\varepsilon)$ where A_ε is a closed linear operator on X with the dense domain $D(A_\varepsilon) \subset X$, for every $\varepsilon \in (0, 1)$. We introduce an equivalence relation in A

$$\left((A_\varepsilon)_\varepsilon, \left(D(A_\varepsilon)\right)_\varepsilon\right) \sim \left((\tilde{A}_\varepsilon)_\varepsilon, \left(D(\tilde{A}_\varepsilon)\right)_\varepsilon\right).$$

If there exist $\varepsilon_0 \in (0, 1)$ such that $D(A_\varepsilon) = D(\tilde{A}_\varepsilon)$, for every $\varepsilon < \varepsilon_0$, and for every $a \in \mathbb{R}$ there exist $C > 0$ and $\varepsilon_a \leq \varepsilon_0$ such that, for $x \in D(A_\varepsilon)$, $p_i\left(A_\varepsilon - \tilde{A}_\varepsilon)x\right) \leq C\varepsilon^a p_i(x)$, $\forall i \ x \in D(A_\varepsilon)$, $\varepsilon \leq \varepsilon_a$. Since A_ε has a dense domain in X, $R_\varepsilon :=$ $A_\varepsilon - \tilde{A}_\varepsilon$ can be extended to be an operator in $\mathscr{L}_c(X)$ satisfying $p_i\left((A_\varepsilon - \tilde{A}_\varepsilon)x\right) =$ $\mathcal{O}_{\varepsilon \to 0}(\varepsilon^a)$, for every $a \in \mathbb{R}$. such an operator R_ε is called the zero operator.

We denote by A the corresponding element of the quotient space $\mathscr{A}/\sim$. Due to Proposition 3, the following definition makes sense.

Definition 6 $A \in \mathscr{A}/\sim$ is the infinitesimal generator of a Colombeau C_0-semigroup S if there exists a representative $(A_\varepsilon)_\varepsilon$ of A such that A_ε is the infinitesimal generator of S_ε, for ε small enough.

By Pazy [9] we have the following proposition

Proposition 4 *Let S be a Colombeau C_0-semigroup with the infinitesimal generator A. Then there exists $\varepsilon_0 \in (0, 1)$ such that:*

- *Mapping $t \longmapsto S_\varepsilon(t)x : \mathbb{R}_+ \longrightarrow X$ is continuous for every $x \in X$ and $\varepsilon < \varepsilon_0$,*
-

$$\lim_{h \to 0} \int_t^{t+h} S_\varepsilon(s)x ds = S_\varepsilon(t)x, \quad \varepsilon < \varepsilon_0, \quad x \in X.$$

-

$$\int_0^t S_\varepsilon(s)x ds \quad \in D(A_\varepsilon), \quad \varepsilon < \varepsilon_0, \quad x \in X.$$

- *For every $x \in D(A_\varepsilon)$, $t \geq 0$ $S_\varepsilon(t)x \in D(A_\varepsilon)$ and*

$$\frac{d}{dt} S_\varepsilon(t)x = A_\varepsilon S_\varepsilon(t)x = S_\varepsilon(t)A_\varepsilon x, \quad \varepsilon < \varepsilon_0. \tag{11}$$

- Let $\left(S_\varepsilon\right)_\varepsilon$ and $\left(\tilde{S}_\varepsilon\right)_\varepsilon$ be representative of Colombeau C_0-semigroup S, with infinitesimal generators A_ε and $\tilde{A}_\varepsilon$, $\varepsilon < \varepsilon_0$, respectively. Then, for every $a \in \mathbb{R}$ and $t \geq 0$, for all i,

$$p_i\left(\frac{d}{dt} S_\varepsilon(t) - \tilde{A}_\varepsilon S_\varepsilon(t)\right) = \mathcal{O}_{\varepsilon \to 0}(\varepsilon^a). \tag{12}$$

- For every $x \in D\left(A_\varepsilon\right)$ and every $t, s \geq 0$.

$$S_\varepsilon(t)x - S_\varepsilon(s)x = \int_s^t S_\varepsilon(\tau)A_\varepsilon x d\tau = \int_s^t A_\varepsilon S_\varepsilon(\tau)x d\tau.$$

Now we will discuss a condition given the equality between two generalized semigroups.

Theorem 1 *Let S and $\tilde{S}$ be Colombeau C_0-semigroups with infinitesimal generators A and $\tilde{A}$, respectively. If $A = \tilde{A}$ then $S = \tilde{S}$.*

Proof Let ε be small enough and $x \in D\left(A_\varepsilon\right) = D\left(\tilde{A}_\varepsilon\right)$. Proposition 4 property 4 implies that for $t \geq 0$, the mapping $s \longmapsto \tilde{S}_\varepsilon(t - s)S_\varepsilon(s)x, t \geq s \geq 0$ is differentiable and

$$\frac{d}{ds}\left(\tilde{S}_\varepsilon(t - s)S_\varepsilon(s)x\right) = -\tilde{A}_\varepsilon \tilde{S}_\varepsilon(t - s)S_\varepsilon(s)x$$

$$+ \tilde{S}_\varepsilon(t - s)A_\varepsilon S_\varepsilon(s)x, \quad t \geq s \geq 0.$$

The assumption $A = \tilde{A}$ implies that $A_\varepsilon = \tilde{A}_\varepsilon + R_\varepsilon$, where R_ε is a zero operator. Since $\tilde{A}_\varepsilon$ commutes with $\tilde{S}_\varepsilon$, for every $x \in D\left(A_\varepsilon\right)$

$$\frac{d}{ds}\left(\tilde{S}_\varepsilon(t - s)S_\varepsilon(s)x\right) = S_\varepsilon(t - s)R_\varepsilon S_\varepsilon(s)x, \quad t \geq s \geq 0,$$

which implies that

$$\tilde{S}_\varepsilon(t - s)S_\varepsilon(s)x - \tilde{S}_\varepsilon(t)x = \int_0^s \tilde{S}_\varepsilon(t - u)R_\varepsilon S_\varepsilon(u)x du, \quad t \geq s \geq 0. \tag{13}$$

Putting $s = t$ in (13), we obtain

$$S_\varepsilon(t)x - \tilde{S}_\varepsilon(t)x = \int_0^s \tilde{S}_\varepsilon(t - u)R_\varepsilon S_\varepsilon(u)x du, \quad t \geq 0, x \in D(A_\varepsilon). \tag{14}$$

Since $D(A_\varepsilon)$ is dense in X, uniform boundedness of S and $\tilde{S}$ on $[0, t]$ implies that (11) holds for every $y \in X$. Let us prove that $(N_\varepsilon)_\varepsilon = (S_\varepsilon - \tilde{S}_\varepsilon)_\varepsilon \in \mathscr{SN}(\mathbb{R}_+ : \mathscr{L}_c(X))$.

The formula (12) and Proposition 4 imply that for some $C > 0$ and $a, \tilde{a} \in \mathbb{R}$, for all i

$$\sup_{t \in [0,T)} p_i \left(N_\varepsilon(t) x \right) \leq \sup_{t \in [0,T)} \int_0^t p_i \left(\tilde{S}_\varepsilon(t-u) \right) p_i \left(R_\varepsilon \right) (S_\varepsilon(u)) \, p_i \left(x \right) du$$

$$\leq T \, C \, \varepsilon^{a+\tilde{a}} p_i \left(R_\varepsilon x \right), \quad x \in X.$$

Since $p_i \left(R_\varepsilon \right) = \mathcal{O}_{\varepsilon \to 0}(\varepsilon^b)$, for every $b \in \mathbb{R}, \left(N_\varepsilon(t) \right)_\varepsilon$ satisfies condition (3) in Proposition 4. Condition (3) follows from the boundedness of $(\tilde{S}_\varepsilon)_\varepsilon$, $(S_\varepsilon)_\varepsilon$ on bounded domain $[0, t)$, the properties of $(R_\varepsilon)_\varepsilon$ and the following expression:

$$p_i \left(\frac{N_\varepsilon(t)}{t} \right) = p_i \left(\frac{1}{t} \int_0^t \tilde{S}_\varepsilon(t-u) R_\varepsilon S_\varepsilon(u) x \, du \right) \leq p_i \left(\tilde{S}_\varepsilon(t) \right) p_i \left(R_\varepsilon \right) (S_\varepsilon)$$

$$\leq const, x \in X, \ t \leq t_0, \ \forall i,$$

for some $t_0 > 0$. Also,

$$\lim_{t \to 0} \frac{N_\varepsilon(t)}{t} = \lim_{t \to 0} \frac{\tilde{S}_\varepsilon(t) x - x}{t} - \lim_{t \to 0} \frac{S_\varepsilon(t) x - x}{t}$$

$$= R_\varepsilon x, \quad \forall x \in D \left(A_\varepsilon \right).$$

Since it is enough that (5) holds for a dense subset of X see the remark after Definition 4 this concludes the proof.

4   Generalized Fixed Points

We will presented the notion of fixed point in Colombeau algebra, by using this notion in locally convex space.

4.1   Contractions in Locally Convex and Complete Spaces

This subsection is devoted to discuss the contraction map in locally convex spaces, which led us to define the contraction map in type Colombeau's algebra. Through this section X is also a locally convex space.

Definition 7 A map $A_\varepsilon : X \longrightarrow X$ is called a contraction if for all $i \in I$ it exists $k_i < 1$ such that

$$\forall (x_\varepsilon, y_\varepsilon) \in X \times X, \ p_i(A_\varepsilon x_\varepsilon - A_\varepsilon y_\varepsilon) \leq k_i \, p_i(x_\varepsilon - y_\varepsilon).$$

We have the following result

Theorem 2 *Any contraction $A_\varepsilon : X \longrightarrow X$ has a fixed point. If X is Hausdorff, this fixed point is unique.*

Proof Starting from $x_{0\varepsilon} \in X$ define $x_{(n+1)\varepsilon} = A_\varepsilon(x_{n\varepsilon})$ by induction. It is easy to verify that $x_{n\varepsilon}$ is a Cauchy sequence in the complete space X and converges to some $x_\varepsilon \in X$. The contraction property of the map A_ε implies obviously its continuity. Then, passing to the limit in $x_{(n+1)\varepsilon} = A_\varepsilon(x_{n\varepsilon})$, we obtain that x_ε is a fixed point of X. If X is Hausdorff, for all $z_\varepsilon \neq 0$ it exists $V \in V(0)$ such that $z_\varepsilon \notin V$. Then it exists i (depending on z_ε) such that $pi(z_\varepsilon) > 0$. If x_ε and y_ε are two different fixed points of X, it exists j (depending on $x_\varepsilon - y_\varepsilon$) such that

$$0 < p_j\left(x_\varepsilon - y_\varepsilon\right) = p_j\left(A_\varepsilon(x_\varepsilon) - A_\varepsilon(y_\varepsilon)\right) \le k_j \; p_j\left(x_\varepsilon - y_\varepsilon\right) < pi\left(x_\varepsilon - y_\varepsilon\right).$$

4.2 Contraction Operator in $\tilde{X}$

We will give a notion of contraction map in type Colombeau algebra.

Definition 8 The following hypotheses permit to well define a map $A : \tilde{X} \longrightarrow \tilde{X}$ and to call it a contraction.

(a) for each $(x_\varepsilon)_\varepsilon \in \mathcal{E}_M(X)$, $(A_\varepsilon x_\varepsilon)_\varepsilon \in \mathcal{E}_M(X)$.
(b) Each A_ε is a contraction in $\left(X, \tau_\varepsilon\right)$ endowed with the family $Q_\varepsilon = (q_{\varepsilon,i})_{i\in I}$ and the corresponding contraction constants are denoted by $l_{\varepsilon,i} < 1$.
(c) For each $i \in I$ and $\varepsilon \in (0, 1]$, $\exists a_{\varepsilon,i} > 0$ and $b_{\varepsilon,i} > 0$ such that

$$a_{\varepsilon,i} \; p_i \le q_{\varepsilon,i} \le b_{\varepsilon,i} \; p_i.$$

(d) For each $i \in I$, $\forall \varepsilon \in (0, 1]$, $(\frac{b_{\varepsilon,i}}{a_{\varepsilon,i}})_\varepsilon$ and $(\frac{1}{1-l_{\varepsilon,i}})_\varepsilon \in |\mathcal{E}_M(\mathbb{R})|$.

The essential result given in this theorem

Theorem 3 *Any contraction $A : \tilde{X} \longrightarrow \tilde{X}$ has a fixed point in $\tilde{X}$.*

Proof Remark that condition **(a)** which is **(i)** in Lemma 2.
Let $(i_\varepsilon)_\varepsilon \in \mathcal{N}(X)$ and $(x_\varepsilon)_\varepsilon \in \mathcal{E}_M(X)$, we have

$$p_i(A_\varepsilon(x_\varepsilon + i_\varepsilon) - A_\varepsilon x_\varepsilon) = p_i(A_\varepsilon(x_\varepsilon + i_\varepsilon - x_\varepsilon))$$
$$p_i(A_\varepsilon(x_\varepsilon + i_\varepsilon) - A_\varepsilon x_\varepsilon) = p_i(A_\varepsilon i_\varepsilon)$$
$$p_i(A_\varepsilon(x_\varepsilon + i_\varepsilon) - A_\varepsilon x_\varepsilon) \le C \; p_i(i_\varepsilon).$$

Then $(A_\varepsilon(x_\varepsilon + i_\varepsilon) - A_\varepsilon x_\varepsilon)_\varepsilon \in \mathcal{N}(X)$ and the condition **(ii)** in Lemma 2 is verified. Then A is well defined.

From Theorem 2 we know that each A_ε has a fixed point z_ε obtained as limit of the Cauchy sequence $z_{n\varepsilon}$ defined by induction by $z_{(n+1)\varepsilon} = A_\varepsilon(z_{n\varepsilon})$. Starting from $z_0 = [z_{0\varepsilon}] \in \tilde{X}$, we deduce that $z_1 = [A_\varepsilon(z_{0\varepsilon})] \in \tilde{X}$, and $z_1 - z_0 \in \tilde{X}$. That is to say $p_i(z_{1\varepsilon} - z_{0\varepsilon})_\varepsilon \in |\mathscr{E}_M(\mathbb{R})|$. By induction we can compute for all $n, p \in \mathbb{N}$

$$q_{\varepsilon,i}(z_{n+p,\varepsilon} - z_{n,\varepsilon}) \leq \frac{l_{\varepsilon,i}^n}{1 - l_{\varepsilon,i}} q_{\varepsilon,i}(z_{1,\varepsilon} - z_{0,\varepsilon}).$$

then

$$q_{\varepsilon,i}(z_{p,\varepsilon} - z_{0,\varepsilon}) \leq \frac{1}{1 - l_{\varepsilon,i}} q_{\varepsilon,i}(z_{1,\varepsilon} - z_{0,\varepsilon}).$$

When taking the limit z_ε of $z_{p\varepsilon}$ in (X, τ_ε) when $p \longrightarrow +\infty$, we get

$$q_{\varepsilon,i}(z_\varepsilon - z_{0,\varepsilon}) \leq \frac{1}{1 - l_{\varepsilon,i}} q_{\varepsilon,i}(z_{1,\varepsilon} - z_{0,\varepsilon}),$$

Writing now

$$q_{\varepsilon,i}(z_\varepsilon) \leq q_{\varepsilon,i}(z_\varepsilon - z_{0,\varepsilon}) + q_{\varepsilon,i}(z_{0,\varepsilon}),$$

we have

$$p_i(z_\varepsilon) \leq \frac{1}{a_{\varepsilon,i}} q_{\varepsilon,i}(z_\varepsilon) \leq \frac{b_{\varepsilon,i}}{a_{\varepsilon,i}} \left[\frac{1}{1 - l_{\varepsilon,i}} (p_i(z_\varepsilon - z_{0,\varepsilon}) + p_i(z_{0,\varepsilon})) \right].$$

Then, from the hypotheses $(p_i(z_\varepsilon))_\varepsilon \in |\mathscr{E}_M(\mathbb{R})|$, that is to say $(z_\varepsilon))_\varepsilon \in \mathscr{E}_M(X)$. If $z = [z_\varepsilon]$ then we have

$$Az = [A_\varepsilon z_\varepsilon] = [z_\varepsilon] = z.$$

Then z is a fixed point of A.

5 Mains Results

Our goal in this section is to prove that the existence and uniqueness of the problem (1). First we get the following proposition

Proposition 5 *Let $A_\varepsilon = (V_\varepsilon - \Delta)$ is the infinitesimal generators of C_0-semigroups (S_ε). Moreover*

$$S \in SG([0, +\infty[, \mathscr{L}(L^2)) \quad \text{with} \quad S = [S_\varepsilon]$$

Proof The C_0-semigroups $S_\varepsilon : [0, +\infty[\to L(\mathscr{L}^2)$, defined by the Heat kernel formula

$$S_\varepsilon(.)\psi(x) = (E(t, x) * \psi(x)) * \varphi_\varepsilon(t)$$

where $E(t, x) = \frac{1}{2\sqrt{\pi t}} \exp \frac{i|x|^2}{4t}$. That is

$$S_\varepsilon(t)\psi(x) = \int_0^t \int_{\mathbb{R}} E(t - \tau, y - x) \psi(y)\varphi_\varepsilon(\tau)d\tau$$

Here $X = L^2(\mathbb{R})$. By Nakumara [9] the mapping

$$S(t) : \begin{cases} L^2(\mathbb{R}) \to L^2(\mathbb{R}) \\ \psi \to E(t, .) * \psi(.) \end{cases}$$

is a semigroup.

But

$$S_\varepsilon(t) = S * \varphi_\varepsilon(t)$$

which completes the proof. Therefore

$$S = [S_\varepsilon] \in SG([0, +\infty[, L(\mathscr{L}^2(\mathbb{R})))$$

Now we consider the existence and uniqueness result for a fractional differential equation given by:

$$\begin{cases} D_c^q u(t) = g(t, u(t)) = Au(t) + f(t), \ t \in [0, b], \\ u(0) = u_0 \in \tilde{\mathbb{R}} \end{cases} \tag{15}$$

where D_c^q is the Caputo derivative of order $0 < q < 1$, $u_0 \in \tilde{R}$, $g \in \mathscr{C}(J \times \tilde{R}; \tilde{R})$, $J = [0, b]$ and $u \in \mathscr{C}(J; \tilde{R})$, $I^q u \in D(A)$, $f : J \longrightarrow X$ is continuous. In the forthcoming analysis, we need the following hypothesis:

H1: the linear operator $A_\varepsilon : D(A_\varepsilon) \subset X \longrightarrow X$ (X banach space) satisfies the Hille–Yosida condition, that is, there exist two constant $\omega \in \mathbb{R}$ and M_1 such that $]w, +\infty[\subset \rho(A_\varepsilon)$ and

$$\| (\lambda I - A_\varepsilon)^{-k} \|_{\mathscr{L}(X)} \le \frac{M_1}{(\lambda - w)^k}, \ for \ all \ \lambda > \omega, k \ge 1.$$

H2: $Q_\varepsilon(t)$ is continuous in the uniform operator topology for $t > 0$ and $\{Q_\varepsilon(t)\}_{t \ge 0}$ is uniformly bounded, that is, there exists $M_2 > 1$ such that $\sup_{t \ge 0} |Q_\varepsilon(t)| < M_2$.

By Hille–Yosida theorem A generate a generalized C_0-semigroup S. We need the following definition before we proceed further.

Definition 9 Let $g \in \tilde{\mathbb{R}}$, We tell that g is globally Lipschitz if, $\forall t \in J, \forall \varepsilon \in]0, 1], \exists k_\varepsilon(t) > 0, \forall (y, z) \in \tilde{\mathbb{R}} \times \tilde{\mathbb{R}}$ we have

$$\mid g_\varepsilon(t, y_\varepsilon) - g_\varepsilon(t, z_\varepsilon) \mid \leq k_\varepsilon(t) \mid y\varepsilon - z_\varepsilon \mid$$

where $\sup_{t \in J} k_\varepsilon(t) = M_{T,\varepsilon} < +\infty$.

Now we will presented the existence and uniqueness result of our problem.

Theorem 4 *Assume that the hypotheses H_1 and H_2 hold and g satisfied a global lipschitz, then (15) admit unique solution.*

Proof for $u_0 \in \tilde{\mathbb{R}}, g \in \mathscr{C}(J \times \tilde{\mathbb{R}}; \tilde{\mathbb{R}}), u \in C(J; X)$, and $I^q u \in D(A)$.
The problem reduces to finding a fixed point of the map

$$\phi : \tilde{\mathbb{R}} \longrightarrow \tilde{\mathbb{R}}$$

such that

$$\forall t \in J \quad \phi(x)(t) = u_0 + SI^q u(t) + I^q f(t)$$

In order to prove the result, we will check the assumptions (a), (b), (c) and (d) of Definition 8, and applying Theorem 3.

(a) We pose

$$\phi_\varepsilon(u)(t) = u_{\varepsilon 0} + S_\varepsilon I^q u_\varepsilon(t) + I^q f_\varepsilon(t), \quad \forall t \in J.$$

Form what it is clear that $\phi_\varepsilon : \mathscr{C}^\infty(J, X) \longrightarrow \mathscr{C}^\infty(J, X)$, here the space $(\mathscr{C}^\infty(J, X, \tau)$ is a topological space where τ is given by the family of norms $(p_T)_{T \in J}$, such that $p_T(u_\varepsilon) = \sup_{t \in [0,T]} \left| u_\varepsilon(t) \right|$, for all $u_\varepsilon \in C^\infty(J, X)$, let $(u_\varepsilon)_\varepsilon \in \mathscr{E}_M^s(J)$ and $(v_\varepsilon)_\varepsilon \in \mathscr{N}^s(J)$ we have

$$\phi_\varepsilon(u_\varepsilon)(t) = u_{\varepsilon 0} + S_\varepsilon I^q u_\varepsilon(t) + I^q f_\varepsilon(t),$$

Since g is Lipchitz then

$$\left| g_\varepsilon(2u_\varepsilon(t)) - g_\varepsilon(u_\varepsilon(t)) \right| \leq k_\varepsilon(t) \left| u_\varepsilon(t) \right|,$$

and

$$\left| S_\varepsilon(u_\varepsilon(t)) \right| \leq k_\varepsilon(t) \left| u_\varepsilon(t) \right|,$$

then $(S_\varepsilon u_\varepsilon(t))_\varepsilon \in \mathscr{E}_M^s(J)$, we have

$$I^q u_\varepsilon(t) = \int_0^t \frac{(t - s)^{q-1}}{\Gamma(q)} u_\varepsilon(s) ds$$

$$\left| I^q u_\varepsilon(t) \right| \le \frac{b^q}{\Gamma(q)} \left| u_\varepsilon(t) \right|,$$

then $I^q u_\varepsilon(t) \in \mathscr{E}_M^s(J)$. So,

$$\left| \phi_\varepsilon(u_\varepsilon)(t) \right| \le \left| u_{\varepsilon 0} \right| + \left| S_\varepsilon I^q u_\varepsilon(t) \right| + \left| I^q f_\varepsilon(t) \right|,$$

then

$$p_T(\phi_\varepsilon(u_\varepsilon)) \in |\mathscr{E}_M^r| \Rightarrow (\phi_\varepsilon(u_\varepsilon))_\varepsilon \in \mathscr{E}_M^s(J).$$

(b) First we have to write (1) in term of representatives

$$\begin{cases} D_c^q u_\varepsilon(t) = g_\varepsilon(t, u_\varepsilon(t)) = S_\varepsilon \, u_\varepsilon(t) + f_\varepsilon(t), \\ u_\varepsilon(0) = u_{0\varepsilon} \in \mathbb{R} \end{cases} \tag{16}$$

from what it is clear that $\phi_\varepsilon : \mathscr{C}^\infty(J, X) \longrightarrow \mathscr{C}^\infty(J, X)$

Denote by $(\mathscr{C}^\infty(J, X), \tau_\varepsilon)$ is here a topological space where τ_ε is given by the family of norms $(q_{T,\varepsilon})_{T \in \mathbb{R}^+}$ such that

$$\forall u_\varepsilon \in \mathscr{C}^\infty(J, X), q_{T,\varepsilon}(u_\varepsilon) = \sup_{t \in [0,T]} \left| u_\varepsilon(t) \right| \exp\left(-t \frac{b^{q-1}}{\Gamma(q)} M_{T,\varepsilon} \right),$$

we have

$$\phi_\varepsilon(u_\varepsilon)(t) - \phi_\varepsilon(v_\varepsilon)(t) = I^q (g_\varepsilon(t, u_\varepsilon(t)) - g_\varepsilon(t, v_\varepsilon(t)))$$

$$= \int_0^t \frac{(t-s)^{q-1}}{\Gamma(q)} (g_\varepsilon(s, u_\varepsilon(s)) - g_\varepsilon(s, v_\varepsilon(s))) ds.$$

which implies that

$$\left| \phi_\varepsilon(u_\varepsilon)(t) - \phi_\varepsilon(v_\varepsilon)(t) \right| \le \int_0^t \frac{(b)^{q-1}}{\Gamma(q)} M_{T,\varepsilon} \left| u_\varepsilon(s) - v_\varepsilon(s) \right| ds,$$

and

$$e^{-t \frac{b^{q-1}}{\Gamma(q)} M_{T,\varepsilon}} \left| \phi_\varepsilon(u_\varepsilon)(t) - \phi_\varepsilon(v_\varepsilon)(t) \right| \le e^{-t \frac{b^{q-1}}{\Gamma(q)} M_{T,\varepsilon}} \int_0^t \frac{(b)^{q-1}}{\Gamma(q)} M_{T,\varepsilon} \left| u_\varepsilon(s) - v_\varepsilon(s) \right| ds.$$

Writing now

$$e^{-t \frac{b^{q-1}}{\Gamma(q)} M_{T,\varepsilon}} \int_0^t \frac{(b)^{q-1}}{\Gamma(q)} M_{T,\varepsilon} \left| u_\varepsilon(s) - v_\varepsilon(s) \right| ds$$

$$= e^{-t \frac{b^{q-1}}{\Gamma(q)} M_{T,\varepsilon}} \int_0^t \frac{b^{q-1}}{\Gamma(q)} M_{T,\varepsilon} e^{-s \frac{b^{q-1}}{\Gamma(q)} M_{T,\varepsilon}} e^{s \frac{b^{q-1}}{\Gamma(q)} M_{T,\varepsilon}} \left| u_\varepsilon(s) - v_\varepsilon(s) \right| ds.$$

Implies that

$$e^{-t\frac{b^{q-1}}{\Gamma(q)}M_{T,\varepsilon}} \int_0^t \frac{(b)^{q-1}}{\Gamma(q)} M_{T,\varepsilon} |u_\varepsilon(s) - v_\varepsilon(s)| ds$$

$$\leq e^{-t\frac{b^{q-1}}{\Gamma(q)}M_{T,\varepsilon}} q_{T,\varepsilon}(u_\varepsilon - v_\varepsilon) \int_0^t \frac{b^{q-1}}{\Gamma(q)} M_{T,\varepsilon} e^{-t\frac{b^{q-1}}{\Gamma(q)}M_{T,\varepsilon}} ds.$$

Thus

$$e^{-t\frac{b^{q-1}}{\Gamma(q)}M_{T,\varepsilon}} \int_0^t \frac{(b)^{q-1}}{\Gamma(q)} M_{T,\varepsilon} |u_\varepsilon(s) - v_\varepsilon(s)| ds \leq q_{T,\varepsilon}(u_\varepsilon - v_\varepsilon)(1 - e^{-t\frac{b^{q-1}}{\Gamma(q)}M_{T,\varepsilon}}).$$

As consequence

$$q_{T,\varepsilon}\big(\phi_\varepsilon(u_\varepsilon) - \phi_\varepsilon(v_\varepsilon)\big) \leq q_{T,\varepsilon}(u_\varepsilon - v_\varepsilon)(1 - e^{-b\frac{b^{q-1}}{\Gamma(q)}M_{T,\varepsilon}}).$$

So, ϕ_ε is a contraction in $(\mathscr{C}^\infty(J, \mathbb{R}), \tau_\varepsilon)$.

(c) We can write for all $T \in J$ and $u_\varepsilon \in C^\infty(J, X)$

$$\sup_{t\in[0,T]} \left\{ |u_\varepsilon(t)| \, e^{-\frac{b^q}{\Gamma(q)}M_{T,\varepsilon}} \right\} \leq \sup_{t\in[0,T]} \left\{ |u_\varepsilon(t)| \, e^{-t\frac{b^{q-1}}{\Gamma(q)}M_{T,\varepsilon}} \right\}$$

$$\leq \sup_{t\in[0,T]} |u_\varepsilon(t)|,$$

then

$$e^{-\frac{b^q}{\Gamma(q)}M_{T,\varepsilon}} p_T \leq q_{T,\varepsilon} \leq p_T.$$

(d) Assume now that for each $T \in J$ we have

$$\left(e^{\frac{b^q}{\Gamma(q)}M_{T,\varepsilon}}\right)_\varepsilon \in \mathscr{E}_M^{or},$$

and

$$\left(\frac{1}{1 - (1 - e^{-b\frac{b^{q-1}}{\Gamma(q)}M_{T,\varepsilon}})}\right)_\varepsilon = (e^{\frac{b^q}{\Gamma(q)}M_{T,\varepsilon}})_\varepsilon \in \mathscr{E}_M^{or}.$$

Finally, from definition contraction map on generalized function of Colombeau the following map

$$\phi : \begin{cases} \tilde{\mathbb{R}} \longrightarrow \tilde{\mathbb{R}}, \\ u(t) = [u_\varepsilon(t)] \longmapsto \phi(u)(t) = [\phi_\varepsilon(u_\varepsilon)(t)] \end{cases}$$

is a contraction, with $w = [w_\varepsilon]$ as fixed point from Theorem 3 z_ε being the unique fixed point of ϕ_ε.

We are going to prove that z is the unique fixed point of ϕ, and therefore the unique solution of (1).

If $v = [v_\varepsilon]$ is another fixed point of ϕ, we have

$$v_\varepsilon = \phi_\varepsilon(v_\varepsilon) + \rho_\varepsilon \ with \ \rho_\varepsilon \in \mathcal{N}^s(\mathbb{R})$$

then $(p_T(i_\varepsilon))_\varepsilon \in |\mathcal{N}^r|$

$$w_\varepsilon(t) - v_\varepsilon(t) = I^q g_\varepsilon(t, w_\varepsilon(t)) - I^q g_\varepsilon(t, v_\varepsilon(t)) - \rho_\varepsilon(t),$$

then

$$w_\varepsilon(t) - v_\varepsilon(t) = S_\varepsilon I^q (w_\varepsilon(t) - v_\varepsilon(t)) - \rho_\varepsilon(t),$$

and

$$w_\varepsilon(t) - v_\varepsilon(t) = \rho_\varepsilon + S_\varepsilon \int_0^t \frac{(t-s)^{q-1}}{\Gamma(q)}(w_\varepsilon(s) - v_\varepsilon(s))ds,$$

implies that

$$p_i(w_\varepsilon(t) - v_\varepsilon(t)) \le p_i(\rho_\varepsilon) + \int_0^t M_{T,\varepsilon}\frac{b^{q-1}}{\Gamma(q)}p_i(w_\varepsilon(s) - v_\varepsilon(s))\,ds, \quad \forall i \in I.$$

Since $\rho_\varepsilon \in \mathcal{N}(\mathbb{R})$, then $p_i(\rho_\varepsilon) \le \varepsilon^a$ for all $a \in \mathbb{R}$, then

$$p_i(w_\varepsilon(t) - v_\varepsilon(t)) \le \varepsilon^a + \int_0^t M_{T,\varepsilon}\frac{b^{q-1}}{\Gamma(q)}p_i(w_\varepsilon(s) - v_\varepsilon(s))\,ds, \quad \forall i.$$

Now by Grönwall Lemma 1, we get

$$p_i\left(w_\varepsilon(t) - v_\varepsilon(t)\right) \le \varepsilon^a\, e^{M_{T,\varepsilon}\frac{b^{q-1}}{\Gamma(q)}t}, \quad \forall i \in I.$$

We have $e^{M_{T,\varepsilon}\frac{b^q}{\Gamma(q)}} \in |\mathcal{E}_M^r|$ and $(p_I(i_\varepsilon))_\varepsilon \in |\mathcal{N}^r|$, then

$$(p_T(w_\varepsilon - v_\varepsilon))_\varepsilon \in |\mathcal{N}^r|.$$

So,

$$w = v.$$

Finally we find the results of the fractional problem of Schrodinger

Theorem 5 *We put* $A = \Delta - V$. *By the previous result the problem* (1) *has unique solution.*

Proof By Proposition 5, $\Delta - V$ generate a generalized C_0-semigroup S, The operator $\Delta - V_\varepsilon$ verified the first condition **H-1**, and S defined in Proposition 5 verified the condition **H-2**. Thus by Theorem 4 the problem has unique solution.

References

1. Branciari, A.: A fixed point theorem for mappings satisfying a general contractive condition of integral type. Int. J. Math. Math. Sci. **29**(9), 531–536 (2002)
2. Colombeau, J.F.: New Generalized Function and Multiplication of Distribution. North Holland, Amsterdam (1984)
3. Colombeau, J.F.: Elementary Introduction to New Generalized Function. North Holland, Amsterdam (1985)
4. Grosser, M., Kunzinger, M., Oberguggenberger, M., Steinbauer, R.: Geometric Theory of Generalized Functions with Applications to General Relativity, Mathematics and its Applications, vol. 537. Kluwer Academic Publishers, Dordrecht (2001)
5. Hermann, R., Oberguggenberger, M.: Ordinary differential equations and generalized functions. Non- linear Theory of Generalized Functions, pp. 85–98. Chapman & Hall, Boca Raton (1999)
6. Kilbas, A.A., Srivastava, H.M., Trujillo, J.J.: Theory and Applications of Fractional Differential Equations. Elsevier B.V, Netherlands (2006)
7. Oberguggenberger, M.: Multiplication of Distributions and Applications to Partial Differential Equations. Pitman Research Notes in Mathematics. Longman Scientific & Technical, Harlow (1992)
8. Marti, J.A.: Fixed points in algebras of generalized functions and applications, HAL Id: hal-01231272
9. Pazy, A.: Semigroups of Linear Operators and Applications to Partial Differential Equations. Applied Mathematical Sciences, vol. 44. Springer, New York (1983)

Relaxed Controllability for Parabolic Linear Systems Using RHUM Approach

Layla Ezzahri, Imad El Harraki and Ali Boutoulout

Abstract In this paper, we treat the problem of relaxed controllability, so called enlarged controllability, for linear systems described by parabolic PDEs. We characterize the optimal control using two approaches; the first one is the Reversible Hilbert Uniqueness Method (RHUM) which is based on HUM approach, and the second one is the penalty method which characterizes the control of minimum energy.

1 Introduction

For distributed parameter systems, the notion of controllability has played a central role throughout the history of modern control theory. The controllability concept consists at steering a system from an initial state to a prescribed one defined on a spatial domain Ω of the evolution system. This concept has often been studied and widely developed [1, 2], and several publications especially for linear parabolic systems, which describe various chemical, physical and biological phenomena such as problems of air pollution, reaction-diffusion or flame propagation [3], have been concerned [4–6].

A situation that is very important in control theory and in practical applications is that of controllability with hard constraints on controls and states [7]. In the case that input constraints are presented on the linear systems, the controllability property has been characterized by Lions for the wave equation case [8]. Also in [8], Lions treated the problem of enlarged controllability for the wave equation with a boundary

L. Ezzahri (✉) · A. Boutoulout
Faculty of Sciences, Meknes, Morocco
e-mail: lailaezzahri@gmail.com

A. Boutoulout
e-mail: boutouloutali@yahoo.fr

I. El Harraki
Rabat Superior National School of Mines, Rabat, Morocco
e-mail: imadharraki@gmail.com

© Springer Nature Switzerland AG 2020
E. H. Zerrik et al. (eds.), *Recent Advances in Modeling, Analysis and Systems Control: Theoretical Aspects and Applications*, Studies in Systems, Decision and Control 243, https://doi.org/10.1007/978-3-030-26149-8_8

103

control. Parabolic problems are one of the fields of mathematics which undergoes a detailed investigation, due to the many problems which rely on this theory.

In this paper, we will develop the concept of enlarged controllability for systems described by parabolic partial differential equations, with internal control and using zone actuators [9, 10]; where the aim is to steer the system from an initial state into a space of constraints G in the considered domain.

This paper is organized as follows. The enlarged internal controllability for a linear parabolic systems shall be presented in Sect. 2. Then, in Sect. 3, the existence of a control that ensures the enlarged controllability will be established using RHUM approach, and based on this, in Sect. 4 a control with minimum energy shall be presented using the penalty method.

2 Enlarged Internal Controllability

Let Ω be an open bounded subset of $\mathbb{R}^n$ with regular boundary $\partial\Omega$. For $T > 0$, let $Q = \Omega \times]0, T[$ and $\Sigma = \partial\Omega \times]0, T[$. We consider the following parabolic system:

$$\begin{cases} y'(x, t) - Ay(x, t) = \chi_D f(x)u(t) & \text{in } Q \\ y(x, 0) = y^0(x) & \text{in } \Omega \\ y(\xi, t) = 0 & \text{on } \Sigma, \end{cases} \tag{1}$$

where A is a second-order elliptic linear symmetric operator, which generates a strongly continuous semi-group $(S(t))_{t \geq 0}$ and excited by a zone actuator (D, f) where $D \subset \Omega$ and $f \in L^2(\Omega)$.

Let us consider a control u in $U = L^2(0, T, \mathbb{R}^p)$ (p depends on the number of the considered actuators) and y^0 the initial condition in the state space $L^2(\Omega)$.

We design the solution of (1) by $y_u(.) \in L^2(\Omega)$ [11]. Let us recall the following definitions:

Definition 1 System (1) is said to be null controllable if for all initial state $y_0 \in L^2(\Omega)$ there exists a control $u \in U$ such that

$$y_u(T) = 0.$$

Definition 2 System (1) is said to be exactly controllable in $L^2(\Omega)$ at time T, that is, for any initial state y_0 and any terminal state $y_d \in L^2(\Omega)$, there exists a control $u \in U$ such that the solution of (1) with $\omega = \Omega$ satisfies

$$y_u(T) = y_d,$$

where ω is an open and nonempty subset of Ω.

Now, let us set the following definition of the concept of enlarged controllability.

Definition 3 We have enlarged controllability for system (1) if, for all $y^0 \in L^2(\Omega)$, there exists a control $u \in U$ such that

$$y_u(T) \in G,$$

where G is a closed vectorial subset of $L^2(\Omega)$.

Remark 1 Let H be the operator from $U \rightarrow L^2(\Omega)$, for $u \in U$, defined by

$$Hu = \int_0^T S(T - s)Bu(s)ds,$$

where $B : u \rightarrow \chi_D f(x)u(t)$. Then the previous definition is equivalent to say that we have enlarged controllability for system (1) if

$$(Im H) \cap G \neq \emptyset.$$

Remark 2

- The notion of enlarged controllability depends on the subset G.
- If $G = \{0\}$, we find the classical notion of null controllability.

Example

If for $T > 0$, we find a control u such that

$$y_u(T) = 0 \text{ in } \omega$$

and

$$G = \{y \in L^2(\Omega)/y = 0 \text{ in } \omega\},$$

then we have the enlarged controllability problem for G.

3 RHUM Approach

The purpose of this section is to explore the Reversible Hilbert Uniqueness Method (RHUM), which is an extension of HUM approach [12] devoted to the computation of the optimal control problem for system (1). First, we will prove under what conditions we can find the enlarged controllability, then we will find the optimal control that steers our system into G.

Let us consider $\phi_0 \in G^0$, where G^0 is the polar set of G in $L^2(\Omega)$, such that

$$\phi_0 \in G^0 \Leftrightarrow \langle \phi_0, g_0 \rangle_{L^2(\Omega)} = 0 \quad \forall \, g_0 \in G.$$

Let $w_j(x)$ be the eigenfunctions of A associated with the eigenvalues λ_j. Let us consider the following retrograde problem

$$\begin{cases} \phi'(x,t) - A\phi(x,t) = 0 & \text{in} \quad Q \\ \phi(x,T) = \phi_0 & \text{in} \quad \Omega \\ \phi(\xi,t) = 0 & \text{on} \quad \Sigma, \end{cases} \tag{2}$$

which admits a unique solution ϕ (see [13]).

Theorem 1 *If* $\langle w_j, f \rangle_{L^2(D)} \neq 0$, *then*

$$\|\phi_0\|_{G^0} = \left(\int_0^T (\langle f, \phi \rangle)^2_{L^2(D)} dt \right)^{\frac{1}{2}}, \tag{3}$$

defines a norm in G^0 *and we have the enlarged controllability for* G *at time* T.

Proof We consider the following problem:

$$\begin{cases} \psi'(x,t) = A\psi(x,t) + \chi_D f(x)u(t) & \text{in} \quad Q \\ \psi(0) = y^0 & \text{in} \quad \Omega \\ \psi(\xi,t) = 0 & \text{on} \quad \Sigma. \end{cases} \tag{4}$$

We obtain the solution ψ of system (4), which is continuous from $[0, T]$ to $L^2(\Omega)$.

The problem of enlarged controllability returns to find $\phi_0 \in G^0$ such that $\psi(T) \in G$ and then

$$u = \langle f, \phi \rangle,$$

is the control that ensures the enlarged controllability and $y_u = \psi$.

To achieve this, let us introduce π the orthogonal projection on $G^\perp$ (the orthogonal of G), and an affine operator Λ from G^0 to $G^\perp$ such that

$$\Lambda\phi_0 = \pi\psi(T). \tag{5}$$

Henceforth the problem returns to solve

$$\Lambda\phi_0 = 0. \tag{6}$$

Now, let us decompose ψ such that $\psi = \psi_1 + \psi_2$, where ψ_1 and ψ_2 are, respectively, the solutions of

$$\begin{cases} \psi_1'(x,t) - A\psi_1(x,t) = 0 & \text{in} \quad Q \\ \psi_1(x,0) = y^0(x) & \text{in} \quad \Omega \\ \psi_1(\xi,t) = 0 & \text{on} \quad \Sigma \end{cases} \tag{7}$$

and

$$\begin{cases} \psi_2'(x, t) = A\psi_2(x, t) + \chi_{_D} f(x)u(t) & \text{in } Q \\ \psi_2(x, 0) = 0 & \text{in } \Omega \\ \psi_2(\xi, t) = 0 & \text{on } \Sigma, \end{cases} \tag{8}$$

then

$$\pi\psi(T) = \pi\psi_1(T) + \pi\psi_2(T). \tag{9}$$

Let us set

$$\pi\psi_2(T) = \Lambda_0\phi_0, \tag{10}$$

where $\Lambda_0 \in \mathcal{L}(G^0; G^\perp)$, then we have

$$\Lambda\phi_0 = \Lambda_0\phi_0 + \pi\psi_1(T). \tag{11}$$

From (6), (11) returns to solve

$$\Lambda_0\phi_0 = -\pi\psi_1(T). \tag{12}$$

We set

$$\mu = \langle\Lambda_0\phi_0, \phi_0\rangle. \tag{13}$$

By definition, we have

$$\langle\pi y - y, z\rangle = 0 \text{ for all } z \in G^0. \tag{14}$$

From (14) and for $z = \phi_0$, $y = \psi_2(T)$ we have

$$\langle\pi\psi_2(T) - \psi_2(T), \phi_0\rangle = 0,$$

then

$$\mu = \langle\psi_2(T), \phi_0\rangle. \tag{15}$$

The computation of (15) is obtained by multiplying system (8) by ϕ and integrating over Q and using the Green's formula. Hence

$$\int_Q\int_0^T \phi\frac{\partial\psi_2}{\partial t}(x, t)dtdx = \int_Q\int_0^T \phi A\psi_2(x, t)dtdx + \int_Q\int_0^T \phi\langle f, \phi\rangle_{L^2(D)} f(x)\chi_{_D} dtdx, \tag{16}$$

which gives

$$\langle\psi_2(T), \phi(T)\rangle = \int_Q\int_0^T \phi(t)\langle f, \phi(t)\rangle_{L^2(D)} f(x)\chi_{_D} dtdx. \tag{17}$$

Or

$$\langle \Lambda_0 \phi_0, \phi_0 \rangle = \int_0^T \langle f, \phi(t) \rangle_{L^2(D)}^2 dt = \|\phi_0\|_{G^0}^2,$$

then

$$\mu = \int_0^T \left(\langle f, \phi \rangle_{L^2(D)} \right)^2 dt. \qquad (18)$$

Now we prove that if $\langle w_j, f \rangle_{L^2(D)} \neq 0$, then the mapping (3) is a norm which is equivalent to this of G^0.

The mapping (3) defines a norm in G^0, indeed

$$\|\phi_0\|_{G^0} = 0 \text{ gives } \langle \phi, f \rangle_{L^2(D)} = 0 \text{ a.e in } [0, T],$$

which is equivalent to

$$\sum_n e^{(\lambda_n t)} \sum_{j=1}^n \langle \phi_0, w_j \rangle w_j \langle w_j, f \rangle = 0. \qquad (19)$$

Thus, for all $t \in]0, T[$ where time T is arbitrarily chosen large enough, then (19) gives

$$\langle \phi_0, w_j \rangle \langle w_j, f \rangle_{L^2(D)} = 0.$$

Using the assumption that $\langle w_j, f \rangle_{L^2(D)} \neq 0$, we deduce that $\langle \phi_0, w_j \rangle = 0$.

It follows that $\phi_0 = 0$ and (3) is a norm, then μ is an isomorphism from G^0 to $G^\perp$, and the equality (12) admits a unique solution. Hence we have the enlarged controllability (relative to G).

4 Minimum Energy Control

Now the problem of minimum energy control will be considered, more precisely, we are interested in the following minimization problem:

$$\begin{cases} \inf \mathscr{J}(u) = \dfrac{1}{2} \int_0^T \|u\|_{\mathbb{R}^p}^2 dt \\ u \in U_{ad}(G), \end{cases} \qquad (20)$$

where

$$U_{ad}(G) = \{u \in U \mid y_u(T) \in G\}.$$

Proposition 1 *We suppose that we have enlarged controllability for G at time T, then (20) has a unique solution given by*

$$u^* = \langle f, \phi \rangle, \qquad (21)$$

which ensures the transfer of system (1) into G, where ϕ is the solution of system (2).

Proof We suppose that we have enlarged controllability and we consider the following non empty set C of couples (u, y)

$$\begin{cases} y'(x, t) - Ay(x, t) - \chi_D f(x)u(t) \in L^2(Q) \\ y(0) = y^0 \quad \text{in} \quad \Omega \\ y(\xi, t) = 0 \quad \text{on} \quad \Sigma \\ y_u(T) \in G, \end{cases} \qquad (22)$$

and, for $\epsilon > 0$, we consider the auxiliary problem of (20)

$$\begin{cases} \inf \mathscr{J}_\epsilon(u, y) \\ (u, y) \in C, \end{cases} \qquad (23)$$

where

$$\mathscr{J}_\epsilon(u, y) = \frac{1}{2} \int_0^T u(t)^2 dt + \frac{1}{2\epsilon} \int_0^T \int_\Omega (y'(t) - Ay(t) - \chi_D f(x)u(t))^2 dxdt.$$

Let us consider $\{u_\epsilon, y_\epsilon\}$ the solution of (23). Then we have

$$\mathscr{J}_\epsilon(u_\epsilon, y_\epsilon) = \inf \mathscr{J}_\epsilon(u, y) \le \inf \mathscr{J}_\epsilon(u), u \in U_{ad}(G) \qquad (24)$$

where

$$\mathscr{J}_\epsilon(u) = \frac{1}{2} \int_0^T u(t)^2 dt, \qquad (25)$$

when $\epsilon \to 0$, we have

$$\begin{cases} u_\epsilon \text{ is bounded } \text{ i.e } \|u_\epsilon\| < k \\ y'_\epsilon(t) - Ay_\epsilon(t) - \chi_D f(x)u_\epsilon(t) = \omega_\epsilon \text{ where } \|\omega_\epsilon\|_{L^2(Q)} \le k\sqrt{\epsilon}, \\ \text{and } k \text{ is a constant independent of } \epsilon. \end{cases} \qquad (26)$$

We can extract a sequence such that

$$\begin{aligned} u_\epsilon &\longrightarrow \tilde{u} \text{ in } U \text{ weakly,} \\ y_\epsilon &\longrightarrow z \text{ in } L^2(Q) \text{ weakly.} \end{aligned} \qquad (27)$$

By the semi continuity of $\mathscr{J}$, we have

$$\mathscr{J}(u^*) \le \liminf \mathscr{J}_\epsilon(u_\epsilon) \le \liminf \mathscr{J}_\epsilon(u_\epsilon, y_\epsilon).$$

Then

$$\mathscr{J}(u^*) = \inf \mathscr{J}(u), u^* \in U_{ad}(G), \qquad (28)$$

and

$$\tilde{u} = u^*. \tag{29}$$

To introduce the optimal system for system (23), let us define

$$p_\epsilon = -\frac{1}{\epsilon}(y'_\epsilon(t) - Ay_\epsilon(t) - \chi_D f(x)u_\epsilon(t)), \tag{30}$$

then

$$\int_0^T u_\epsilon(t)u(t)dt + \int_0^T \langle p_\epsilon, \xi'(t) - A\xi(t)\rangle dt = -\int_0^T \langle p_\epsilon, f\rangle u(t)dt \tag{31}$$

for $u \in U_{ad}(G)$ and ξ such that

$$\begin{cases} \xi'(x,t) - A\xi(x,t) = \chi_D f(x)u(t) \text{ in } Q \\ \xi(0) = 0 & \text{in } \Omega \\ \xi(v,t) = 0 & \text{on } \Sigma \\ \xi(T) \in G. \end{cases} \tag{32}$$

We deduce that p_ϵ verifies

$$\begin{cases} p'_\epsilon(x,t) - Ap_\epsilon(x,t) = \chi_D f(x)\langle p_\epsilon, f\rangle_{L^2(D)} \text{ in } Q \\ p_\epsilon(x,0) = 0 & \text{in } \Omega \\ p_\epsilon = 0 & \text{on } \Sigma, \end{cases} \tag{33}$$

and

$$\langle p_\epsilon(T), \xi(T)\rangle = 0 \quad (\forall \xi(T) \in G), \tag{34}$$

then $p_\epsilon \in G^0$.

If we suppose that

$$\int_0^T \langle p_\epsilon, f\rangle^2 dt \geq k \parallel p_\epsilon \parallel^2_{L^2(\Omega)}, \tag{35}$$

then we can switch to the limit (when $\epsilon \to 0$) and if we have enlarged controllability relative to G, we obtain the following system:

$$\begin{cases} y'(x,t) - Ay(x,t) = \chi_D f(x)u(t) & \text{in } Q \\ y(x,0) = y^0 & \text{in } \Omega \\ y = 0 & \text{on } \Sigma \\ p'(x,t) - Ap(x,t) = \chi_D f(x)\langle p, f\rangle_{L^2(D)} \text{ in } Q \\ p'(T) \in G^0 \\ p = 0 & \text{on } \Sigma. \end{cases} \tag{36}$$

We take $p(T) \in G^0$, we introduce ϕ solution of (2), we define ψ by (4) then we have $\psi = y$ if

$$\psi(T) \in G, \tag{37}$$

which prove that the Eq. (12) admits a unique solution for $\phi_0 \in G^0$.

5 Conclusion

This paper treated the problem of enlarged controllability for parabolic systems using two approaches. Future works aim to extend this notion to the case of systems described by non linear PDEs .

References

1. Curtain, R.F., Zwart, H.: An Intoduction to Infinite Dimentional Linear Systems Theory. Springer, Berlin (1995)
2. Curtain, R.F., Pritchard, A.J.: Infinite Dimensional Linear Systems Theory. Springer, Berlin (1978)
3. Henry, D.: Geometric Theory of Semilinear Parabolic Equations. Springer, New York (1981). https://doi.org/10.1007/BFb0089647
4. Lions, J.-L.: Exact controllability: stabilization and perturbations for distributed systems. SIAM Rev. **30**, 1–68 (1988)
5. Lasiecka, I., Triggiani, R.: Control theory for partial differential equations: continuous and approximation theories. Encycl. Math. Appl. **1**(2), 74–75 (2000)
6. Li, X., Yong, J.: Optimal Control Theory for Infinite Dimensional Systems. Birkhäauser, Boston (1995)
7. Heemels, W.P.M.H., Camlibel, M.K.: Controllability of linear systems with input and state constraints. In: Proceedings of the 46th IEEE Conference on Decision and Control New Orleans, USA, 12-14 (2007)
8. Lions, J.L.: Partial differential equations and the calculus of variations. Progress in Nonlinear Differential Equations and Their Applications, vol. II (1989)
9. El Jai, A., Pritchard, A.J.: Sensors and Actuators in Distributed Systems Analysis. Ellis Horwood Series in Applied Mathematics. Wiley, New York (1988)
10. Zerrik, E., El Jai, A., Boutoulout, A.: Actuators and regional boundary controllability of parabolic system. Int. J. Syst. Sci. **30**, 73–82 (2000)
11. Lions, J.L., Magenes, E.: Problèmes aux limites non homogènes et applications. Dunod, Paris (1968)
12. Lions, J.L.: Contrôlabilité Exacte Perturbation et Stabilisation des systèmes distribués. Masson (1988)
13. Lions, J.L.: Problèmes aux limites non classiques pour equation devolution, Revista Mathematica de universidad computense de Madrid (1988)

A New Mathematical Model for the Efficiency Calculation

Aníbal Galindro, Micael Santos, Delfim F. M. Torres
and Ana Marta-Costa

Abstract During the past sixty years, a lot of effort has been made regarding the productive efficiency. Such endeavours provided an extensive bibliography on this subject, culminating in two main methods, named the Stochastic Frontier Analysis (parametric) and Data Envelopment Analysis (non-parametric). The literature states this methodology also as the benchmark approach, since the techniques compare the sample upon a chosen "more-efficient" reference. This article intends to disrupt such premise, suggesting a mathematical model that relies on the optimal input combination, provided by a differential equation system instead of an observable sample. A numerical example is given, illustrating the application of our model's features.

Keywords Frontier estimation models · Efficiency analysis · Technical inefficiency · From data to differential equations

2010 Mathematics Subject Classification 90B50 · 90C08

A. Galindro · M. Santos · A. Marta-Costa
Centre for Transdisciplinary Development Studies, University of Trás-os-Montes
and Alto Douro, Polo II–ECHS, Quinta de Prados, 5000-801 Vila Real, Portugal
e-mail: anibalg@utad.pt

M. Santos
e-mail: micaels@utad.pt

A. Marta-Costa
e-mail: amarta@utad.pt

D. F. M. Torres (✉)
Center for Research and Development in Mathematics and Applications (CIDMA),
Department of Mathematics, University of Aveiro, 3810-193 Aveiro, Portugal
e-mail: delfim@ua.pt

© Springer Nature Switzerland AG 2020
E. H. Zerrik et al. (eds.), *Recent Advances in Modeling, Analysis and Systems Control:
Theoretical Aspects and Applications*, Studies in Systems, Decision and Control 243,
https://doi.org/10.1007/978-3-030-26149-8_9

1   Introduction

The word *efficiency* may acquire several meanings, but in economics such word is closely related to the general premise of the field, since the available resources are limited and how should we use them to attain the maximum level of output or the maximum utility is a natural question. The productivity efficiency literature began in 1957 with Farrell's work and is a rising theme applied to several sectors. Farrell [1] defined two concepts of efficiency: Technical Efficiency (TE) and Allocative Efficiency (AE). The first is evident when a certain level of inputs is given, the Decision Making Unit (DMU) is able to produce the maximum level of output or, fixing a certain level of output, the DMU is able to minimize the level of input [2, 3]. The AE reflects the ability of a firm to use the inputs in their optimal proportions (given their respective prices), to minimize the cost or maximize the revenue [2, 4]. Another type of efficiency is the scale efficiency, which tells us if a DMU is operating on an optimal scale [5]. Over time, several methodologies have been developed to estimate the productive efficiency. Those methodologies can be categorized mainly as parametric or non-parametric. The most or the main methods used in each category are the Stochastic Frontier Analysis (SFA) and Data Envelopment Analysis, respectively [2, 6–9]. However, with the development of the methodologies, such separation is not so clear nowadays. For example, Johnson and Kuosmanen [10] have developed a semi-nonparametric one-stage DEA and many efforts have been made to develop non-parametric or semiparametric SFA, as documented on the work of Kumbhakar, Parmeter, and Zelenyuk [11]. However, all these methodologies are benchmarking approaches, in other words, they compare the productive efficiency of a DMU against a reference performance (the most-efficient DMU's). In contrast, the goal of this work is to create a new methodology that does not follow this benchmark premise and looks beyond the self-proclaimed better or more efficient observation. Our differential equation model relies on overall optimality per delivered output, which endures a different interpretation of the problem.

2   Literature Review

The productive efficiency methodologies lay on the benchmarking approach, since all the techniques rely on a continuous and systematic process of comparing a certain chosen sample upon a reference (benchmark) performance [12, 13]. The reference performance is normally the "best-practice", i.e., the methodologies identify the most efficient DMUs to build a frontier and then compare the least efficient DMUs against this frontier. The greater the distance from a DMU performance to the best-practice, the greater its level of inefficiency. The benchmark approach in the productive efficiency can be categorized mainly in two groups: non-parametric and parametric. Non-parametric techniques are normally based on programming techniques and do not require a production, cost, or profit function, calculating the relation of the inputs with the outputs without an econometric estimation [2, 9, 13]. Within the non-parametric approaches, we have the DEA and the Free Disposal Hull (FDH). Usually, the productive efficiency methods assume the convexity of the production set. However, FDH was created by Deprins, Simar, and Tulkens [14], who proposed the elimination of the convexity assumption [15]. FDH is a variation of DEA, based on a programing technique, and the frontier estimated by this method may coincide or be below the DEA frontier, so the efficiency scores estimated by FDH tend to be higher than those estimated by DEA [16]. The DEA is the most used non-parametric TE model. It is based on mathematical programming techniques and does not require the specification of a functional form for the technology. It should be noted that the majority of the non-parametric methods are deterministic, therefore they do not allow any random noises or measurement errors [2, 9]. Another characteristic of the DEA methods is their potential sensitivity of efficiency scores to the number of observations, as well as to the dimensionality of the frontier and to the number of outputs and inputs [2, 9, 17]. However, some developments have been made in the traditional DEA. For example, Simar and Wilson developed a stochastic DEA using bootstrapping techniques in [18, 19]. Johnson and Kuosmanen [10] developed a semi-nonparametric one-stage DEA based on the critics of the two-stage DEA and on the work of Banker and Natarajan [20], who incorporate a noise term that has a truncated distribution. SFA normally is a parametric and stochastic method, which was introduced by Aigner, Lovell, and Schmidt [21] and Meeusen and van Den Broeck [22]. This method imposes more structure on the shape of the frontier by specifying a functional form for the production function, such as the Cobb–Douglas or the Translog form. Moreover, the SFA was developed to allow random errors, but neither the random error nor inefficiency can be observed, so separating them requires an assumption on the distribution of the efficiencies scores and on the random error. Although the SFA is traditionally parametric, some developments have given it some degree of convergence with non-parametric models, which are referred as non-parametric or semiparametric SFA [11, 23–27]. Essentially, Fan et al. [24] and Kneip and Simar [28] provide the baseline studies for these methodologies. More recently, the advances in stochastic frontier models were documented in Kumbhakar et al. [11].

3 A New Theoretical Model

The main goal of this article is to overcome the limitations of the benchmark approach, extensively used in the previously presented bibliography, appended to several efficiency calculation methods. The generalized method of obtaining a sample (at least one) based totally on efficient benchmark does not guarantee a proper or correct solution. Therefore, the subsequent process of evaluating the efficiency/inefficiency level of the other observations might be biased. We propose a differential equation based method to solve a single output and multiple input efficiency problem. Let m be the sample size containing the data points for n input variables $X = \{x_1, x_2, x_3, \ldots, x_n\}$

and Y the output associated to each data point such that $Y = \{y_1, y_2, y_3, \ldots, y_m\}$. The explicit matricial form is stated on Eq. (1):

$$
\begin{bmatrix} y^1 \\ y^2 \\ \vdots \\ y^m \end{bmatrix} = \begin{bmatrix} x_1{}^1 & x_2{}^1 & \cdots & x_n{}^1 \\ x_1{}^2 & x_2{}^2 & \cdots & x_n{}^2 \\ \vdots & \vdots & \ddots & \vdots \\ x_1{}^m & x_2{}^m & \cdots & x_n{}^m \end{bmatrix}.
\tag{1}
$$

The main idea is to interpret the input/output dynamics as a differential equation system that retrieves the optimal input combination X^* per each output level $Y \in \mathbb{R}^+$. Therefore, the system compiles n differential equations, one per each input variable, and it is defined by Eq. (2), where $\mathbf{A_i} = \{\beta_i{}^0, \beta_i{}^1, \ldots, \beta_i{}^n\}$, $i = 1, 2, \ldots, n$, represents the whole inner input trade-off combinations that forces optimal input allocation per each output level. Such parameters can be numerically estimated for both linear and non-linear differential equations using methods developed by Ramsay [29, 30]:

$$
\begin{cases}
\dfrac{dx_1(y)}{dy} = \beta_1^0 + \beta_1^1 x_1(y) + \beta_1^2 x_2(y) + \cdots + \beta_1^n x_n(y), \\[2mm]
\dfrac{dx_2(y)}{dy} = \beta_2^0 + \beta_2^1 x_1(y) + \beta_2^2 x_2(y) + \cdots + \beta_2^n x_n(y), \\[2mm]
\vdots \\[2mm]
\dfrac{dx_n(y)}{dy} = \beta_n^0 + \beta_n^1 x_1(y) + \beta_n^2 x_2(y) + \cdots + \beta_n^n x_n(y).
\end{cases}
\tag{2}
$$

We settle that the set X of input variables is expressed in similar values (Euro, for example), to avoid the concern of different weights or costs in the model development. The optimality of A^* does not guarantee the feasibility of the process with non-negative inputs across every output value. Therefore, the condition on Eq. (3) should be met:

$$
0 \le x_1, x_2, \ldots, x_n \ \forall \ y.
\tag{3}
$$

The model initial conditions should follow Eq. (4), where the null output is generated by zero inputs:

$$
x_1(0) = 0, \quad x_2(0) = 0, \quad \ldots, \quad x_n(0) = 0.
\tag{4}
$$

Nonetheless, we intend to also obtain the optimal output level y^* for the given problem. Assuming increasing returns to scale until some single point (which would be y^*) and decreasing returns to scale afterwards, per each $y^d > y^*$, the point y^* guarantees profit maximization and solution singleness. The profit function given by Eq. (5) acquaints that the resulting output is sold at an arbitrary price c. Each level of output y is bundled with an aggregate variation rate of every input on Eq. (5). The integration extracts the numerical input for each level of production y. The maximizing profit stretched by the optimal y^*, is obtained when the derivative of $J(y)$,

$$J(y) = cy - \int_0^y \frac{dx_i(y)}{dy},$$ (5)

equals zero:

$$\frac{dJ(y)}{dy} = 0.$$ (6)

The newly discovered y^* eases straightforwardly the computation of the ideal input set $X^* = \{x_1^*, x_2^*, x_3^*, \ldots, x_n^*\}$, where each real value can be obtained by the integration of Eq. (7):

$$x_i^* = \int_0^{y^*} \frac{dx_i(y)}{dy}.$$ (7)

The optimal X^* can be interpreted as a point in a n-dimensional space, each non-optimal observation $X_j^{no} = \{x_{1j}^{no}, x_{2j}^{no}, x_{3j}^{no}, \ldots, x_{nj}^{no}\} \in \mathbb{R}^n$, $j = 1, 2, 3, \ldots, m$, from the sample m outlies a distance vector from the optimal solution on Eq. (8):

$$d_j = \sqrt{(x_{1j}^{no} - x_1^*)^2 + (x_{2j}^{no} - x_2^*)^2 + \cdots + (x_{nj}^{no} - x_n^*)^2}.$$ (8)

Let Ω be the set of feasible input combinations such that $f : \Omega \subset \mathbb{R}^n \to \mathbb{R}$ is a function with domain Ω with values in $\mathbb{R}$ (the output level). Analogously, we settle a subset function using a restricted domain $\Omega' \subseteq \Omega$, which contains the feasible data points that obey a certain restriction. A given point from Ω' is extracted from the Ω set only if $\forall \Omega'$, $0 \le \frac{dy}{dx_i}$, and the condition on Eq. (9) is met:

$$\sum_{i=0}^n \frac{dy}{dx_i} > \varepsilon.$$ (9)

Naturally, the selection of ε acquaints a certain level of parsimony, since larger values may absorb and exclude valid and empirical observations. On the other hand, choosing an ε that is too small may expand unnecessarily the feasible region Ω'. Considering each element from the subset Ω', obtained with a certain level of ε, it is possible to generate an analogous set Ω'' that contains the distance given by Eq. (8) of each feasible input combination from Ω'. Let X_w be the input combination from Ω' that obtains the maximum value from the Ω'' set (d_w), which is collinear with the observation that we intend to attain information about their efficiency. Let it be X_m^{no}. Literally, X_m^w work as the most inefficient points and as a general reference for the efficiency calculations. Being $X_j^{no} = \{x_{1j}^{no}, x_{2j}^{no}, x_{3j}^{no}, \ldots, x_{nj}^{no}\} \in \mathbb{R}^n$, $j = 1, 2, \ldots, m$, the sampled non-optimal data points, with distances to X^* given by Eq. (8), their efficiency level is given by Eq. (10):

$$I_d = 1 - \frac{d_j}{d_w},$$ (10)

for all $d_j \leq d_w$. Therefore, the efficiency level I_d of each sampled observation is bounded between 0 and 1. Since the weights or costs are acquainted, since the beginning, within the input variables, the distances on the generated n-dimensional space provide the same effort throughout the efficient point. Let X_a and X_b be to sampled data points such that $I_a = I_b$ is the equidistant measurement provided in the n-dimensional space embedded with the pre-accounted weights, ensuring that the effort through the optimal point (efficiency improvement) of both samples X_a and X_b is the same, validating the fairness when we measure their efficiency levels. Nonetheless, we can stretch forward our model to acquaint singularly technical and allocative efficiency, instead of the previous general effort throughout the optimal point y^*. If we reduce the domain of f even further, and assume subsets Ω^m, $m \in \mathbb{N}$, we get m functions f_m such that $f_m : \Omega^m \subset \mathbb{R}^n \rightarrow \mathbb{R}$ satisfies $f_m(x) = y_m$ for all $x \in \Omega^m$. Each given function f_m with domain Ω^m acquaints a fixed output level y^m. Therefore, the resulting optimal points, considering that there is only an optimal set of inputs, is now given by X^{m*} bundled with y_{m*}, extracted from the optimal movement of the differential equations on Eq. (2). Now, each sampled observation lays on a different f_m for each output level y_m. Assuming a sampled non-optimal input combination X_m^{no}, entangled with a certain level of output y_m, the productive efficiency is now given by the ratio of the distance to the optimal point X^{m*} (Eq. (10)) alongside the distance from the worst feasible point X_m^w considered to X^{m*}. It is also possible to analogously shift the formulation to attain information about the technical efficiency of each observation. Instead of working with a subset that compiles similar levels of output, such subset should only feature similar levels of input given by the explicit sum of the input combination.

Summarizing, we have just obtained the following result.

Theorem 1 *Let $\{\beta_i{}^0, \beta_i{}^1, \ldots, \beta_i{}^n\} \in \mathbb{R}, i = 1, 2, \ldots, n$, compile the minimal input differential equation system (2) per each y. Settle the initial conditions as (4), while assuring that the non-negativity condition (3) holds. A feasible Ω region is obtained with (9). Using the optimal X^* per level of output y, the frontier points of Ω yield the worst efficiency levels X_m^w. Sampling an observation X_j with the associated collinear most inefficient point X_w, the distance vector is calculated according to (8) using the optimal point X^*. Finally, the efficiency levels are given by (10).*

4   An Illustrative Numerical Application

In order to illustrate our method, we give here a simple numerical example. Following the differential equation system premise on Eq. (2), a two input/single output system is created according to Eq. (11):

Fig. 1 The dashed/
continuous line represents
the optimal input level x_1/x_2
per each level of output y,
respectively

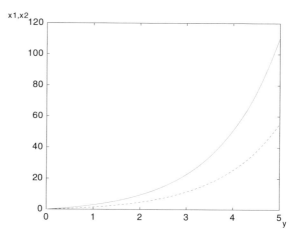

$$\begin{cases} \dfrac{dx_1(y)}{dy} = 1 + 0.25x_1(y) + 0.25x_2(y), \\ \dfrac{dx_2(y)}{dy} = 2 + 0.50x_1(y) + 0.50x_2(y). \end{cases} \tag{11}$$

The aforementioned system describes a limited capacity productive system with decreasing returns to scale. Indeed, the input needed to increment the output values grows exponentially to bulky and unrealistic combinations. Such system does not violate the non-negativity condition on Eq. (3) for any $y > 0$, considering also the initial conditions stated on Eq. (4) for both x_1 and x_2. The Fig. 1 displays the differential equation development considering different levels of output y. The c value is chosen such that the Eqs. (5) and (6) of the maximizing profit hold. To obtain a specific example, we chose the optimal input combination for $y^1 = 1$, which using the integral on Eq. (7) yields $x_1 = 1.25$ and $x_2 = 2.5$. The next step is to settle the acceptable domain Ω, which does not violate the Eq. (9), for an almost neglectable ε. Such condition of Ω lay on the Eq. (12) restrictions:

$$\begin{aligned} x_1 &\geq 1.25, \\ x_2 &\geq 2.50, \\ x_1^2 + x_2^2 &\leq 100. \end{aligned} \tag{12}$$

With the admissible space settled, it is now possible to proceed with concrete computations. Let us take the example of a company producing $y = 1$ with input combinations settled on the point $X^d = [x_1, x_2] = [1.25, 6.21]$ with the worst directional choice X_w defined as $X^w = [x_1^w, x_2^w] = [1.25, 9.92]$ (see Fig. 2). The distance vectors d_j and d_w are easily obtained from Eq. (8), yielding the values 3.71 and 7.42, respectively, with d_w retrieving the worst case scenario considering that the chosen point belongs to it's vector. The efficiency level I_d of our example is delivered on Eq. (13):

Fig. 2 The grey area represents the admissible input combinations Ω while X^*, X^d, and X^w represent the optimal, sampled and worst collinear input levels, respectively. The output is settled in $y = 1$

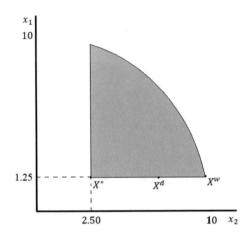

$$I_d = 1 - \frac{3.71}{7.42} = 0.5. \tag{13}$$

Since the point is equidistant to the optimal point, as the worst case scenario we can state that both the inefficiency and efficiency levels are 0.5.

5 Conclusions

We presented a literature review of the state-of-the-art efficiency methods on benchmark settling techniques based on existing data. In order to overcome the sensibility of the existing data, we proposed a new non-sample method based on differential equations that mimics the target single-output/multiple-output productive system. A few assumptions were theorized in order to obtain a differential equation system that retrieves the optimal input level per each output. Such generalizations allow us to settle an optimality condition to obtain the utopic input/output level that maximizes the overall profit. Using this result as a reference, the sample efficiency level is calculated in a generated sub-space that acquaints both feasible and rational data points in $\mathbb{R}^n$. Since the inputs already acquaint their inner weights, the n-dimensional space (considering n inputs) distance measuring, assures proportionality among the sample. The same premise can be followed to attain productive and technical efficiency/inefficiency levels, considering alternative endeavours of the selected model subset. Our theoretical approach relies on unbiased calculations of the coefficients β_i^j on Eq. (2). Introducing error with real data may demote the global optimum output y^* to an estimated one. Nonetheless, as long as the weights or costs are directly observable, the sample efficiency level should withstand even though the optimal point is biased. The presented numerical simulation displays a simple two input and one output example, where the efficiency level of a sampled company was calculated. Nonetheless, data based numerical simulations can also be computationally expensive when the number of inputs grows. Our model can be expanded to multiple output approaches using the same premises.

Acknowledgements This work was supported by the R&D Project INNOVINE & WINE Vineyard and Wine Innovation Platform Operation NORTE-01-0145-FEDER-000038, co-funded by the European and Structural Investment Funds (FEDER) and by Norte 2020 (Programa Operacional Regional do Norte 2014/2020). Torres was supported by FCT through CIDMA, project UID/MAT/04106/2019.

References

1. Farrell, M.J.: The measurement of productive efficiency. J. R. Stat. Soc. Ser. A (General) **120**(3), 253–290 (1957)
2. Bravo-Ureta, B.E., Solís, D., Moreira López, V.H., Maripani, J.F., Thiam, A., Rivas, T.: Technical efficiency in farming: a meta-regression analysis. J. Product. Anal. **27**(1), 57–72 (2007)
3. Fleming, E., Mounter, S., Grant, B., Griffith, G., Villano, R.: The new world challenge: performance trends in wine production in major wine-exporting countries in the 2000s and their implications for the Australian wine industry. Wine Econ. Policy **3**(2), 115–126 (2014)
4. Aparicio, J., Borras, F., Pastor, J.T., Vidal, F.: Accounting for slacks to measure and decompose revenue efficiency in the Spanish Designation of Origin wines with DEA. Eur. J. Oper. Res. **231**(2), 443–451 (2013)
5. Anang, B.T., Bäckman, S., Rezitis, A.: Does farm size matter? Investigating scale efficiency of peasant rice farmers in Northern Ghana. Econ. Bull. **36**(4), 2275–2290 (2016)
6. Djokoto, J.G.: Technical efficiency of organic agriculture: a quantitative review. Stud. Agric. Econ. **117**(2), 61–71 (2015)
7. Djokoto, J.G., Srofenyoh, F.Y., Arthur, A.A.A.: Technical inefficiency effects in agriculture - a meta-regression. J. Agric. Sci. **8**(2), 109–121 (2016)
8. Mareth, T., Thomé, A.M.T., Cyrino Oliveira, F.L., Scavarda, L.F.: Systematic review and meta-regression analysis of technical efficiency in dairy farms. Int. J. Product. Perform. Manag. **65**(3), 279–301 (2016)
9. Thiam, A., Bravo-Ureta, B.E., Rivas, T.E.: Technical Efficiency in developing country agriculture: a meta-analysis. Agric. Econ. **25**(2–3), 235–243 (2001)
10. Johnson, A.L., Kuosmanen, T.: One-stage and two-stage DEA estimation of the effects of contextual variables. Eur. J. Oper. Res. **220**(2), 559–570 (2012)
11. Kumbhakar, S.C., Parmeter, C.F., Zelenyuk, V.: Stochastic Frontier Analysis: Foundations and Advances, pp. 1–103 (2017)
12. Jamasb, T., Pollitt, M.: Benchmarking and regulation: international electricity experience. Util. Policy **9**(3), 107–130 (2000)
13. Khetrapal, P., Thakur, T.: A review of benchmarking approaches for productivity and efficiency measurement in electricity distribution sector. Int. J. Electron. Electr. Eng. **2**(3), 214–221 (2014)
14. Deprins, D., Simar, L., Tulkens, H.: Measuring labor efficiency in post office. The Performance of Public Enterprises: Concepts and Measurement, pp. 243–268. Elsevier Science Ltd, North-Holland (1984)
15. Leleu, H.: A linear programming framework for free disposal hull technologies and cost functions: primal and dual models. Eur. J. Oper. Res. **168**, 340–344 (2006)
16. Tulkens, H.: On FDH efficiency analysis: some methodological issues and applications to retail banking, courts, and urban transit. J. Product. Anal. **4**(1–2), 183–210 (1993)

17. Ramanathan, R.: An Introduction to Data Envelopment Analysis: A Tool for Performance Measurement. Sage Publications, London (2003)
18. Simar, L., Wilson, P.W.: Estimation and inference in two-stage, semi-parametric models of production processes. J. Econom. **136**(1), 31–64 (2007)
19. Simar, L., Wilson, P.W.: Statistical inference in nonparametric frontier models: the state of the Art. J. Product. Anal. **13**(1), 49–78 (2000)
20. Banker, R.D., Natarajan, R.: Evaluating contextual variables affecting productivity using data envelopment analysis. Oper. Res. **56**(1), 48–58 (2008)
21. Aigner, D., Lovell, C.A.K., Schmidt, P.: Formulation and estimation of stochastic frontier production function models. J. Econom. **6**(1), 21–37 (1977)
22. Meeusen, W., van Den Broeck, J.: Efficiency estimation from Cobb-Douglas production functions with composed error. Int. Econ. Rev. **18**(2), 435–444 (1977)
23. Banker, R.D., Maindiratta, A.: Maximum likelihood estimation of monotone and concave production frontiers. J. Product. Anal. **3**(4), 401–415 (1992)
24. Fan, Y., Li, Q., Weersink, A.: Semiparametric estimation of stochastic production frontier models. J. Bus. & Econ. Stat. **14**(4), 460–468 (1996)
25. Kuosmanen, T., Kortelainen, M.: Stochastic non-smooth envelopment of data: semi-parametric frontier estimation subject to shape constraints. J. Product. Anal. **38**(1), 11–28 (2012)
26. Noh, H.: Frontier estimation using kernel smoothing estimators with data transformation. J. Korean Stat. Soc. **43**(4), 503–512 (2014)
27. Parmeter, C.F., Racine, J.S.: Smooth constrained frontier analysis. Recent Advances and Future Directions in Causality, Prediction, and Specification Analysis, pp. 463–488. Springer, New York (2013)
28. Kneip, A., Simar, L.: A general framework for frontier estimation with panel data. J. Product. Anal. **7**(2–3), 187–212 (1996)
29. Ramsay, J.O.: From Data to Differential Equations. In: Antoch, J. (ed.) COMPSTAT 2004 - Proceedings in Computational Statistics. Physica, Heidelberg (2004)
30. Ramsay, J.O., Hooker, G., Campbell, D., Cao, J.: Parameter estimation for differential equations: a generalized smoothing approach. Stat. Methodol. **69**(5), 741–796 (2007)

Minimum Energy Control of Fractional Linear Systems Using Caputo-Fabrizio Derivative

Touria Karite, Ali Boutoulout and Amir Khan

Abstract We investigate regional controllability for time fractional diffusion systems of Caputo–Fabrizio type. The problem is studied via Hilbert uniqueness method (HUM) introduced by J. L. Lions in 1988. The used approach allow us to characterize the control of minimum energy.

1 Introduction

Fractional calculus owes its origin to a question of whether the meaning of a derivative to an integer order n could be extended to still be valid when n is not an integer. This question was first raised by L'Hôpital on September 30th, 1695. On that day, in a letter to Leibniz, he posed a question about $\dfrac{d^n x}{dx^n}$, Leibniz's notation for the nth derivative of the linear function $f(x) = x$. L'Hôpital curiously asked what the result would be if $n = \dfrac{1}{2}$. Leibniz responded that it would be "an apparent paradox, from which one day useful consequences will be drawn," [1].

Following this unprecedented discussion, the subject of fractional calculus caught the attention of other great mathematicians, many of whom directly or indirectly contributed to its development. They included Euler, Laplace, Fourier, Lacroix, Abel, Riemann and Liouville.

Researchers consider fractional order calculus as a generalization of the integer order calculus to a real or complex numbers and, nowadays, it plays an important

T. Karite (✉) · A. Boutoulout
TSI Team, MACS Laboratory, Department of Mathematics and Computer Science,
Faculty of Sciences, Moulay Ismail University, Meknes, Morocco
e-mail: touria.karite@gmail.com

A. Boutoulout
e-mail: boutouloutali@yahoo.fr

A. Khan
Department of Mathematics and Statistics, University of Swat, Swat, Pakistan
e-mail: amir.maths@gmail.com

© Springer Nature Switzerland AG 2020
E. H. Zerrik et al. (eds.), *Recent Advances in Modeling, Analysis and Systems Control:*
Theoretical Aspects and Applications, Studies in Systems, Decision and Control 243,
https://doi.org/10.1007/978-3-030-26149-8_10

role in various fields: physics (classic and quantum mechanics, thermodynamics, vescoelasticity, etc.), chemistry, biology, economics, engineering, signal and image processing, and control. Niels Henrik Abel, in 1823, was probably the first to give an application of fractional calculus. Abel applied the fractional calculus in the solution of an integral equation which arises in the formulation of the problem of finding the shape of a frictionless wire lying in a vertical plane such that the time of a bead placed on the wire slides to the lowest point of the wire in the same time regardless of where the bead is placed. This problem is, usually, designated by Tautochrone Problem but sometimes it's also referred as Isochrone Problem. The cycloid is the isochrone as well as the brachistochrone curve. The brachistochrone problem deals with the shortest time of slide and, from mathematical point of view, it marks the begin of the Calculus of Variations.

It is worth to mention that the controllability problem of a fractional order sub-diffusion system can be reformulated as a problem of infinite dimensional control system. Moreover, in the case of diffusion systems, it should be pointed out that, in general, not all the states can be reached [2–4]. That's why a wide range of researchers were interested in the notion of "regional controllability". This notion has been introduced by El Jai et al. (1995) for parabolic systems and by E. Zerrik and R. Larhrissi (2000) for hyperbolic linear ones. It is commonly used to refer to control problems in which the target of our interest is not fully specified as a state, but refers only to a smaller internal region ω of the system domain Ω.

Over the years, many mathematicians, using their own notation and approach, have found various definitions that fit the idea of a non-integer order integral or derivative. This paper deals with the controllability properties of a fractional diffusion equation in the sense of Caputo–Fabrizio. We define the regional (exact and weak) controllability and we compute the control minimizing the cost functional.

The remainder of this paper is organized as follows: some definitions and preliminaries on fractional calculus are given in the next section. In Sect. 3, we define the regional fractional controllability of the system. Our main results on the regional controllability are proved in Sect. 4, where we prove that the control steering our systems to the final desired state is of minimum energy. And we end with Sect. 5 of conclusions and some interesting open questions that deserve further investigations.

2 Preliminaries

Let us recall the usual Caputo fractional time derivative of order α, given by

$$^C D^\alpha f(t) = \frac{1}{\Gamma(1-\alpha)} \int_a^t \frac{\dot{f}(\tau)}{(t-\tau)^\alpha} d\tau, \tag{1}$$

with $\alpha \in (0, 1)$ and $a \in [-\infty, t)$, $f \in H^1(a, b)$, $b > a$. By changing the kernel $(t - \tau)^a$ with the function $exp(\frac{-\alpha}{1-\alpha} t)$ and $\frac{1}{\Gamma(1-\alpha)}$ with $\frac{M(\alpha)}{1-\alpha}$, we obtain the following new definition of fractional time derivative (Caputo–Fabrizio derivative [5])

$$^{CF}D^\alpha f(t) = \frac{M(\alpha)}{1-\alpha} \int_a^t \dot{f}(\tau) \exp\left(-\frac{\alpha(t-\tau)}{1-\alpha}\right) d\tau. \qquad (2)$$

where $M(\alpha)$ is a normalization function such that $M(0) = M(1) = 1$. According to (2), the new fractional derivative is equal to zero when the function $f(t)$ is a constant similarly to (1), but, the kernel does not have singularity for $t = \tau$. However, if the function does not belongs to $H^1(a, b)$ then, the derivative can be reformulated as

$$^{CF}D^\alpha f(t) = \frac{\alpha M(\alpha)}{1-\alpha} \int_a^t (f(t) - f(\tau)) \exp\left(-\frac{\alpha(t-\tau)}{1-\alpha}\right) d\tau.$$

We define the fractional integral of order α of a function f by the following.

Definition 1 ([6]) Let $0 < \alpha < 1$. The fractional integral of order α of a function f is defined by

$$I_t^\alpha (f(t)) = \frac{2(1-\alpha)}{(2-\alpha)M(\alpha)} f(t) + \frac{2\alpha}{(2-\alpha)M(\alpha)} \int_0^t f(s)ds, \ t \geq 0 \qquad (3)$$

where $M(\alpha)$ is a normalization function such that $M(0) = M(1) = 1$.

In order to prove our main results, the following lemma is needed.

Lemma 1 ([7]) *Let the reflection operator $\mathcal{Q}$ on the interval $[0, T]$ be defined as follows:*

$$\mathcal{Q}h(t) := h(T - t),$$

for some function h which is differentiable and integrable. Then the following relations hold:

$$\mathcal{Q}_0 I_t^\alpha h(t) = {}_t I_T^\alpha \mathcal{Q}h(t), \qquad \mathcal{Q}_0 D_t^\alpha h(t) = {}_t D_T^\alpha \mathcal{Q}h(t)$$

and

$$_0 I_t^\alpha \mathcal{Q}h(t) = \mathcal{Q}_t I_T^\alpha h(t), \qquad {}_0 D_t^\alpha \mathcal{Q}h(t) = \mathcal{Q}_t D_T^\alpha h(t).$$

3 Regional Fractional Controllability

Let Ω be an open regular bounded set of $\mathbb{R}^n$ and $Z = L^2(\Omega)$. Consider the state-space fractional system of order $\alpha \in (0, 1)$

$$\begin{cases} ^{CF}D^\alpha y(t) = Ay(t) + Bu(t) & t \in [0, T] \\ y(0) = y_0 & \text{in } \mathcal{D}(A), \end{cases} \qquad (4)$$

where $^{CF}D^\alpha$ denotes the Caputo-Fabrizio fractional order derivative (for more details see references [5, 8]). The second order operator A is linear with dense domain such that the coefficients do not depend on t and generates a C_0-semi-group $(T(t))_{t\geq 0}$ on the Hilbert space Z and we refer the readers to Engel and Nagel [9] and Renardy and Rogers [10] for more properties on operator A. $\mathcal{D}(A)$ holds for the domain of the operator A. $y \in L^2(0, T; Z)$ and $u \in U = L^2(0, T; \mathbb{R}^m)$ (where m is the number of actuators). The initial datum y_0 is in Z, the control operator $B : \mathbb{R}^m \longrightarrow Z$ is linear (possibly unbounded) depending on the number and structure of actuators.

Theorem 1 ([11]) *The solution $y(t, u)$ of system (4) for a given initial condition $y(0) = y_0$ and input $u(t)$ has the form*

$$y(t, u) = e^{\hat{A}t}\left(\hat{y}_0 + \hat{B}u_0\right) + \int_0^t e^{\hat{A}(t-\tau)}\hat{B}\left[\beta u(\tau) + \dot{u}(\tau)\right]d\tau, \qquad (5)$$

where

$$\hat{A} = \alpha\left[I_n - (1-\alpha)A\right]^{-1}A,$$

$$\hat{B} = \left[I_n - (1-\alpha)A\right]^{-1}(1-\alpha)B, \quad \beta = \frac{\alpha}{1-\alpha},$$

$$\hat{y}_0 = \left[I_n - (1-\alpha)A\right]^{-1}y_0, \quad e^{\hat{A}t} = \mathcal{L}^{-1}\left\{\left[I_n s - \hat{A}\right]^{-1}\right\}, \qquad (6)$$

$$\dot{u}(\tau) = \frac{du(\tau)}{d\tau}, \quad u(0) = u_0.$$

Let $\omega \subseteq \Omega$ be a given region of positive Lebesgue measure. We denote the projection operator on ω by the following restriction mapping

$$\chi_\omega : L^2(\Omega) \longrightarrow L^2(\omega)$$
$$y \longmapsto \chi_\omega y = y|_\omega$$

Definition 2 • System (4) is said to be exactly regionally controllable on ω at time T if, for any $z_d \in L^2(\omega)$, there exists a control $u \in U$ such that

$$\chi_\omega y(T, u) = z_d. \qquad (7)$$

• System (4) is said to be approximately regionally controllable on ω at time T if, for any $z_d \in L^2(\omega)$, given $\epsilon > 0$, there exists a control $u \in U$ such that

$$\|\chi_\omega y(T, u) - z_d\|_{L^2(\omega)} \leq \epsilon \qquad (8)$$

4 Minimum Energy Control

The study of fractional optimal control problems is a subject under strong development: see [12, 13] and references therein. In this section, inspired by the results of [14–19], we show that the steering controls are minimizers of a suitable optimal control problem.

Let G be a closed subspace of Z. The extended controllability problem consists in finding a minimum norm control that steers the system to G at time T

$$\begin{cases} \int_0^T \frac{1}{2} \|u(t)\|^2 dt \\ u \in U_{ad}, \end{cases} \tag{9}$$

where $U_{ad} = \{u \in U \mid y(T, u) - z_d \in G\}$.

Identifying $L^2(\Omega)$ with its dual we have $Z \subset L^2(\Omega) \subset Z^*$. Let

$$G = \{g \in Z^* \mid g = 0 \text{ in } \Omega \setminus \omega\}$$

For $\varphi_0 \in G_0$, we consider the system in Z^*

$$\begin{cases} {}^{CF}D^\alpha Q\varphi(t) = -A^*Q\varphi(t) \\ \lim_{t \to T^-} {}_t I_T^{1-\alpha} Q\varphi(t) = \varphi_0 \in \mathcal{D}(A^*) \subseteq L^2(\Omega), \end{cases} \tag{10}$$

and let

$$\|\varphi_0\|_{G_0}^2 = \int_0^T \|B^*\varphi(t)\|^2 dt. \tag{11}$$

be a semi-norm on G_0.

It follows from Lemma 1 that (10) can be rewritten as

$$\begin{cases} {}^{CF}_0 D_t^\alpha \varphi(t) = -A^*\varphi(t) \\ \lim_{t \to 0^+} {}_0 I_t^{1-\alpha} \varphi(t) = \varphi_0 \in \mathcal{D}(A^*) \subseteq L^2(\Omega), \end{cases} \tag{12}$$

We also consider

$$\begin{cases} {}^{CF}_0 D_t^\alpha \psi(t) = A\psi(t) + BB^*\varphi(t) \quad t \in [0, T] \\ \psi(0) = y_0, \end{cases} \tag{13}$$

and define the operator M by

$$M\varphi_0 = \mathcal{P}(\psi(T)), \tag{14}$$

where $\mathcal{P} = \chi_\omega^* \chi_\omega$.

M is affine and can be decomposed as

$$M\varphi_0 = \mathcal{P}\left(\psi_0(T) + \psi_1(T)\right),$$

where ψ_0 solution of the following system

$$\begin{cases} {}^{CF}_0 D_t^\alpha \psi_0(t) = A\psi_0(t) & t \in [0, T] \\ \psi_0(0) = 0, \end{cases} \tag{15}$$

and ψ_1 solution of

$$\begin{cases} {}^{CF}_0 D_t^\alpha \psi_1(t) = A\psi_1(t) + BB^*\varphi(t) & t \in [0, T] \\ \psi_1(0) = y_0 \in \mathcal{D}(A). \end{cases} \tag{16}$$

Let

$$\Lambda\varphi_0 = \mathcal{P}\left(\psi_1(T)\right). \tag{17}$$

With these notations, the regional controllability problem leads to the resolution of equation

$$\Lambda\varphi_0 = \chi_\omega^* z_d - \mathcal{P}\left(\psi_0(T)\right). \tag{18}$$

We have then the following result.

Theorem 2 *If system* (1) *is weakly regionally controllable then Eq.* (18) *has a unique solution* $\varphi_0 \in G_0$ *and the control*

$$u^*(t) = B^*\varphi(t) \tag{19}$$

steers system (1) *to* z_d *at time* T *in* ω

$$y(T, u^*) - z_d \in G.$$

Moreover, this control minimizes the cost function

$$J(u) = \frac{1}{2}\|u(t)\|^2 dt \quad on \quad U_{ad}$$

Proof First, we prove that if system (4) is weakly regionally controllable, then (11) defines a norm.

$$\|\varphi_0\|_{G_0}^2 = 0 \iff \int_0^T \|B^*\varphi(t)\|^2 dt = 0$$

$$\Longleftrightarrow B^*\varphi(t) = 0$$
$$\Longleftrightarrow -B^* t^{\alpha-1} e^{\hat{A}(T-t)} \varphi_0 = 0$$

$$\Longleftrightarrow \varphi_0 = 0$$

We denote the completion of the set G_0 with respect to the norm (11) again by G_0. For $\varphi_0 = \chi_\omega^* \varphi$ let $\|\varphi\|_F = \|\chi_\omega^* \varphi\|_{G_0}$, then there exists k, $\bar{k}$ such that $\|\varphi\|_F \leq k\|\varphi_0\|_{Z^*} \leq \bar{k}\|\varphi_0\|_{L^2(\omega)}$.

If F is the completion of $L^2(\omega)$ with respect to $\|\cdot\|_F$ then $L^2(\omega) \subset Z^* \subset F$ and by duality $F^* \subset Z \subset L^2(\omega)$. Moreover, the set G_0^* is given by $G_0^* = \{z \in Z : z|_\omega \in F^*\}$. We have

$$\langle \Lambda\varphi_0, \varphi_0 \rangle_{G_0^*, G_0} = \langle \mathcal{P}(\psi_1(T)), \varphi_0 \rangle = \langle \psi_1(T), \varphi_0 \rangle.$$

But $\psi_1(T) = \int_0^T S(T-t)BB^*\varphi(s)ds$, so

$$\langle \Lambda\varphi_0, \varphi_0 \rangle_{G_0^*, G_0} = \left\langle \int_0^T S(T-t)BB^*\varphi(s)ds, \varphi_0 \right\rangle$$
$$= \int_0^T \langle S(T-t)BB^*\varphi(s), \varphi_0 \rangle ds$$
$$= \int_0^T \|B^*\varphi(s)\|^2 ds = \|\varphi_0\|_{G_0}$$

Hence $\Lambda : G_0 \longrightarrow G_0^*$ is one to one. By the assumption $U_{ad} \neq 0$ we have $z_d - \psi_0(T) \in F^*$ and so there is a unique φ_0 solving (18).

Further, let $u = u^*$ in (4), we see that $\chi_\omega y(T, u^*) = z_d$.

For any $u_1 \in U$ avec $\chi_\omega y(T, u_1^*) = z_d$, we obtain that

$$\chi_\omega \left[y(T, u^*) - y(T, u_1) \right] = 0,$$

and for any $\varphi_0 \in G_0$ we have

$$\langle \varphi_0, \chi_\omega \left[y(T, u^*) - y(T, u_1) \right] \rangle = 0.$$

It follows that

$$\int_0^T \langle u^*(s) - u_1(s), B^*\varphi(t) \rangle ds = 0.$$

Moreover, since

$$J'(u^*)(u^* - u_1) = 2 \int_0^T \langle u^*(s), u^*(s) - u_1(s) \rangle \, ds$$
$$= 2 \int_0^T \langle B^* \varphi(t), u^*(s) - u_1(s) \rangle \, ds$$
$$= 0.$$

by Theorem 1.3 in [20], we conclude that u^* solves the minimum energy problem (9) and the proof is complete.

5 Conclusion and Perspectives

In this work we considered fractional diffusion equation using Caputo–Fabrizio operator. We studied the regional fractional controllability using an extention of HUM approach. The formulation presented and the resulting equations are very similar to those for classical optimal control problems. Also it is evident that by modifying:

- the boundary conditions (Dirichlet, Neumann, mixed),
- the nature of the control (distributed, boundary),
- boundedness of the control operator B,
- the type of the equation and the order of the studied equation,

we can have many other open problems that need further investigations and which can be developed in future works.

Acknowledgements This work has been carried out with a grant from Hassan II Academy of Sciences and Technology project N° 630/2016.

References

1. Miller, K., Ross, B.: An Introduction to the Fractional Calculus and Fractional Differential Equations. Wiley, New York (1993)
2. El Jai, A., Pritchard, A.J.: Sensors and Controls in the Analysis of Distributed Systems. Halsted Press (1988)
3. Ge, F., Chen, Y.Q., Kou, C.: Regional gradient controllability of sub-diffusion processes. J. Math. Anal. Appl. **440**(2), 865–884 (2016)
4. Ge, F., Chen, Y.Q., Kou, C.: Regional boundary controllability of time fractional diffusion processes. IMA J. Math. Control Inf. **34**(3), 871–888 (2016)
5. Caputo, M., Fabrizio, M.: A new definition of fractional derivative without singular kernel. Prog. Fract. Differ. Appl. **1**(2), 73–85 (2015)
6. Julaighim Algahtani, O.J.: Comparing the Atangana–Baleanu and Caputo–Fabrizio derivative with fractional order: Allen Cahn model. Chaos, Solitons and Fractals **89**, 552–559 (2016)
7. Małgorzata, K.: On Solutions of Linear Fractional Differential Equations of a Variational Type. Czestochowa University of Technology, Czestochowa (2009)
8. Atanacković, T.M., Pilipović, S., Zorica, D.: Properties of the Caputo-Fabrizio fractional derivative and its distributional settings. Fract. Calc. Appl. Anal. **21**(1), 29–44 (2018)

9. Engel, K.J., Nagel, R.: A Short Course on Operator Semigroups. Springer, New York (2006)
10. Renardy, M., Rogers, R.C.: An Introduction to Partial Differential Equations. Springer, New York (2004)
11. Kaczorek, T.: Analysis of positive and stable fractional continuous-time linear systems by the use of Caputo-Fabrizio derivative. Control Cybern. **45**(3), 289–299 (2016)
12. Mozyrska, D., Torres, D.F.M.: Minimal modified energy control for fractional linear control systems with the Caputo derivative. Carpathian J. Math. **26**(2), 210–221 (2010)
13. Mozyrska, D., Torres, D.F.M.: Modified optimal energy and initial memory of fractional continuous-time linear systems. Signal Process. **91**(3), 379–385 (2011)
14. El Jai, A., Pritchard, A.J.: Sensors and Actuators in Distributed Systems Analysis. Wiley, New York (1988)
15. Zerrik, E.: Regional Analysis of Distributed Parameter Systems. University of Rabat, Morocco (1993). PhD Thesis
16. El Jai, A., Pritchard, A.J., Simon, M.C., Zerrik, E.: Regional controllability of distributed systems. Int. J. Control **62**, 1351–1365 (1995)
17. Karite, T., Boutoulout, A.: Regional enlarged controllability for parabolic semilinear systems. Int. J. Appl. Pure Math. **113**(1), 113–129 (2017)
18. Karite, T., Boutoulout, A.: Regional boundary controllability of semilinear parabolic systems with state constraints. Int. J. Dyn. Syst. Differ. Equ. **8**(1/2), 150–159 (2018)
19. Karite, T., Boutoulout, A., Torres, D.F.M.: Enlarged controllability of Riemann–Liouville fractional differential equations. J. Comput. Nonlinear Dynam. **13**(9), 090907 (2018). 6 pp
20. Lions, J.L.: Optimal Control of Systems Governed by Partial Differential Equations. Springer, New York (1971)

Parametric Identification of the Dynamics of Inter-Sectoral Balance: Modelling and Forecasting

Olena Kostylenko, Helena Sofia Rodrigues and Delfim F. M. Torres

Abstract This work is devoted to modelling and identification of the dynamics of the inter-sectoral balance of a macroeconomic system. An approach to the problem of specification and identification of a weakly formalized dynamical system is developed. A matching procedure for parameters of a linear stationary Cauchy problem with a decomposition of its upshot trend and a periodic component, is proposed. Moreover, an approach for detection of significant harmonic waves, which are inherent to real macroeconomic dynamical systems, is developed.

Keywords Leontief's model · Cyclical processes · Harmonic waves · Prediction

2010 Mathematics Subject Classification 91B02 · 91B84

1 Introduction

For any dynamical object (technical, economic, environmental, etc.), the most critical is the problem of limited resources and their optimal use. Thereby, there is a need to build mathematical models that adequately describe the existing trends and provide high-precision predictive properties of the dynamical systems. Therefore, it is necessary to evaluate the quality of mathematical models of dynamical processes through the prism of imitation and predictive properties.

O. Kostylenko · H. S. Rodrigues · D. F. M. Torres (✉)
Department of Mathematics, Center for Research and Development in Mathematics and Applications (CIDMA), University of Aveiro, 3810-193 Aveiro, Portugal
e-mail: delfim@ua.pt

O. Kostylenko
e-mail: o.kostylenko@ua.pt

H. S. Rodrigues
School of Business Studies, Polytechnic Institute of Viana do Castelo,
4930-678 Valença, Portugal
e-mail: sofiarodrigues@esce.ipvc.pt

© Springer Nature Switzerland AG 2020
E. H. Zerrik et al. (eds.), *Recent Advances in Modeling, Analysis and Systems Control: Theoretical Aspects and Applications*, Studies in Systems, Decision and Control 243,
https://doi.org/10.1007/978-3-030-26149-8_11

The meaning of mathematical modelling, as a research method, is determined by the fact that a model is a conceptual tool focused on the analysis and forecasting of dynamical processes using differential or differential-algebraic equations. However, the parameters of these equations are usually unknown in advance. Therefore, in practice, any direct problem (imitation, prediction, and optimization) is always preceded by an inverse problem (model specification and identification of parameters and variables included in it).

To construct dynamical models in economics, it is necessary to formulate principles, based on economic theory, and consider equations that adequately determine the evolution of the investigated process. The functioning of a market economy, as well as any economic system, is not uniform and uninterrupted. An important tool for its prediction is the dynamical model of inter-industry equilibrium balance developed by Leontief (the so called "Leontief's input-output model"), which became the basis of mathematical economics [1]. Leontief's input-output model is described by a system of linear differential equations in which the input is the non-productive consumption of sectors, and the output is the issue of these sectors. In practice, the matrices of the system of differential equations are unknown in advance. Therefore, there is a need to evaluate them based on statistical information of the macroeconomic system under study, for a certain period of time. The problem of parametric identification of macroeconomic dynamics is discussed in the papers [2–5].

One feature of the dynamical Leontief model is that it can be considered as taking into account the cyclical nature of the macroeconomic processes, typical for developed countries of Western Europe, USA, Canada, Japan, and others [6, 7]. According to statistical analysis, their economic growth (upturn phase) alternates with the processes of stagnation and decline in production volumes (downturn phase), i.e., a decline in all the economic activity. After a decline phase, again, the economic cycle continues, similarly with ups and downs. Such periodic fluctuations indicate a cyclical nature of the economic development.

The analysis of the input-output model is critical for developing government regulatory programs, that are an integral part of the economic and environmental policy for any developed market system. Such an analysis is a useful tool for studying the strategic directions of economic development, the expansion of dynamical interindustry relations, and the interaction between economic sectors and the natural environment [8, 9].

This paper is organized as follows. Section 2 is devoted to a detailed economic and mathematical description of the Leontief macroeconomic model. Sect. 3 focuses on the problem of specification and identification of this model and Sect. 4 to the testing of the Leontief model with time series data of a real macroeconomic dynamics. In Sect. 5, the main conclusions are carried out.

2 Problem Statement

The static model of inter-industry balance is obtained when one equates the net production of branches to the final demand for the products of these industries:

$$\mathbf{x} - A\mathbf{x} = \mathbf{y} \,, \tag{1}$$

where $\mathbf{x} = (x_1, x_2, \ldots, x_n)'$ is the column vector of gross output by industries; $\mathbf{y} = (y_1, y_2, \ldots, y_n)'$ is the column vector of final demand; and A is the matrix of direct costs. The vector of final demand $\mathbf{y}$ can be represented as a sum of two vectors: investments, $\mathbf{y}_1(t) = B\dot{\mathbf{x}}(t)$, and consumer products, $\mathbf{y}_2(t) = \mathbf{u}(t)$. Then, the model of dynamical inter-industry balance is

$$\mathbf{x}(t) = A\mathbf{x}(t) + B\dot{\mathbf{x}}(t) + \mathbf{u}(t) \,, \quad \dot{\mathbf{x}}(t) = \frac{d\mathbf{x}(t)}{dt} \,, \tag{2}$$

where $\mathbf{u}(t)$ is the vector-column of final (non-productive) consumption and B is the matrix of capital coefficients. A discrete-time dynamic balance model can be reduced to a model with continuous time as follows:

$$B\dot{\mathbf{x}}(t) = (E - A)\mathbf{x}(t) - \mathbf{u}(t) \,. \tag{3}$$

If the inverse matrix B^{-1} exists, then the Leontief's model can be rewritten as a linear multiply-connected system, with the final consumption vector $\mathbf{u}(t)$ as the input and with the gross output vector $\mathbf{x}(t)$ as the output:

$$\dot{\mathbf{x}}(t) = B^{-1}(E - A)\mathbf{x}(t) - B^{-1}\mathbf{u}(t) \,, \quad t \in [t_0, t_f] \,. \tag{4}$$

The differential equation (4) must be supplemented with the boundary condition

$$\mathbf{x}(t_*) = \mathbf{x}_* \,, \tag{5}$$

where t_* is the point of the segment $[t_0, t_f]$ at which the boundary state is specified. To determine the unknown parameters A, B, and $\mathbf{x}_*$ of the Cauchy problem (4) (5), in accordance with the methodology proposed in [10], we divide the interval $[t_0, t_f]$ into two intervals: the identification period $[t_0, t_*)$ and the forecasting period $[t_*, t_f]$. We assume that the base period $[t_0, t_*)$ has statistical information $\mathbf{x}_t$ and $\mathbf{y}_t$ regarding industry's outputs and consumption at its discrete-times $t = 1, 2, \ldots, N$, and the model will be adjusted for the vector of unknown coefficients θ. For the interval $[t_*, t_f]$, we assume that $t_f - t_* \ll N$. Through this approach, the boundary condition (5) is satisfied in the first integer point of the forecast period. Then, if the model is stationary, and due to the inertia of the dynamical system, the vector θ can be translated for the forecast period. The stationariness of the model is characterized by the high quality of approximation and forecasting and robustness [11]. However, for modelling and prediction of linear stationary models, the identification period N should be large enough to stabilize the interrelations between the elements of the system.

3   Algorithm

First of all, for convenience, we make the data dimensionless and then, we proceed by solving the Cauchy problem (4)–(5). To do this, we need to specify the vector of phase coordinates $\mathbf{x}(t)$ and estimate the unknown parameters that will appear in the specification process [10]. The vector $\mathbf{x}(t)$ is searched by decomposing the trajectories of the phase coordinates motion into components [10]. If these trajectories are defined, then the vector $\mathbf{u}(t)$ can be found. It will be supplied to the input of the dynamical system and it can be used to solve the problem of specifying the sectors' output and manage their movement. The vector $\mathbf{u}$ will be considered as a control vector and it can be found using the inverse relationship

$$\mathbf{u}(t) = P\mathbf{x}(t) - B\dot{\mathbf{x}}(t) , \quad P = E - A . \tag{6}$$

At each time step, the vector $\mathbf{u}$ is a function of the phase coordinates and their derivatives. The regulator implements the critical idea of control theory: the principle of inverse connection, being used for identifying the model (6). In our research, the regulator consists of two devices. At any time t, they form a total value of phase coordinates (issues sectors) and controls (non-productive consumption sectors):

$$\mathbf{x}(t) = \sum_{m=1}^{n} \mathbf{x}_m(t) , \quad \mathbf{u}(t) = \sum_{m=1}^{n} \mathbf{u}_m(t) . \tag{7}$$

If the modelling trajectories $\mathbf{x}(t)$ and $\mathbf{u}(t)$, where $\mathbf{x}(t) = (x_1(t), x_2(t), \ldots, x_n(t))'$ and $\mathbf{u}(t) = (u_1(t), u_2(t), \ldots, u_n(t))'$, are adjusted to high imitation and prediction properties, then the total trajectories (7) should have the same properties. Following [10], we assume that the trajectory of the dynamical system is represented by an additive combination of its components:

$$\mathbf{x}_t = \mathbf{x}_*(t) + \varepsilon_t . \tag{8}$$

The development tendency is characterized by a linear trend,

$$\mathbf{x}_*(t) = \bar{\mathbf{x}} + \mathbf{b}(t - \bar{t}) , \quad \bar{\mathbf{x}}_* = \bar{\mathbf{x}} + \mathbf{b}(t - \bar{t}) = \bar{\mathbf{x}} , \tag{9}$$

while the fluctuation process is described by a linear combination of harmonics with some frequencies over the time interval $[1, N]$:

$$\varepsilon_t = \sum_{k} (\mathbf{a}_k \cos \omega_k t + \mathbf{b}_k \sin \omega_k t) + \nu_t , \tag{10}$$

where ω_k is the frequency of the harmonic k; $\mathbf{a}_k, \mathbf{b}_k$, $k = 1, \ldots, n-1$, are the unknown coefficients of the decomposition in the truncated Fourier series; and v_t is the vector of random residuals. We decompose the process of random oscillations as a truncated Fourier series in order to reflect the imitation and predict the real properties of the oscillations. Since the trend and periodic components correlate with each other, then the phase trajectories can be found after evaluating the regression model:

$$\mathbf{x}_t - \mathbf{x} = \mathbf{b}(t - \bar{t}) + \sum_{k=1}^{n-1} (\mathbf{a}_k \cos \omega_k t + \mathbf{b}_k \sin \omega_k t) + v_t , \quad t = 1, \ldots, N . \quad (11)$$

It is assumed that in regression models the average value of residuals is zero. Therefore, it is necessary to choose the frequency fluctuation, so that the average harmonics values would be equal to zero. Harmonic fluctuations are adjusted in the spectrum of frequencies:

$$\begin{cases} \sin \omega_k t = 0 \\ \cos \omega_k t = 0 \end{cases} \Rightarrow \omega_k = \frac{2\pi}{N} k, \quad k = 1, 2, \ldots \quad (12)$$

Determination of frequencies from the spectrum given in (12), as well as the period T of oscillations of this system, can be done using the first regulator device (6), which calculates the output of the entire macroeconomic system. If the fluctuation frequencies belong to the spectrum (12), where the variable N is the sample size, then, from the mathematical point of view, we need the fluctuation period T instead of the sample size. So, we assume that $N = T$. The optimal value N, as well as the fluctuation period, can be identified using the back-extrapolation method. Minimization of the residual sum of squares

$$S = \sum_{t=1}^{N} v_t^2 = \sum_{t=1}^{N} (x_t - \bar{x} - b(t - \bar{t}) - \sum_k (a_k \cos \omega_k t + b_k \sin \omega_k t))^2 \quad (13)$$

provides the following Ordinary Least Squares (OLS) estimates:

$$\hat{b} = \frac{\frac{2}{N} \sum_{t=1}^{N} (x_t - \bar{x})(t + \sum_k (\cot \frac{\omega_k}{2} \sin \omega_k t - \cos \omega_k t))}{\frac{N^2-1}{6} - \sum_k \frac{1}{\sin^2 \frac{\omega_k}{2}}} ,$$

$$\varepsilon_t = x_t - \bar{x} - \hat{b}(t - \bar{t}) , \quad \hat{a}_k = \frac{2}{N} \sum_{t=1}^{N} x_t \cos \omega_k t - \hat{b} , \quad (14)$$

$$\hat{b}_k = \frac{2}{N} \sum_{t=1}^{N} x_t \sin \omega_k t + \hat{b} \cot \frac{\omega_k}{2} , \quad \hat{c}_k = \sqrt{\hat{a}_k^2 + \hat{b}_k^2} .$$

According to (14), the trend correlates with harmonics and, vice versa, harmonic waves depend on the trend around which they fluctuate. Harmonics of the regression model of fluctuations, which have different spectra (12), do not correlate with each other. If the estimates $\hat{a}_k$ and $\hat{b}_k$, $k = 1, 2, \ldots$, are optimal (14), then the minimum sum of squares of deviations is equal to

$$\min S = \sum_{t=1}^{N} \hat{v}^2 = \sum_{t=1}^{N} \varepsilon_t^2 - \frac{N}{2} \sum_{k} \hat{c}_k^2 . \tag{15}$$

The significance of the OLS-estimates (14) was tested using Student's t-test. The optimal dimension n of the phase space is determined when the $n - 1$ significant harmonics are found. The component specification $\mathbf{x}$ is executed with a specified value n. If the phase coordinates are chosen, then we assume that the inherent harmonic fluctuations are tuned to the frequency (12). The number of significant harmonics in models of different phase coordinates may differ, due to the fact that each subset $x_1, x_2, \ldots, x_n$ of the set $\mathbf{x}$ has its own specificity of functioning. If the phase coordinate responds quickly to qualitative changes in a given dynamical system, then this coordinate will have the maximum number, that is, $n - 1$ harmonics. The minimum number of harmonics will be in the decomposition of those phase coordinates that are weakly responsive to changes in other subsets of the system. If the insignificant OLS-estimates of the decomposition coefficients are discarded, then the model of the phase coordinate fluctuations around the corresponding trends is back-extrapolated for the period $t \leq 0$ and checked if matches the statistical value ε_t, $t = 0, -1, \ldots$ If we are satisfied with the test, then we model the trajectories of the phase coordinate movement as

$$\mathbf{x}(t) = \bar{\mathbf{x}} + \mathbf{b}(t - \bar{t}) + \sum_{k} (\hat{\mathbf{a}}_k \cos \omega_k t + \hat{\mathbf{b}}_k \sin \omega_k t) . \tag{16}$$

The approximation properties of the obtained model curves are described by using the coefficients of determination R^2. The coefficients of determination linear trends of the phase coordinates are calculated by the formula

$$\mathbf{R}_{tr}^2 = \frac{\hat{\mathbf{b}}^2 \frac{(N^2-1)}{12} N}{\sum_{t=1}^{N} (\mathbf{x}_t - \bar{\mathbf{x}})^2} . \tag{17}$$

Taking into account the periodic component in the model motion paths (16), we get

$$\mathbf{R}^2 = 1 - \frac{\sum_{t=1}^{N} v_t^2}{\sum_{t=1}^{N} (\mathbf{x}_t - \bar{\mathbf{x}})^2} , \quad v_t^2 = \sum_{t=1}^{N} \varepsilon_t^2 - \frac{N}{2} \sum_{k} \mathbf{c}_k^2 . \tag{18}$$

For the fluctuation model, one gets

$$\mathbf{R}_{fl}^2 = \frac{\frac{N}{2}\sum\limits_{k}\hat{\mathbf{c}}_k^2}{\sum\limits_{t=1}^{N}\varepsilon_t^2}. \tag{19}$$

Particles of harmonic dispersions, in the general dispersion fluctuations of each sector $m = 1, \ldots, n$, are calculated by using respective coefficients of determination. The proportion of dispersion of the kth harmonic, in the total dispersion of the fluctuation of the phase coordinates, is

$$\mathbf{R}_k^2 = \frac{\frac{N}{2}\hat{\mathbf{c}}_k^2}{\sum\limits_{t=1}^{N}\varepsilon_t^2} = \frac{\hat{\mathbf{c}}_k^2}{\sum\limits_{k}\hat{\mathbf{c}}_k^2}\mathbf{R}_{fl}^2. \tag{20}$$

The specification of the control vector $\mathbf{u}$ is carried out at the given phase vector $\mathbf{x}(t)$ and its derivative $\dot{\mathbf{x}}(t)$. To identify the control vector $\mathbf{u}(t)$ taking into account (6), we construct the following regression model:

$$\mathbf{u}(t) - \bar{\mathbf{u}} = P\left(\mathbf{x}(t) - \bar{\mathbf{x}}\right) - B\left(\dot{\mathbf{x}} - \hat{\mathbf{b}}\right) + \mathbf{r}_t, \tag{21}$$

where $\mathbf{r}_t$ is the vector of random perturbations. The adequacy of the curves is checked using the coefficients of determination, as well as the second controller device, which models the trajectory of total non-production consumption, according to the second balance equation (7).

4 Implementation

The approbation of the constructed algorithm was carried out for real macroeconomic dynamics. The research was done using the statistical data of France [12], which includes information from 1949 to 2017. It was found that the period of fluctuation for France is fifty years ($T = 50$). Consequently, $N = 50$, while 1966–2015 is the identification period and 2016–2017 the forecast period. The next and main step is to establish the significant harmonics. For this, we have used the Student t-test with a significance level of $\alpha = 0.005$. If the k-harmonic appears insignificant by Student's t-test, then all insignificant harmonics are sequentially turned over, and we need to remove them from the regression. Then, coming back to the previous step, we start recalculations, as long as all the harmonics become significant. When we find four significant harmonics, Kondratieff's wave ($k = 1$), Kuznets' wave ($k = 3$), Juglar's wave ($k = 6$), and the wave that equals half the period of the Kondratieff wave ($k = 2$), then we conclude that the number of all sectors is five, that is, $n = 4$ harmonics

Table 1 The coefficients of trends determination

Sector	1	2	3	4	5	Σ
R^2	0.8076	0.6871	0.7773	0.7233	0.8151	0.7883

plus 1 trend [7]. Thus, the economy must be divided into five sectors. Empirically, we set the optimum division of the economy into sectors: Industry and Agriculture (Sector 1); Construction and Transport (Sector 2); Finance and Real estate (Sector 3); Communication and Science (Sector 4); and Service Industries (Sector 5). In each sector, the four harmonics are present. Parametric identification of the regression model of the sectors outputs gives the following coefficients of trends determination R^2 (Table 1), around which fluctuation occurs. Results in Table 1 show that, for the economy of France, the fluctuation of outputs around the corresponding trend is perceptible. The harmonic waves interact with the trend, complicating their analysis. But if we consider the pure oscillatory process, then the harmonics of Fourier's series become uncorrelated, simplifying the analysis of the influence of individual harmonics on the general oscillation process. Particles of harmonic dispersions in the general dispersion fluctuations of each sector $m = 1, \ldots, n$ are calculated using the respective coefficients of determination (20), the values of which are given in Table 2, where the "—" means that the harmonic is insignificant in the sector according to Student's t-test. The analysis of the share of harmonics in the general dispersions fluctuations of each sector, calculated by appropriate coefficients of determination, shows that:

- The Kondratieff wave $k = 1$ (long wave) substantially influences the 1st and 4th sectors.
- The wave with period of 25 years ($k = 2$) prevails in the 2nd and 3rd sectors.
- The Kuznec wave $k = 3$ (rhythms with period 15–20 years), especially manifests itself in the 1st and 2nd sectors.
- The contribution of the Juglar wave ($k = 6$) is less significant in comparison with other waves, but it makes significant adjustments to the gross output function for the Sectors 2.
- The total contribution of harmonics, into the dispersion of fluctuations of the sectors, ranges from 93.64% (1st sector) to 98.75% (5th sector).

Therefore, regression models of fluctuations have the qualitative approximation properties and we can expect a significant contribution into the dispersions of issues. The values of the coefficients of determination of modelling trajectories of issues are given in Table 3. Comparing the results shown in Table 1 and Table 3, we can make a conclusion about the significant influence of harmonic waves on trajectories of sectors issues (the quality of modelling trajectories exceeds 99%).

In Fig. 1, the graphics of modelling curves of gross issues and the trajectories corresponding to the fluctuations of the macroeconomic system are shown. The graphics show points that give statistical information and the solid line is the trajectory. Comparison of the predicted values with the real data (the last two points,

Table 2 Contribution of harmonics into the oscillatory process

Sector	$k = 1$	$k = 2$	$k = 3$	$k = 6$	$\sum$
1	0,8202	0,0254	0,0908	–	0,9364
2	0,3719	0,4622	0,0982	0,0173	0,9496
3	0,6460	0,2916	0,0208	–	0,9583
4	0,8855	0,0884	0,0073	–	0,9812
5	0,7354	0,2521	–	–	0,9875
$\sum$	0,7429	0,2186	0,0094	0,0089	0,9798

Table 3 The quality of modelling trajectories of issues

Sector	1	2	3	4	5	$\sum$
R^2	0,9983	0,9972	0,9987	0,9991	0,9998	0,9995

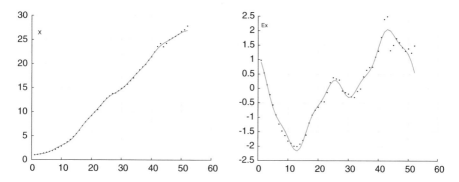

Fig. 1 Modelling curves of gross outputs (left) and fluctuation of gross output (right) of all economy

which correspond to years 2016 and 2017), testifies that the model has high accuracy predictive properties.

Analysis of modelling curves shows that they give a qualitative approximation of statistical data. Therefore, the proposed method can be used for effective forecasting in real macroeconomic systems. Investigating the economy of France, we have received the following modelling trajectory:

$$x(t) = -7.10705 + 3.3429t + 7.2722 \cos(\omega_1 t) - 8.898 \sin(\omega_1 t)$$
$$+ 4.4042 \cos(\omega_2 t) - 4.1148 \sin(\omega_2 t) - 1.3436 \cos(\omega_3 t)$$
$$- 0.1238 \sin(\omega_3 t) + 0.2233 \cos(\omega_6 t) + 0.0976 \sin(\omega_6 t) .$$

Let us analyse the harmonics that are present in this modelling trajectory and, therefore, have an influence on the macroeconomic system of France. Figure 2 shows the graphics of the first, second, third, and sixth harmonics within the interval $t \in [35, 80]$, that is, from years 2000 to 2045. The analysis shows that since 2000,

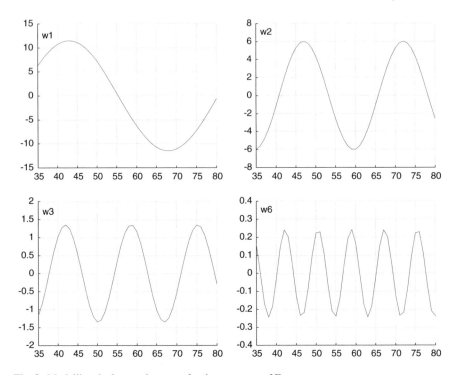

Fig. 2 Modelling the harmonic waves for the economy of France

the first three harmonics are in the lifting phase. Since 2003, it is supplemented by the 6th harmonic lifting phase. So this period can be considered as the "golden era" of France. But in 2008, the 1st, 3rd, and 4th harmonics entered a phase of decline. As a result, at the end of 2008, the economic crisis has begun. The cause of the crisis was the fact that all the basic harmonics entered into the phase of decline. It is supplemented by the 2nd harmonic since 2012. Further analysis shows the 6th harmonic will enter the growth phase in 2020. The 3rd harmonic is in lifting phase now and until 2024, when the wave starts to decline. The analysis of the 2nd wave shows that it is now in a phase of decline. Changes are expected to begin in 2025, when the 2nd wave enters the growing phase until 2040. However, significant changes should be expected starting with the year 2034, when the Kondratieff wave will enter the phase of lifting. In 2034, it will be supplemented with the 3rd harmonic, in 2037 with the 6th harmonics. Therefore, after that time, we observe a gradual growth of the economic development of France.

5   Conclusions

The constructed mechanism, for simulation and forecasting dynamics of a macroeconomic system, allows to establish interconnections between individual economy sectors. The identification algorithm of the structural model of the inter-branch balance can be used for efficient allocation of resources in the formation of relationships between different branches. The resulting trajectories of issues and non-productive consumption have high imitation and forecast properties. The developed method can also be used for the analysis and forecasting development of real macroeconomic systems.

Acknowledgements The approach used in this paper is based on previous work carried out by O. Kostylenko under the supervision of Nazarenko [4, 5], which was awarded a diploma at "All-Ukrainian competition of students' research papers". The research was supported by the Portuguese national funding agency for science, research and technology (FCT), within the Center for Research and Development in Mathematics and Applications (CIDMA), project UID/MAT/04106/2019. Kostylenko is also supported by the Ph.D. fellowship PD/BD/114188/2016.

References

1. Leontief, W.: Input-Output Economics. Oxford University Press, New York (1986)
2. Ramsay, J.O., Hooker, G., Campbell, D.: Parameter estimation for differential equations: a generalized smoothing approach. J. R. Stat. Soc. Ser. B Stat. Methodol. **69**(5), 741–796 (2007)
3. Nazarenko, O.M., Filchenko, D.V.: Parametric identification of state-space dynamic systems: a time-domain perspective, Int. J. Innov. Comput. Inf. Control **4**(7), 1553–1566 (2008)
4. Kostylenko, O.O., Nazarenko, O.M: Modelling and identification of macroeconomic system dynamic interindustry balance. Mech. Econ. Regul. **1**(63), 76–86 (2014)
5. Kostylenko, O.O., Nazarenko, O.M: Identification of dynamic "input-output" model and forecasting of macroeconomic system. Nov. Econ. Cybern. **3**, 50–64 (2013)
6. Diebolt, C., Doliger, C.: Kondratieff waves, warfare and world security. In: Economic Cycles Under Test: A Spectral Analysis, pp. 39–47. IOS Press, Amsterdam, 2006
7. Korotayev, A.V., Tsirel, S.V.: Spectral analysis of world GDP dynamics: Kondratieff waves, Kuznets swings, Juglar and Kitchin cycles in global economic development, and the 2008–2009 economic crisis. Struct. Dyn. **4**(1), 3–57 (2010)
8. Fied, B.C.: Environmental Economics: An Introduction. McGraw-Hill, New York (1997)
9. Dobos, I., Floriska, A.: A dynamic Leontief model with non-renewable resources. Econ. Syst. Res. **17**, 319–328 (2005)
10. Nazarenko, O.M: Construction and identification of linear-quadratic models of weakly formalized dynamic systems (in Ukrainian), Math. Modelling Inform. Tech. Aut. Control Systems **10**(833), 185–192 (2008)
11. Greene, W.H.: Econometric Analysis, 5th edn. Pearson Education International, New York (2003)
12. INSEE, http://www.bdm.insee.fr/bdm2/index.action

Ekeland's Variational Principle for the Fractional $p(x)$-Laplacian Operator

Elhoussine Azroul, Abdelmoujib Benkirane and Mohammed Shimi

Abstract In this paper, using adequate variational techniques, mainly based on Ekeland's variational principle, we will establish the existence of a continuous family of eigenvalues for problems driven by the fractional $p(x)$-Laplacian operator $(-\Delta_{p(x)})^s$, with homogenous Dirichlet boundary conditions. More precisely, we show that there exists $\lambda^* > 0$, such that for all $\lambda \in (0, \lambda^*)$ is an eigenvalue of the problem

$$(\mathscr{P}_s) \quad \begin{cases} (-\Delta_{p(x)})^s u(x) = \lambda |u(x)|^{r(x)-2} u(x) \ in \ \Omega, \\ \\ u = 0 \qquad\qquad\qquad in \ \mathbb{R}^N \setminus \Omega, \end{cases}$$

where Ω is a bounded open set of $\mathbb{R}^N$, $\lambda > 0$, p and r are a continuous variable exponents.

1 Introduction

In recent years, problems involving fractional p-Laplacian have been studied intensively. These topics has become the center of studying PDEs because of its mathematical challenges and real applications. Indeed, this type of operator arises in a quite natural way in different contexts, such as the description of several physical phenomena, optimization, population dynamics, and mathematical finance. The fractional Laplacian operator $(-\Delta)^s$, $0 < s < 1$, also provides a simple model to describe some jump Lévy processes in probability theory (see [1, 4–6] and the references therein).

E. Azroul · A. Benkirane · M. Shimi (✉)
Laboratory of Mathematical Analysis and Applications, Faculty of Sciences Dhar Al Mahraz,
Sidi Mohamed Ben Abdellah University, Fez, Morocco
e-mail: mohammed.shimi2@usmba.ac.ma

E. Azroul
e-mail: elhoussine.azroul@gmail.com

A. Benkirane
e-mail: abd.benkirane@gmail.com

© Springer Nature Switzerland AG 2020   145
E. H. Zerrik et al. (eds.), *Recent Advances in Modeling, Analysis and Systems Control:*
Theoretical Aspects and Applications, Studies in Systems, Decision and Control 243,
https://doi.org/10.1007/978-3-030-26149-8_12

The literature on fractional Sobolev spaces, nonlocal problems driven by these operators and their applications is very interesting and quite large, we refer the interested readers to (see, for instance [5, 6, 8, 20–22] for further details). Specifically, we refer to Di Nezza, Palatucci and Valdinoci [8], for a full introduction to the study of the fractional Sobolev spaces and the fractional Laplacian operators.

On the other hand, attention has been paid to the study of partial differential equations involving the $p(x)$-Laplacian operators; see [9–12, 14, 15, 19] and the references therein.

So the natural question that arises is to see which result we will obtain, if we replace the $p(x)$-Laplacian operator by its fractional version (the fractional $p(x)$-Laplacian operator).

Currently, as far as we know, the only results for fractional Sobolev spaces with variable exponents and fractional $p(x)$-Laplacian operator are obtained by [2, 3, 7, 13]. In particular, the authors generalized the last operator to the fractional case. Then, they introduced an appropriate functional space to study problems in which a fractional variable exponent operator is present.

Now, let us introduce the fractional Sobolev space with the variable exponent as it is defined in [7].

Let Ω be a smooth bounded open set in $\mathbb{R}^N$. we start by fixing $s \in (0, 1)$ and let $p : \overline{\Omega} \times \overline{\Omega} \longrightarrow (1, +\infty)$ be a continuous bounded function. We assume that

$$1 < p^- = \min_{(x,y)\in\overline{\Omega}\times\overline{\Omega}} p(x, y) \leqslant p(x, y) \leqslant p^+ = \max_{(x,y)\in\overline{\Omega}\times\overline{\Omega}} p(x, y) < +\infty \quad (1)$$

and

$$p \text{ is symmetric, that is, } p(x, y) = p(y, x) \text{ for all } (x, y) \in \overline{\Omega} \times \overline{\Omega}. \quad (2)$$

We set

$$\bar{p}(x) = p(x, x) \text{ for all } x \in \overline{\Omega}.$$

We define the fractional Sobolev space with variable exponent via the Gagliardo approach as follows:

$$W = W^{s,p(x,y)}(\Omega)$$
$$= \left\{ u \in L^{\bar{p}(x)}(\Omega) : \int_{\Omega\times\Omega} \frac{|u(x) - u(y)|^{p(x,y)}}{\lambda^{p(x,y)}|x - y|^{sp(x,y)+N}} \, dxdy < +\infty, \text{ for some } \lambda > 0 \right\}$$

where $L^{\bar{p}(x)}(\Omega)$ is the Lebesgue space with variable exponent (see Sect. 2).

The space $W^{s,p(x,y)}(\Omega)$ is a Banach space (see [13]) if it is equipped with the norm

$$\|u\|_W = \|u\|_{L^{\bar{p}(x)}(\Omega)} + [u]_{s,p(x,y)},$$

where $[.]_{s,p(x,y)}$ is a Gagliardo seminorm with variable exponent, which is defined by

$$[u]_{s,p(x,y)} = [u]_{s,p(x,y)}(\Omega) = inf\left\{\lambda > 0 : \int_{\Omega \times \Omega} \frac{|u(x) - u(y)|^{p(x,y)}}{\lambda^{p(x,y)}|x - y|^{sp(x,y)+N}}\, dxdy \leqslant 1\right\}.$$

$(W, \|.\|_W)$ is a separable reflexive space, see [3, Lemma 3.1].

The fractional $p(x)$-Laplacian operator is given by

$$(-\Delta_{p(x)})^s u(x) = p.v. \int_\Omega \frac{|u(x) - u(y)|^{p(x,y)-2}(u(x) - u(y))}{|x - y|^{N+sp(x,y)}}\, dy \quad \text{for all } x \in \Omega,$$

where $p.v.$ is a commonly used abbreviation in the principal value sense.

Remark 1 Note that $(-\Delta_{p(x)})^s$ is a generalized operator of the fractional p-Laplacian operator $(-\Delta_p)^s$ (i.e., when $p(x, y) = p = constant$) and is the fractional version of the $p(x)$-Laplacian operator $\Delta_{p(x)}u(x) = div(|\nabla u(x)|^{p(x)-2}u(x))$, which is associated with the variable exponent Sobolev space.

In this paper, we are concerned with the study of the eigenvalue problem,

$$(\mathscr{P}_s) \quad \begin{cases} (-\Delta_{p(x)})^s u(x) = \lambda|u(x)|^{r(x)-2}u(x) \text{ in} & \Omega, \\ \\ u = 0 & \text{in } \mathbb{R}^N \setminus \Omega, \end{cases}$$

where Ω is a smooth open and bounded set in $\mathbb{R}^N$ ($N \geqslant 3$), $\lambda > 0$ is a real number, $p : \overline{\Omega} \times \overline{\Omega} \longrightarrow (1, +\infty)$ is a continuous function satisfying (1) and (2) and $r : \overline{\Omega} \longrightarrow (1, +\infty)$ is a continuous function such that

$$1 < r^- = \min_{x \in \overline{\Omega}} r(x) \leqslant r(x) \leqslant r^+ = \max_{x \in \overline{\Omega}} r(x) < p^- \quad \text{for all } x \in \overline{\Omega}. \quad (3)$$

We will show that any $\lambda > 0$ sufficiently small is an eigenvalue of the above non-local nonhomogeneous problem. The proof relies on simple variational arguments based on Ekeland's variational principle.

Our main result generalizes the work of Mihăilescu and Rădulescu [16], in the fractional case. More precisely, we replace $\Delta_{p(x)}$ which is a local operator, by the nonlocal operator $(-\Delta_{p(x)})^s$.

In the context of eigenvalue, problems involving variable exponent represent a starting point in analyzing more complicated equations. An eigenvalue problem involving variable exponent growth conditions intensively studied is the fallowing

$$(\mathscr{P}_1) \quad \begin{cases} -div(|\nabla u(x)|^{p(x)-2}\nabla u(x)) = \lambda|u(x)|^{q(x)-2}u(x) \text{ in } \Omega, \\ \\ u = 0 & \text{on } \partial\Omega, \end{cases}$$

where $p, q : \overline{\Omega} \longrightarrow (1, +\infty)$ are two continuous functions and $\lambda > 0$ is a real number.

In the case $p(x) = q(x)$ Fan, Zhang and Zhao in [11] establish the existence of infinitely many eigenvalues for problem $(\mathscr{P}_1)$ by using an argument based on the Ljusternik–Schnirelmann critical point theory.

Note that when $p(x) \neq q(x)$, the competition between the growth rates involved in problem $(\mathscr{P}_1)$ is essential in describing the set of eigenvalues of this problem and we cite the following:

- In the case when $\min_{x \in \overline{\Omega}} q(x) < \min_{x \in \overline{\Omega}} p(x)$ and $q(x)$ has a subcritical growth, Mihăilescu and Rădulescu [16] used *Ekeland's variational principle* in order to prove the existence of a continuous family of eigenvalues which lies in a neighborhood of the origin. This result is later extended by Fan in [9].
- In the case when $\max_{x \in \overline{\Omega}} p(x) < \min_{x \in \overline{\Omega}} q(x)$ and $q(x)$ has a subcritical growth, a mountain pass argument, similar with that used by Fan and Zhang [10], can be applied in order to show that any $\lambda > 0$ is an eigenvalue of problem $(\mathscr{P}_2)$.
- Finally, in the case when $\max_{x \in \overline{\Omega}} q(x) < \min_{x \in \overline{\Omega}} p(x)$ it can be proved that the energetic functional which can be associated with the eigenvalue problem has a nontrivial minimum for any positive λ large enough (see, [10]). Clearly, in this case, the result of Mihăilescu and Rădulescu [16] can be also applied. Consequently, in this situation there exist two positive constants λ^* and λ^{**} such that any $\lambda \in (0, \lambda^*) \cup (\lambda^{**}, +\infty)$ is an eigenvalue of the problem.

In an appropriate context, we also point out the study of the fallowing eigenvalue problem involving variable exponent growth conditions and a nonlocal term

$$(\mathscr{P}_2) \quad \begin{cases} -\eta[u]div\big(|\nabla u(x)|^{p(x)-2})\nabla u(x)\big) = \lambda f(x, u(x)) & in \quad \Omega, \\ \\ u \quad = \quad 0 & on \quad \partial\Omega, \end{cases}$$

where $p : \overline{\Omega} \longrightarrow (1, +\infty)$ is continuous function, $\eta[u]$ is a non-local term defined by the fallowing relation

$$\eta[u] = 2 + \left(\int_{\Omega} \frac{1}{p(x)}|\nabla u(x)|^{p(x)}dx\right)^{\frac{p^+}{p^-}-1} + \left(\int_{\Omega} \frac{1}{p(x)}|\nabla u(x)|^{p(x)}dx\right)^{\frac{p^-}{p^+}-1},$$

λ is a real number and $f : \overline{\Omega} \times \mathbb{R} \to \mathbb{R}$ is given by

$$f(x, t) = \begin{cases} |t|^{p(x)-2}t & if \ |t| < 1 \\ \\ |t|^{r(x)-2}t & if \ |t| \geqslant 1 \end{cases}$$

with $r : \overline{\Omega} \longrightarrow (1, +\infty)$ is a continuous function satisfying

$$\frac{(p^+)^2}{p^-} < r^- \leqslant r^+ < \frac{Np^-}{N - p^-}$$

For this problem Mihăilescu, Stancu-Dumitru [17] proved the existence of a continuous set of eigenvalues in a neighborhood at the right of the origin by using as main argument the mountain-pass theorem.

This paper is organized as follows. In Sect. 2, we give some definitions and fundamental properties of the spaces $L^{q(x)}$ and W. In Sect. 3, we introduce some important lemmas which shows that the functional $\mathscr{I}_\lambda$ (see Sect. 3) satisfies the geometrical conditions of the Mountain Pass theorem. Finally, using the Ekeland's variational principle we prove that the problem $(\mathscr{P}_s)$ has a continuous family of eigenvalues lying in a neighborhood at the right of the origin.

2 Some Preliminary Results

In this section, we recall some necessary properties of variable exponent spaces. For more details we refer the reader to [12, 14, 19], and the references therein.

Consider the set

$$C_+(\overline{\Omega}) = \left\{ q \in C(\overline{\Omega}) : q(x) > 1 \quad \text{for all } x \in \overline{\Omega} \right\}.$$

For all $q \in C_+(\overline{\Omega})$, we define

$$q^+ = \sup_{x \in \overline{\Omega}} q(x) \quad and \quad q^- = \inf_{x \in \overline{\Omega}} q(x).$$

For any $q \in C_+(\overline{\Omega})$, we define the variable exponent Lebesgue space as

$$L^{q(x)}(\Omega) = \left\{ u : \Omega \longrightarrow \mathbb{R} \text{ measurable} : \int_\Omega |u(x)|^{q(x)} dx < +\infty \right\}.$$

This vector space endowed with the *Luxemburg norm*, which is defined by

$$\|u\|_{L^{q(x)}(\Omega)} = \inf \left\{ \lambda > 0 : \int_\Omega \left| \frac{u(x)}{\lambda} \right|^{q(x)} dx \leqslant 1 \right\}$$

is a separable reflexive Banach space.

Let $\hat{q} \in C_+(\overline{\Omega})$ be the conjugate exponent of q, i.e., $\frac{1}{q(x)} + \frac{1}{\hat{q}(x)} = 1$. Then we have the following Hölder-type inequality

Lemma 1 (Hölder inequality) *If $u \in L^{q(x)}(\Omega)$ and $v \in L^{\hat{q}(x)}(\Omega)$, then*

$$\left| \int_{\Omega} uv dx \right| \leqslant \left(\frac{1}{q^-} + \frac{1}{\hat{q}^-} \right) \|u\|_{L^{q(x)}(\Omega)} \|v\|_{L^{\hat{q}(x)}(\Omega)} \leqslant 2 \|u\|_{L^{q(x)}(\Omega)} \|v\|_{L^{\hat{q}(x)}(\Omega)}$$

A very important role in manipulating the generalized Lebesgue spaces with variable exponent is played by the modular of the $L^{q(x)}(\Omega)$ space, which defined by

$$\rho_{q(.)} : L^{q(x)}(\Omega) \longrightarrow \mathbb{R}$$

$$u \longrightarrow \rho_{q(.)}(u) = \int_{\Omega} |u(x)|^{q(x)} dx.$$

Proposition 1 *Let $u \in L^{q(x)}(\Omega)$; then we have*

(i) $\|u\|_{L^{q(x)}(\Omega)} < 1$ *(resp.* $= 1, > 1) \Leftrightarrow \rho_{q(.)}(u) < 1$ *(resp.* $= 1, > 1$),
(ii) $\|u\|_{L^{q(x)}(\Omega)} < 1 \Rightarrow \|u\|_{L^{q(x)}(\Omega)}^{q^+} \leqslant \rho_{q(.)}(u) \leqslant \|u\|_{L^{q(x)}(\Omega)}^{q^-}$,
(iii) $\|u\|_{L^{q(x)}(\Omega)} > 1 \Rightarrow \|u\|_{L^{q(x)}(\Omega)}^{q^-} \leqslant \rho_{q(.)}(u) \leqslant \|u\|_{L^{q(x)}(\Omega)}^{q^+}$.

Proposition 2 *If $u, u_k \in L^{q(x)}(\Omega)$ and $k \in \mathbb{N}$, then the following assertions are equivalent*

(i) $\lim_{k \to +\infty} \|u_k - u\|_{L^{q(x)}(\Omega)} = 0$,
(ii) $\lim_{k \to +\infty} \rho_{q(.)}(u_k - u) = 0$,
(iii) $u_k \longrightarrow u$ *in measure in Ω and* $\lim_{k \to +\infty} \rho_{q(.)}(u_k) = \rho_{q(.)}(u)$.

In [13], the authors introduced the variable exponent Sobolev fractional space as follows

$$E = W^{s,q(x),p(x,y)}(\Omega)$$

$$= \left\{ u \in L^{q(x)}(\Omega) : \int_{\Omega \times \Omega} \frac{|u(x) - u(y)|^{p(x,y)}}{\lambda^{p(x,y)} |x - y|^{sp(x,y)+N}} \, dx dy < +\infty, \text{ for some } \lambda > 0 \right\},$$

where $q : \overline{\Omega} \longrightarrow (1, +\infty)$ is a continuous function, such that

$$1 < q^- = \min_{(x,y) \in \overline{\Omega} \times \overline{\Omega}} q(x) \leqslant q(x) \leqslant q^+ = \max_{(x,y) \in \overline{\Omega} \times \overline{\Omega}} q(x) < +\infty$$

We would like to mention that the continuous and compact embedding theorem is proved in [13] under the assumption $q(x) > \bar{p}(x) = p(x, x)$. Here, we give a slightly different version of compact embedding theorem assuming that $q(x) = \bar{p}(x)$ which can be obtained by following the same discussions in [13].

Theorem 1 *Let Ω be a smooth bounded domain in $\mathbb{R}^N$ and let $s \in (0, 1)$. Let $p : \overline{\Omega} \times \overline{\Omega} \longrightarrow (1, +\infty)$ be a continuous variable exponent with $sp(x, y) < N$ for all*

$(x, y) \in \overline{\Omega} \times \overline{\Omega}$. Let (1) and (2) be satisfied. Let $r : \overline{\Omega} \longrightarrow (1, +\infty)$ be a continuous variable exponent such that

$$p_s^*(x) = \frac{N\bar{p}(x)}{N - s\bar{p}(x)} > r(x) \geq r^- = \min_{x \in \overline{\Omega}} r(x) > 1 \ \text{ for all } x \in \overline{\Omega}.$$

Then, there exists a constant $C = C(N, s, p, r, \Omega) > 0$ such that for any $u \in W$,

$$\|u\|_{L^{r(x)}(\Omega)} \leq C\|u\|_W.$$

Thus, the space W is continuously embedded in $L^{r(x)}(\Omega)$ with $1 < r(x) < p_s^*(x)$ for all $x \in \overline{\Omega}$. Moreover, this embedding is compact.

Remark 2 Let W_0 denote the closure of $C_0^\infty(\Omega)$ in W, that is,

$$W_0 = \overline{C_0^\infty(\Omega)}^{\|.\|_W}$$

(i) Theorem 1 remains true if we replace W by W_0.
(ii) Since $p_s^*(x) > \bar{p}(x) \geq p^- > 1$, then Theorem 1 implies that $[.]_{s, p(x, y)}$ is a norm on W_0, which is equivalent to the norm $\|.\|_W$. So $(W_0, [.]_{s, p(x, y)})$ is a Banach space.

Definition 1 Let $p : \overline{\Omega} \times \overline{\Omega} \longrightarrow (1, +\infty)$, be a continuous variable exponent and $s \in (0, 1)$. For any $u \in W_0$, we define the modular $\rho_{p(.,.)} : W_0 \longrightarrow \mathbb{R}$, by

$$\rho_{p(.,.)}(u) = \int_{\Omega \times \Omega} \frac{|u(x) - u(y)|^{p(x, y)}}{|x - y|^{N+sp(x, y)}} \, dxdy$$

and

$$\|u\|_{\rho_{p(.,.)}} = \inf\left\{\lambda > 0 : \rho_{p(.,.)}\left(\frac{u}{\lambda}\right) \leq 1\right\} = [u]_{s, p(x, y)} = \|.\|_{W_0}.$$

Remark 3 The modular $\rho_{p(.,.)}$ also check the results of Propositions 1 and 2.

We could also get the following properties:

Lemma 2 Let $p : \overline{\Omega} \times \overline{\Omega} \longrightarrow (1, +\infty)$, be a continuous variable exponent and $s \in]0, 1[$. For any $u \in W_0$, we have

(i) $1 \leq [u]_{s, p(x, y)} \Rightarrow [u]_{s, p(x, y)}^{p^-} \leq \int_{\Omega \times \Omega} \frac{|u(x) - u(y)|^{p(x, y)}}{|x - y|^{N+sp(x, y)}} \, dxdy \leq [u]_{s, p(x, y)}^{p^+}$,

(ii) $[u]_{s, p(x, y)} \leq 1 \Rightarrow [u]_{s, p(x, y)}^{p^+} \leq \int_{\Omega \times \Omega} \frac{|u(x) - u(y)|^{p(x, y)}}{|x - y|^{N+sp(x, y)}} \, dxdy \leq [u]_{s, p(x, y)}^{p^-}$.

Let denote by $\mathcal{L}$ the operator associated to the $(-\Delta_{p(x)})^s$ defined as

$$\mathcal{L} : W_0 \longrightarrow W_0^*$$

$$u \longrightarrow \mathcal{L}(u) : W_0 \longrightarrow \mathbb{R}$$

$$\varphi \longrightarrow <\mathcal{L}(u), \varphi >$$

such that

$$< \mathcal{L}(u), \varphi >= \int_{\Omega \times \Omega} \frac{|u(x) - u(y)|^{p(x,y)-2}(u(x) - u(y))(\varphi(x) - \varphi(y))}{|x - y|^{N+sp(x,y)}} \, dxdy,$$

where W_0^* is the dual space of W_0.

Lemma 3 ([3]) *Assume that assumptions (1) and (2) are satisfied and $0 < s < 1$. Then, the following assertions hold:*

- $\mathcal{L}$ *is a bounded and strictly monotone operator.*
- $\mathcal{L}$ *is a mapping of type (S_+), i.e.*
  if $u_k \rightharpoonup u$ in W_0 and $\lim\sup\limits_{k \longrightarrow +\infty} < \mathcal{L}(u_k) - \mathcal{L}(u), u_k - u > \leqslant 0$, then $u_k \longrightarrow u$ in
  W_0.
- $\mathcal{L}$ *is a homeomorphism.*

3   Main Results

Definition 2 We say that $u \in W_0$ is a weak solution of problem $(\mathcal{P}_s)$, if for all $\varphi \in W_0$ we have

$$\int_{\Omega \times \Omega} \frac{|u(x) - u(y)|^{p(x,y)-2}(u(x) - u(y))(\varphi(x) - \varphi(y))}{|x - y|^{N+sp(x,y)}} \, dxdy - \lambda \int_{\Omega} |u(x)|^{r(x)-2}u(x)\varphi(x)dx = 0. \quad (4)$$

Moreover, we say that λ is an eigenvalue of problem $(\mathcal{P}_s)$, if there exists $u \in W_0 \setminus \{0\}$ which satisfies (4), i.e. u is the corresponding eigenfunction to λ.

Let us consider the energy functional $\mathcal{J}_\lambda$ corresponding to the problem $(\mathcal{P}_s)$, defined by

$$\mathcal{J}_\lambda : W_0 \longrightarrow \mathbb{R}$$

$$\mathcal{J}_\lambda(u) = \int_{\Omega \times \Omega} \frac{1}{p(x, y)} \frac{|u(x) - u(y)|^{p(x,y)}}{|x - y|^{N+sp(x,y)}} \, dxdy - \lambda \int_{\Omega} \frac{1}{r(x)} |u(x)|^{r(x)}dx$$

for any $\lambda > 0$,

   Now, we introduce some important lemmas that show that the functional $\mathcal{J}_\lambda$ satisfies the geometrical conditions of mountain pass theorem that are necessary to establish the proof of the existence result.

Lemma 4 *Let Ω be a smooth bounded open set in $\mathbb{R}^N$ and $s \in (0, 1)$. Let $p : \overline{\Omega} \times \overline{\Omega} \longrightarrow (1, +\infty)$, be a continuous variable exponent satisfied (1) and (2) with $sp(x, y) < N$ for all $(x, y) \in \overline{\Omega} \times \overline{\Omega}$ and let $r : \overline{\Omega} \longrightarrow (1, +\infty)$ be a continuous variable exponent such that $1 < r(x) < p^-$ for any $x \in \overline{\Omega}$. Then,*

(i) $\mathscr{J}_\lambda$ is well defined.

(ii) $\mathscr{J}_\lambda \in C^1(W_0, \mathbb{R})$ and for all $u, \varphi \in W_0$, its Gâteaux derivative is given by:

$$< \mathscr{J}'_\lambda(u), \varphi >= \int_{\Omega \times \Omega} \frac{|u(x) - u(y)|^{p(x,y)-2}(u(x) - u(y))(\varphi(x) - \varphi(y))}{|x - y|^{N+sp(x,y)}} \, dxdy$$

$$- \lambda \int_\Omega |u(x)|^{r(x)-2} u(x) \varphi(x) dx.$$

Proof (i)- Let $u \in W_0$; then

$$\mathscr{J}_\lambda(u) = \int_{\Omega \times \Omega} \frac{1}{p(x,y)} \frac{|u(x)-u(y)|^{p(x,y)}}{|x-y|^{N+sp(x,y)}} \, dxdy \; - \lambda \int_\Omega \frac{1}{r(x)} |u(x)|^{r(x)} dx$$

$$\leqslant \frac{1}{p^-} \int_{\Omega \times \Omega} \frac{|u(x)-u(y)|^{p(x,y)}}{|x-y|^{N+sp(x,y)}} \, dxdy - \frac{\lambda}{r^+} \int_\Omega |u(x)|^{r(x)} dx$$

$$= \frac{1}{p^-} \rho_{p(.,.)}(u) - \frac{\lambda}{r^+} \rho_{r(.)}(u)$$

By Proposition 1 and Remark 3 we get

$$\mathscr{J}_\lambda(u) \leqslant \frac{1}{p^-} \left[\|u\|_{W_0}^{p+} + \|u\|_{W_0}^{p-} \right] - \frac{\lambda}{r^+} \left[\|u\|_{L^{r(x)}(\Omega)}^{r+} + \|u\|_{L^{r(x)}(\Omega)}^{r-} \right]$$

Using Theorem 1 and Remark 2-(i), we obtain

$$\mathscr{J}_\lambda(u) \leqslant \frac{1}{p^-} \left[\|u\|_{W_0}^{p+} + \|u\|_{W_0}^{p-} \right] - \frac{\lambda}{r^+} \left[C^{r+} \|u\|_{W_0}^{r+} + C^{r-} \|u\|_{W_0}^{r-} \right]$$

$$\leqslant \frac{1}{p^-} \left[\|u\|_{W_0}^{p+} + \|u\|_{W_0}^{p-} \right] - \frac{\lambda}{r^+} \, max \, \{C^{r+}, C^{r-}\} \left[\|u\|_{W_0}^{r+} + \|u\|_{W_0}^{r-} \right]$$

$$\leqslant \left(\frac{1}{p^-} - \frac{\lambda}{r^+} \, max \, \{C^{r+}, C^{r-}\} \right) \left[\|u\|_{W_0}^{p+} + \|u\|_{W_0}^{r+} \right] < +\infty.$$

(ii)- **Existence of the Gâteaux derivative**. We define

$$\Psi(u) = \int_{\Omega \times \Omega} \frac{1}{p(x,y)} \frac{|u(x) - u(y)|^{p(x,y)}}{|x - y|^{N+sp(x,y)}} \, dxdy \quad and \quad \Phi_\lambda(u) = \lambda \int_\Omega \frac{1}{r(x)} |u(x)|^{r(x)} dx.$$

Then

$$\mathscr{J}_\lambda(u) = \Psi(u) - \Phi_\lambda(u) \quad and \quad \mathscr{J}'_\lambda(u) = \Psi'(u) - \Phi'_\lambda(u) \tag{5}$$

• For any $u, \varphi \in W_0$, we have

$$< \Psi'(u), \varphi > = \int_{\Omega \times \Omega} \frac{|u(x) - u(y)|^{p(x,y)-2}(u(x) - u(y))(\varphi(x) - \varphi(y))}{|x - y|^{N+sp(x,y)}} \, dxdy \tag{6}$$

Indeed,

$$< \Psi'(u), \varphi > = \lim_{t \to 0} \frac{\Psi(u+t\varphi) - \Psi(u)}{t}$$

$$= \lim_{t \to 0} \left\{ \int_{\Omega \times \Omega} \frac{|(u(x)+t\varphi(x))-(u(y)+t\varphi(y))|^{p(x,y)} - |u(x)-u(y)|^{p(x,y)}}{tp(x,y)|x-y|^{N+sp(x,y)}} \, dxdy \right\} \tag{7}$$

Let us consider $M : [0, 1] \longrightarrow \mathbb{R}$

$$\alpha \longmapsto \frac{\left|(u(x) - u(y)) + \alpha t(\varphi(x) - \varphi(y))\right|^{p(x,y)}}{tp(x, y)|x - y|^{N+sp(x,y)}}.$$

M is continuous on $[1, 0]$ and differentiable on $(0, 1)$. Then by the mean value theorem, there exists $\theta \in (0, 1)$ such that

$$M'(\alpha)(\theta) = M(1) - M(0)$$

Then

$$\frac{\left|(u(x) - (u(y)) + \theta t(\varphi(x) - \varphi(y))\right|^{p(x,y)-2}\left[(u(x) - u(y)) + t\theta(\varphi(x) - \varphi(y))\right](\varphi(x) - \varphi(y))}{|x - y|^{N+sp(x,y)}}$$

$$= S_t(u, \varphi) = \frac{|(u(x) - u(y)) + t(\varphi(x) - \varphi(y))|^{p(x,y)} - |u(x) - u(y)|^{p(x,y)}}{tp(x, y)|x - y|^{N+sp(x,y)}} \tag{8}$$

Combining (7) and (8), we get

$$< \Psi'(u), \varphi > = \lim_{t \to 0} \int_{\Omega \times \Omega} S_t(u, \varphi) \, dxdy$$

Since $t, \theta \in [0.1]$, so $t\theta \leqslant 1$, which implies

$$S_t(u, \varphi) \leqslant \frac{\left|(u(x) - (u(y)) + (\varphi(x) - \varphi(y))\right|^{p(x,y)-2}\left[(u(x) - u(y)) + (\varphi(x) - \varphi(y))\right](\varphi(x) - \varphi(y))}{|x - y|^{N+sp(x,y)}}.$$

On the other hand

$$S_t(u, \varphi) \xrightarrow[t \to 0]{} \frac{|u(x) - u(y)|^{p(x,y)-2}(u(x) - u(y))(\varphi(x) - \varphi(y))}{|x - y|^{N+sp(x,y)}}.$$

Hence, by the dominated convergence theorem, we obtain (6).

By the same argument we have

$$< \Phi'_\lambda(u), \varphi > \lambda \int_\Omega |u(x)|^{r(x)-2} u(x)\varphi(x)dx.$$

Then by relation (5), the result is holds.

Continuity of the Gâteaux derivative of $\mathscr{J}_\lambda$. Assume that $u_k \longrightarrow u$ in W_0 and we show that $\Psi'(u_k) \longrightarrow \Psi'(u)$ in W_0^*. Indeed,

$$< \Psi'(u_k) - \Psi'(u), \varphi >=$$

$$\int_{\Omega \times \Omega} \frac{\left[|u_k(x)-u_k(y)|^{p(x,y)-2}(u_k(x)-u_k(y))-|u(x)-u(y)|^{p(x,y)-2}(u(x)-u(y))\right]}{|x-y|^{N+sp(x,y)}}$$

$$\times (\varphi(x) - \varphi(y)) \, dxdy$$

$$= \int_{\Omega \times \Omega} \left[\frac{|u_k(x)-u_k(y)|^{p(x,y)-2}(u_k(x)-u_k(y))}{|x-y|^{(\frac{N}{p(x,y)}+s)(p(x,y)-1)}} - \frac{|u(x)-u(y)|^{p(x,y)-2}(u(x)-u(y))}{|x-y|^{(\frac{N}{p(x,y)}+s)(p(x,y)-1)}} \right]$$

$$\times \frac{(\varphi(x)-\varphi(y))}{|x-y|^{\frac{N}{p(x,y)}+s}} dxdy$$

Let us set

$$F_k(x, y) = \frac{|u_k(x) - u_k(y)|^{p(x,y)-2}(u_k(x) - u_k(y))}{|x - y|^{(\frac{N}{p(x,y)}+s)(p(x,y)-1)}} \in L^{\hat{p}(x,y)}(\Omega \times \Omega),$$

$$F(x, y) = \frac{|u(x) - u(y)|^{p(x,y)-2}(u(x) - u(y))}{|x - y|^{(\frac{N}{p(x,y)}+s)(p(x,y)-1)}} \in L^{\hat{p}(x,y)}(\Omega \times \Omega),$$

$$\overline{\varphi}(x, y) = \frac{\varphi(x) - \varphi(y)}{|x - y|^{\frac{N}{p(x,y)}+s}} \in L^{p(x,y)}(\Omega \times \Omega),$$

where $\frac{1}{p(x,y)} + \frac{1}{\hat{p}(x,y)} = 1$.

Hence, by Hölder inequality (see Lemma 1), we obtain

$$< \Psi'(u_k) - \Psi'(u), \varphi > \leqslant 2\|F_k - F\|_{L^{\hat{p}(x,y)}(\Omega \times \Omega)} \|\overline{\varphi}\|_{L^{p(x,y)}(\Omega \times \Omega)}.$$

Thus

$$\|\Psi'(u_k) - \Psi'(u)\|_{W_0^*} \leqslant 2\|F_k - F\|_{L^{\hat{p}(x,y)}(\Omega \times \Omega)}.$$

Now, let

$$v_k(x, y) = \frac{u_k(x) - u_k(y)}{|x - y|^{\frac{N}{p(x,y)} + s}} \in L^{p(x,y)}(\Omega \times \Omega) \quad and \quad v(x, y) = \frac{u(x) - u(y)}{|x - y|^{\frac{N}{p(x,y)} + s}} \in L^{p(x,y)}(\Omega \times \Omega)$$

Since $u_k \longrightarrow u$ in W_0. Then $v_k \longrightarrow v$ in $L^{p(x,y)}(\Omega \times \Omega)$.
Hence, for a subsequence of $(v_k)_{k \geqslant 0}$, we get

$$v_k(x, y) \longrightarrow v(x, y) \ a.e. \ in \ \Omega \times \Omega \ and \ \exists h \in L^{p(x,y)}(\Omega \times \Omega) \ such \ that \ |v_k(x, y)| \leqslant h(x, y).$$

So we have

$$F_k(x, y) \longrightarrow F(x, y) \ a.e. \ in \ \Omega \times \Omega \quad and \quad |F_k(x, y)| = |v_k(x, y)|^{p(x,y)-1} \leqslant |h(x, y)|^{p(x,y)-1}$$

Then, by the dominated convergence theorem, we deduce that

$$F_k \longrightarrow F \ in \ L^{\hat{p}(x,y)}(\Omega \times \Omega).$$

Consequently

$$\Psi'(u_k) \longrightarrow \Psi'(u) \ in \ W_0^*.$$

By the same argument, we show that

$$\Phi'_\lambda(u_k) \longrightarrow \Phi'_\lambda(u) \ in \ \left(L^{r(x)}(\Omega)\right)^*$$

Then by relation (5), we deduce the continuity of $\mathscr{J}'_\lambda$.
  The proof of Lemma 4 is complete.   □

  The following result shows that the functional $\mathscr{J}_\lambda$ satisfies the first geometrical
condition of the mountain pass theorem;

Lemma 5 *Let Ω be a smooth bounded open set in $\mathbb{R}^N$ and let $s \in (0, 1)$. Let $p :
\overline{\Omega} \times \overline{\Omega} \longrightarrow (1, +\infty)$, be a continuous variable exponent satisfied (1) and (2) with
$sp(x, y) < N$ for all $(x, y) \in \overline{\Omega} \times \overline{\Omega}$ and let $r : \overline{\Omega} \longrightarrow (1, +\infty)$ be a continuous
variable exponent such that $1 < r(x) < p^-$ for all $x \in \overline{\Omega}$. Then, there exists $\lambda^* > 0$
such that for any $\lambda \in (0, \lambda^*)$, there exist $R, a > 0$ such that $\mathscr{J}_\lambda(u) \geqslant a > 0$ for any
$u \in W_0$ with $\|u\|_{W_0} = R$.*

Proof of Lemma 5. Since $r(x) < p_s^*(x)$ for all $x \in \overline{\Omega}$, so by Remark 2-(i) W_0 is
continuously embedded in $L^{r(x)}(\Omega)$. Then there exists a positive constant c_1 such
that

$$\|u\|_{L^{r(x)}(\Omega)} \leqslant c_1 \|u\|_{W_0} \quad for \ all \ u \in W_0. \tag{9}$$

We fix $R \in (0, 1)$ such that $R < \frac{1}{c_1}$. Then relation (9) implies

$$\|u\|_{L^{r(x)}(\Omega)} < 1 \quad for \ all \ u \in W_0 \ with \ R = \|u\|_{W_0}.$$

By Proposition 1-(ii), we get

$$\int_{\Omega} |u(x)|^{r(x)} dx \leqslant \|u\|^{r^-}_{L^{r(x)}(\Omega)} \qquad \text{for all } u \in W_0 \ \ with \ R = \|u\|_{W_0}. \qquad (10)$$

Combining (9) and (10), we get

$$\int_{\Omega} |u(x)|^{r(x)} dx \leqslant c_1^{r^-} \|u\|^{r^-}_{W_0} \quad \text{for all } u \in W_0 \ \ with \ R = \|u\|_{W_0}. \qquad (11)$$

Using the fact that $\|u\|_{W_0} < 1$ and (11), we deduce that for any $u \in W_0$ with $R = \|u\|_{W_0}$ the following inequalities hold true

$$\mathscr{I}_\lambda(u) \geqslant \tfrac{1}{p^+} \int_{\Omega \times \Omega} \tfrac{|u(x)-u(y)|^{p(x,y)}}{|x-y|^{N+sp(x,y)}} \, dxdy - \tfrac{\lambda}{r^-} \int_{\Omega} |u(x)|^{r(x)} dx$$

$$\geqslant \tfrac{1}{p^+} \int_{\Omega \times \Omega} \tfrac{|u(x)-u(y)|^{p(x,y)}}{|x-y|^{N+sp(x,y)}} \, dxdy - \tfrac{\lambda}{r^-} \int_{\Omega} |u(x)|^{r(x)} \, dx$$

$$\geqslant \tfrac{1}{p^+} \|u\|^{p^+}_{W_0} - \tfrac{\lambda}{r^-} c_1^{r^-} \|u\|^{r^-}_{W_0} \qquad (12)$$

$$\geqslant \tfrac{1}{p^+} R^{p^+} - \tfrac{\lambda}{r^-} c_1^{r^-} R^{r^-}$$

$$\geqslant R^{r^-} \left(\tfrac{1}{p^+} R^{p^+-r^-} - \tfrac{\lambda}{r^-} c_1^{r^-} \right).$$

By the inequality (12), we can choose λ^* in order to

$$\frac{1}{p^+} R^{p^+-r^-} - \frac{\lambda}{r^-} c_1^{r^-} > 0.$$

Hence, if we define

$$\lambda^* = \frac{R^{p^+-r^-}}{2p^+} \cdot \frac{r^-}{c_1^{r^-}}, \qquad (13)$$

then for any $\lambda \in (0, \lambda^*)$ and any $u \in W_0$ with $\|u\|_{W_0} = R$ there exists $a = \frac{R^{p^+}}{2p^+} > 0$ such that

$$\mathscr{I}_\lambda(u) \geqslant a > 0,$$

which completes the proof of Lemma 5. □

The following result shows that the functional $\mathscr{I}_\lambda$ satisfies the second geometrical condition of mountain pass theorem;

Lemma 6 *Let Ω be a smooth bounded open set in $\mathbb{R}^N$ and $s \in (0, 1)$. Let $p : \overline{\Omega} \times \overline{\Omega} \longrightarrow (1, +\infty)$, be a continuous variable exponent satisfied (1) and (2) with $sp(x, y) < N$ for all $(x, y) \in \overline{\Omega} \times \overline{\Omega}$ and let $r : \overline{\Omega} \longrightarrow (1, +\infty)$ be a continuous* *variable exponent such that* $1 < r(x) < p^-$ *for all* $x \in \overline{\Omega}$. *Then, there exists* $\varphi \in W_0$ *such that* $\varphi \geqslant 0$, $\varphi \neq 0$ *and* $\mathscr{J}_\lambda(t\varphi) < 0$ *for any* t *small enough.*

Proof of Lemma 6. Assumption (3) implies that $r^- < p^-$. Let $\varepsilon > 0$ be such that $r^- + \varepsilon \leqslant p^-$. Since $r \in C(\overline{\Omega})$, then we can find an open set $\Omega_0 \subset \Omega$ such that

$$|r(x) - r^-| \leqslant \varepsilon, \quad \text{for all } x \in \Omega_0.$$

Consequently

$$r(x) \leqslant r^- + \varepsilon \leqslant p^- \quad \text{for all } x \in \Omega_0.$$

Let $\varphi \in C_0^\infty(\Omega)$ be such that $\overline{\Omega}_0 \subset supp\, \varphi, \varphi(x) = 1$ for all $x \in \Omega_0$, and $0 \leqslant \varphi \leqslant 1$ in $\in \Omega$. Then using the above information for any $t \in (0, 1)$ we have

$$\mathscr{J}_\lambda(t\varphi) = \int_{\Omega \times \Omega} \frac{t^{p(x,y)}}{p(x,y)} \frac{|\varphi(x) - \varphi(y)|^{p(x,y)}}{|x-y|^{N+sp(x,y)}}\, dxdy - \lambda \int_\Omega \frac{t^{r(x)}}{r(x)} |\varphi(x)|^{r(x)} dx$$

$$\leqslant \frac{t^{p^-}}{p^-} \int_{\Omega \times \Omega} \frac{|\varphi(x) - \varphi(y)|^{p(x,y)}}{|x-y|^{N+sp(x,y)}}\, dxdy - \lambda \int_{\Omega_0} \frac{t^{r(x)}}{r(x)} |\varphi(x)|^{r(x)} dx$$

$$\leqslant \frac{t^{p^-}}{p^-} \rho_{p(.,.)}(u) - \frac{\lambda}{r^+} t^{r^- + \varepsilon} \int_{\Omega_0} |\varphi(x)|^{r(x)} dx$$

$$\leqslant t^{r^- + \varepsilon} \left[\frac{\rho_{p(.,.)}(\varphi)}{p^-} t^{p^- - r^- - \varepsilon} - \frac{\lambda}{r^+} \int_{\Omega_0} |\varphi(x)|^{r(x)} dx \right]$$

Thus

$$\mathscr{J}_\lambda(t\varphi) < 0 \quad \text{for any } t < \xi^{\frac{1}{p^- - r^- - \varepsilon}}$$

where

$$0 < \xi < min\left\{ 1, \frac{\frac{\lambda p^-}{r^+} \int_{\Omega_0} |\varphi(x)|^{r(x)} dx}{\rho_{p(.,.)}(\varphi)} \right\}$$

Finally, we point out that $\rho_{p(.,.)}(\varphi) > 0$ (this fact implies that $\varphi \neq 0$). Indeed, since $supp\, \varphi \subset \Omega_0 \subset \Omega$ and $0 \leqslant \varphi \leqslant 1$ in Ω, so we get

$$0 < \int_{\Omega_0} |\varphi(x)|^{r(x)} dx \leqslant \int_\Omega |\varphi(x)|^{r(x)} dx \leqslant \int_\Omega |\varphi(x)|^{r^-} dx \tag{14}$$

On the other hand, since $1 < r^- < p_s^*(x)$ for all $x \in \overline{\Omega}$, then W_0 is continuously embedded in $L^{r^-}(\Omega)$, so there exists $c_2 > 0$ such that

$$\|\varphi\|_{L^{r^-}(\Omega)} \leqslant c_2 \|\varphi\|_{W_0}. \tag{15}$$

Combining (14) and (15), we get

$$0 < \frac{1}{c_2} \|\varphi\|_{L^{r^-}(\Omega)} \leqslant \|\varphi\|_{W_0}.$$

This fact and Proposition 1 ((ii) or (iii)) imply that

$$\rho_{p(.,.)}(\varphi) > 0.$$

The Lemma 6 is proved. $\qquad\qquad\qquad\qquad\qquad\qquad\qquad\qquad\qquad\qquad\qquad$ □

Our main result is given by the following theorem

Theorem 2 *Let Ω be a smooth bounded open set in $\mathbb{R}^N$ and let $s \in (0, 1)$. Let $p : \overline{\Omega} \times \overline{\Omega} \longrightarrow (1, +\infty)$, be a continuous variable exponent satisfied (1) and (2) with $sp(x, y) < N$ for all $(x, y) \in \overline{\Omega} \times \overline{\Omega}$ and let $r : \overline{\Omega} \longrightarrow (1, +\infty)$ be a continuous variable exponent such that $1 < r(x) < p^-$ for all $x \in \overline{\Omega}$. Then there exists $\lambda^* > 0$ such that for all $\lambda \in (0, \lambda^*)$ is an eigenvalue of problem $(\mathscr{P}_s)$.*

The proof of Theorem 2 is based on the Ekeland's variational principle and the mountain pass theorem.

Proof Let $\lambda^* > 0$ be defined as in (13) and let $\lambda \in (0, \lambda^*)$. By Lemma 5, it follows that

$$\inf_{\partial B_R(0)} \mathscr{J}_\lambda > 0, \tag{16}$$

where $\partial B_R(0) = \{u \in B_R(0) : \|u\|_{W_0} = R\}$ and $B_R(0)$ is the ball centered at the origin and of radius R in W_0.

On the other hand, by Lemma 6, there exists $\varphi \in W_0$ such that $\mathscr{J}_\lambda(t\varphi) < 0$ for any t small enough. Moreover, by (12), for all $u \in B_R(0)$, we get

$$\mathscr{J}_\lambda(u) \geqslant \frac{1}{p^+}\|u\|_{W_0}^{p^+} - \frac{\lambda}{r^-}c_1^{r^-}\|u\|_{W_0}^{r^-}. \tag{17}$$

Then, we have

$$-\infty < \bar{c} = \inf_{\overline{B_R(0)}} \mathscr{J}_\lambda < 0. \tag{18}$$

Combining (16) and (18), then we can assume that

$$0 < \varepsilon < \inf_{\partial B_R(0)} \mathscr{J}_\lambda - \inf_{B_R(0)} \mathscr{J}_\lambda.$$

Applying Ekeland's variational principale to the functional $\mathscr{J}_\lambda : \overline{B_R(0)} \longrightarrow \mathbb{R}$, we find $u_\varepsilon \in \overline{B_R(0)}$ such that

$$\begin{cases} \mathscr{J}_\lambda(u_\varepsilon) < \inf_{\overline{B_R(0)}} \mathscr{J}_\lambda + \varepsilon, \\ \mathscr{J}_\lambda(u_\varepsilon) < \mathscr{J}_\lambda(u) + \varepsilon\|u - u_\varepsilon\|_{W_0}, \quad \forall u \neq u_\varepsilon. \end{cases} \tag{19}$$

So

$$\mathscr{J}_\lambda(u_\varepsilon) \leqslant \inf_{\overline{B_R(0)}} \mathscr{J}_\lambda + \varepsilon \leqslant \inf_{B_R(0)} \mathscr{J}_\lambda + \varepsilon < \inf_{\partial B_R(0)} \mathscr{J}_\lambda$$

It follows that $u_\varepsilon \in B_R(0)$.

Now, we consider $\mathscr{I}_\lambda^\varepsilon : \overline{B_R(0)} \longrightarrow \mathbb{R}$

$$u \longrightarrow \mathscr{J}_\lambda(u) + \varepsilon \|u - u_\varepsilon\|_{W_0}.$$

By (19), we get

$$\mathscr{I}_\lambda^\varepsilon(u_\varepsilon) = \mathscr{J}_\lambda(u) < \mathscr{I}_\lambda^\varepsilon(u) \quad \text{for all } u \neq u_\varepsilon.$$

Thus, u_ε is a minimum point of $\mathscr{I}_\lambda^\varepsilon$ on $\overline{B_R(0)}$. It follows that for any $t > 0$ small enough and $v \in B_R(0)$

$$\frac{\mathscr{I}_\lambda^\varepsilon(u_\varepsilon + tv) - \mathscr{I}_\lambda^\varepsilon(u_\varepsilon)}{t} \geqslant 0.$$

By this fact, we claim that

$$\frac{\mathscr{J}_\lambda(u_\varepsilon + tv) - \mathscr{J}_\lambda(u_\varepsilon)}{t} + \varepsilon \|v\|_{W_0} \geqslant 0.$$

When t tends to 0^+, we have that

$$< \mathscr{J}_\lambda'(u_\varepsilon), v > + \varepsilon \|v\|_{W_0} \geqslant 0.$$

This gives

$$\|\mathscr{J}_\lambda(u_\varepsilon)\|_{W_0^*} \leqslant \varepsilon. \tag{20}$$

From (20), we deduce that there exists a sequence $\{w_k\} \subset B_r(0)$ such that

$$\mathscr{J}_\lambda(w_k) \longrightarrow \bar{c} \qquad \text{and} \qquad \mathscr{J}_\lambda'(w_k) \longrightarrow 0. \tag{21}$$

From (17) and (21), we have that $\{w_k\}$ is bounded in W_0. Thus there exists $w \in W_0$ such that $w_k \rightharpoonup w$ in W_0.

By (3) we have that $r(x) < p_s^*(x)$ for all $x \in \overline{\Omega}$, so by Theorem 1 and Remark 2, we deduce that W_0 is compactly embedded in $L^{r(x)}(\Omega)$; then

$$w_k \longrightarrow w \quad \text{in} \quad L^{r(x)}(\Omega). \tag{22}$$

Using Lemma 1, we have

$$\int_\Omega |w_k|^{r(x)-2} w_k(w_k - w)dx \leqslant 2\|w_k\|_{L^{r(x)}(\Omega)} \|w_k - w\|_{L^{r(x)}(\Omega)}.$$

So, by (22), we get

$$\lim_{k \to +\infty} \int_\Omega |w_k|^{r(x)-2} w_k(w_k - w)dx = 0. \tag{23}$$

On the other hand, from (21), we get

$$\lim_{k \to +\infty} < \mathscr{J}'_\lambda(w_k), w_k - w >= 0.$$

Namely,

$$\lim_{k \to +\infty} \left\{ \int_{\Omega \times \Omega} \frac{|w_k(x) - w_k(y)|^{p(x,y)-2}(w_k(x) - w_k(y))((w_k(x) - w_k(y)) - (w(x) - w(y)))}{|x - y|^{N+sp(x,y)}} \, dxdy \right.$$

$$\left. -\lambda \int_{\Omega} |w_k(x)|^{r(x)-2} w_k(x)(w_k(x) - w(x)) dx \right\} = 0.$$

Hence, relation (23), yield

$$\lim_{k \to +\infty} \int_{\Omega \times \Omega} \frac{|w_k(x) - w_k(y)|^{p(x,y)-2}(w_k(x) - w_k(y))((w_k(x) - w_k(y)) - (w(x) - w(y)))}{|x - y|^{N+sp(x,y)}} dxdy = 0.$$

Using the above information, Lemma 3-(ii), and the fact that $w_k \rightharpoonup w$ in W_0, we get

$$\begin{cases} \lim \sup < \mathscr{L}(w_k), w_k - w > \leqslant 0, \\ \quad w_k \rightharpoonup w \text{ in } W_0, \qquad \Rightarrow w_k \longrightarrow w \text{ in } W_0. \\ \mathscr{L} \text{ is a mapping of type } (S_+). \end{cases}$$

Then by (21), we obtain

$$\mathscr{J}_\lambda(w) = \lim_{k \to +\infty} \mathscr{J}_\lambda(w_k) = \bar{c} < 0 \qquad and \qquad \mathscr{J}'_\lambda(w) = 0.$$

We conclude that w is a nontrivial critical point of $\mathscr{J}_\lambda$. Thus w is a nontrivial weak solution for problem $(\mathscr{P}_s)$. Finally any $\lambda \in (0, \lambda^*)$ is an eigenvalue of problem $(\mathscr{P}_s)$. The proof of Theorem 2 is complete. $\qquad \square$

References

1. Applebaum, D.: Lévy processes and stochastic calculus. Cambridge University Press, Cambridge (2009)
2. Bahrouni, A.: Comparison and sub-supersolution principles for the fractional $p(x)$-Laplacian. J. Math. Anal. Appl. **458**, 1363–1372 (2018)
3. Bahrouni, A., Rădulescu, V.: On a new fractional Sobolev space and applications to nonlocal variational problems with variable exponent. Discret. Contin. Dyn. Syst. **11**, 379–389 (2018)
4. Bisci, G.M., Rădulescu, V., Servadi, R.: Variational Methods for Nonlocal Fractional Problems. Encyclopedia of mathematics and its applications. Cambridge University Press, Cambridge (2016)

5. Bucur, C., Valdinoci, E.: Non Local Diffusion and Application. Lecture Notes of the Unione Matematica Italiana. Springer, Switzerland (2016)
6. Caffarelli, L.: Nonlocal diffusions, drifts and games. In: Equations, Nonlinear Partial Differential (ed.) Abel Symposia, vol. 7, pp. 37–52. Springer, Berlin (2012)
7. Del Pezzo, L.M., Rossi, J.D.: Traces for fractional Sobolev spaces with variable exponents. Adv. Oper. Theory **2**, 435–446 (2017)
8. Di Nezza, E., Palatucci, G., Valdinoci, E.: Hitchhiker's guide to the fractional Sobolev spaces. Bull. Sci. Math. **136**(5), 521–573 (2012)
9. Fan, X.L.: Remarks on eigenvalue problems involving the p(x)-Laplacian. J. Math. Anal. Appl. **352**, 85–98 (2009)
10. Fan, X.L., Zhang, Q.H.: Existence of solutions for $p(x)$-Laplacian Dirichlet problem. Nonlinear Anal. **52**, 1843–1852 (2003)
11. Fan, X.L., Zhang, Q., Zhao, D.: Eigenvalues of p(x)-Laplacian Dirichlet problem. J. Math. Anal. Appl. **302**, 306–317 (2005)
12. Fan, X.L., Zhao, D.: On the Spaces $L^{p(x)}(\Omega)$ and $W^{m,p(x)}(\Omega)$. J. Math. Anal. Appl **263**, 424–446 (2001)
13. Kaufmann, U., Rossi, J.D., Vidal, R.: Fractional Sobolev spaces with variable exponents and fractional p(x)-Laplacians. Electr. J. Qual. Theory Differ. Equ. **76**, 1–10 (2017)
14. Kováčik, O., Rákosník, J.: On Spaces $L^{p(x)}(\Omega)$ and $W^{m,p(x)}(\Omega)$. Czechoslov. Math. J. **41**(4), 592–618 (1991)
15. Mihăilescu, M., Rădulescu, V.D.: Continuous spectrum for a class of nonhomogeneous differentials operators. Manuscripta Math. **125**(2), 157–167 (2008)
16. Mihăilescu, M., Rădulescu, V.D.: On a nonhomogeneous quasilinear eigenvalue problem in Sobolev spaces with variable exponent. Proc. Am. Math. Soc. **135**(9), 2929–2937 (2007)
17. Mihăilescu, M., Stancu-Dumitru, D.: On an eigenvalue problem involving the p(x)-Laplace operator plus a nonlocal term. Differ. Equ. Appl. **1**(3), 367–378 (2009)
18. Rabinowitz, P.H.: Minimax Methods in Critical Point Theory with Applications to Differential Equations. CBMS Regional Conference Series in Mathematics, vol. 65. American Mathematical Society, Providence (1986)
19. Rădulescu, V.D., Repovš, D.: Partial Differential Equations with Variable Exponents: Variational Methods and Qualitative Analysis. Monographs and Research Notes in Mathematics. CRC Press, Taylor and Francis Group, Boca Raton (2015)
20. Servadei, R., Valdinoci, E.: Variational methods for nonlocal operators of elliptic type. Discret. Contin. Dyn. Syst. **33**(5), 2105–2137 (2013)
21. Servadei, R., Valdinoci, E.: Mountain pass solutions for nonlocal elliptic operators. J. Math. Anal. Appl. **389**(2), 887–898 (2012)
22. Servadei, R., Valdinoci, E.: Weak and viscosity solutions of the fractional Laplace equation. Publ. Mat. **58**, 133–154 (2014)
23. Shuzhong, S.: Ekeland's variational principle and the mountain pass lemma. Acta Math Sin **1**, 348–355 (1985)

Observer-Based Tracking Control Design for a Class of Nonlinear Discrete-Time Systems

Zakary Omar, Rachik Mostafa, Lhous Mustapha and Bidah Sara

Abstract In this paper, The problem of robust tracking and model following is considered for a class of discrete-time nonlinear systems, where the nonlinearities satisfy the Lipschitz condition. In this paper, it is assumed that the system state is not accessible. A nonlinear observer is designed firstly, and then based on the observed states the controller is designed. Based on Lyapunov stability theory, we prove that the constructed controller can drive the system's output function to the desired output that is generated by a reference model, and the tracking error decreases asymptotically to zero. Simulations on controlling systems are investigated, and the results show that the designed controllers are feasible and efficient.

1 Introduction

During the past decades, the robust tracking and model following problem for nonlinear dynamical systems has been widely investigated [1, 2] and references therein. It has seen several new developments in the design of state feedback-control of linear and nonlinear dynamical systems [3–6]. In [7], the Takagi–Sugeno fuzzy model

Z. Omar (✉) · R. Mostafa · B. Sara
Laboratory of Analysis Modelling and Simulation, Department of mathematics,
Faculty of sciences Ben M'sik, Casablanca, Morocco
e-mail: zakaryma@gmail.com

R. Mostafa
e-mail: m_rachik@yahoo.fr

B. Sara
e-mail: sarabidah@gmail.com

L. Mustapha
Laboratory of Modeling, Analysis, Control and Statistics, Department of Mathematics
and Computer Science, Faculty of Sciences Ain Chock, Hassan II University of Casablanca,
Casablanca, Morocco
e-mail: mlhous17@gmail.com

© Springer Nature Switzerland AG 2020
E. H. Zerrik et al. (eds.), *Recent Advances in Modeling, Analysis and Systems Control:
Theoretical Aspects and Applications*, Studies in Systems, Decision and Control 243,
https://doi.org/10.1007/978-3-030-26149-8_13

approach is extended to the stability analysis and control design for both continuous and discrete-time nonlinear systems with time delay. The problem of state and output feedback nonlinear model predictive control has been addressed in [8]. Wen-Hua Chen in his paper [9] proposes a general framework for the design of controllers for nonlinear systems under disturbances using disturbance observer based control techniques.

Most of these researches are limited to the continuous case, and the results are based on the assumption that system states must be fully accessible, whereas, in practice, this assumption is often unreasonable. This has motivated research in observer-based control for nonlinear systems [10, 11]. Several research have been developed to deal with observer design for nonlinear dynamical systems. Observers design problem is an important problem that has various applications such as output feedback control, process identification and fault detection. This remains one of the challenging and open problems in the area of control theory. The classical useful observer for linear time invariant systems is the well known Luenberger observer [12].

A class of nonlinear systems in triangular form, where nonlinearities are satisfying certain growth conditions was considered in [13, 14]. In [15], the authors show that global stabilization is possible using linear feedback for a class of nonlinear systems which have triangular structure and nonlinearities satisfy some norm bounded growth conditions. A back-stepping design procedure for dynamic feedback stabilization for a class of triangular Lipschitz nonlinear systems with unknown time-varying parameters was given in [16]. Observer design techniques for a class of Lipschitz nonlinear systems were considered in [17–19]. The existence of a stable observer for Lipschitz nonlinear systems was addressed in [20].

In this paper, we consider a class of nonlinear discrete-time systems, especially, discrete-time nonlinear Lipschitz systems, wherein the system states are supposed to be inaccessibles and/or unknowns, thus, an observer is firstly designed to estimate the unknown states, then based on the observed states the controller is designed, such that the corresponding output function tracks the desired reference output. The observer design techniques proposed here are based on quadratic Lyapunov functions and thus depend on the existence of a positive definite solution of a Lyapunov equation. Based on these developments, this work aims mainly to provide a new design of an observer-based control law of Lipschitz nonlinear systems, to ensure that the tracking error decreases asymptotically to zero. Then, our main contribution consists in determining new sufficient conditions of the controlled system augmented by its observer. The proposed control approach, based on the Lyapunov stability theory, was formulated.

The rest of the paper is organized as follows. In Sect. 2, we give the considered class of Lipschitz nonlinear systems. The observer design problem is considered in Sect. 3. Section 4 gives the procedure of the controller design under the sufficient conditions. Illustrative examples are given in Sect. 5. The paper is concluded in Sect. 6.

2 Problem Statement and Preliminaries

In this paper, we consider a class of discrete-time systems given by

$$\begin{cases} x_{i+1} = Ax_i + N(x_i) + Bu_i \\ x_0 \in \mathbb{R}^n \text{ unknown} \end{cases} \tag{1}$$

and the associated output function is

$$y_i = Cx_i \in \mathbb{R}^p \tag{2}$$

where the unknown state variable $x_i \in \mathbb{R}^n$ and A, B, C are respectively $(n \times n)$, $(n \times m)$, $(p \times n)$ matrices, N is a $k - Lipschitz$ nonlinear map and $u_i \in \mathbb{R}^m$ is the control function, which is introduced such that the associated output function (5) tracks a desired output y_i^m generated by a reference system of the form

$$\begin{cases} x_{i+1}^m = A_m x_i^m \\ y_i^m = C_m x_i^m \end{cases} \tag{3}$$

where x_i^m is the state vector of the reference model, and $y_i^m \in \mathbb{R}^p$ has the same dimension as y_i in (5).

Any nonlinear system $x_{i+1} = f(x_i, u_i)$ can be expressed in the form of (1), at least locally, if $f(x, u)$ is continuously differentiable with respect to x. Many nonlinearities are locally Lipschitz. Examples include trigonometric nonlinearities occurring in robotics, nonlinearities which are square or cubic in nature, etc. Nonlinearities can also be considered as a perturbation affecting the system [21].

We use the following notation throughout the paper. $\| M \|$ denotes the Euclidean norm of the matrix or vector M. $\lambda_{min}(M)$ and $\lambda_{max}(M)$ are the minimum and maximum eigenvalues of the symmetric matrix M respectively.

3 Observer Design

Based on the fact that the output function y_i is measurable, an observer is designed to estimate the state variable x_i as follows

$$\begin{cases} z_{i+1} = Fz_i + Nz_i + Ly_i + Bu_i \\ z_0 \end{cases} \tag{4}$$

and the associated output function is

$$y_i^p = Cz_i \in \mathbb{R}^p \tag{5}$$

where F, L are constants matrices with appropriate dimension, such that

$$F + LC = A. \tag{6}$$

Let define the error

$$e_i = z_i - x_i$$

we have

$$e_{i+1} = z_{i+1} - x_{i+1}$$
$$e_{i+1} = Fe_i + (F + LC - A)x_i + Nz_i - Nx_i$$

By (6) we have

$$e_{i+1} = Fe_i + Nz_i - Nx_i = (F + N)e_i. \tag{7}$$

Case 1: N is k-contraction.

Proposition 1 *If L is chosen such that*

$$F + LC = A \tag{8}$$
$$\|F\| < 1 - k \tag{9}$$

where k is the Lipschitz constant of N. Then, the error e_i in (7) is asymptotically stable.

Proof From (7) one deduce that

$$\|e_{i+1}\| \le (\|F\| + k) \|e_i\| .$$

By iteration we have

$$\|e_i\| \le (\|F\| + k)^i \|e_0\| \tag{10}$$

and it follows from (9) and (10) that

$$\|e_i\| \to 0, \text{ when } i \to \infty.$$

Case 2: N is k-Lipschitz function $(k \ge 1)$.

We assume here that N satisfies a Lipschitz condition,

$$\|N(x) - N(y)\| \le k \|x - y\|$$

for some positive constant $k \ge 1$.

Assumption 1 The pair (A, C) given in (1) is completely observable.

It follows from Assumption 1 that there exists a constant matrix L such that $F = A - LC$ is Hurwitz. And for any given symmetric positive definite matrix Q, there exists a unique symmetric positive definite matrix P as the solution of the Lyapunov equation

$$P = F^T P F + Q. \tag{11}$$

Thus, we have the following result

Theorem 2 *If L is chosen such that $k < \sqrt{\|F\|^2 + \frac{\lambda_{min}(Q)}{\lambda_{max}(P)}} - \|F\|$, where P is the unique solution of Lyapunov Equation (11). Then, the error e_i in (7) is asymptotically stable.*

Proof Assume that L has been chosen so that eigenvalues of F all have negative real part. Consider the positive definite function

$$V(e_i) = e_i^T P e_i \tag{12}$$

where P is the unique solution of Lyapunov Equation (11) for a given symmetric positive definite matrix Q. The increment of the Lyapunov function in (12) is given by

$$
\begin{aligned}
\triangle V(e_{i+1}) &= e_{i+1}^T P e_{i+1} - e_i^T P e_i \\
&= e_i^T \left(F^T P F - P \right) e_i + 2 e_i^T F P (N z_i - N x_i) \\
&\quad + (N(x_i + e_i) - N(x_i))^T P \left(N(x_i + e_i) - N(x_i) \right) \\
&= -e_i^T Q e_i + 2 e_i^T F P (N z_i - N x_i) \\
&\quad + (N(x_i + e_i) - N(x_i))^T P \left(N(x_i + e_i) - N(x_i) \right)
\end{aligned}
$$

Since

$$\|N(x_i + e_i) - N(x_i)\| \le k \|e_i\|$$

and

$$\lambda_{min}(Q) \le \|Q\|$$
$$\|P\| \le \lambda_{max}(P)$$

then

$$\triangle V(e_{i+1}) \le \left(-\lambda_{min}(Q) + k^2 \lambda_{max}(P) + 2k \lambda_{max}(P) \|F\| \right) \|e_i\|^2 .$$

Thus $\triangle V(e_{i+1}) = (k - k_1)(k - k_2)$, $\forall k \in [k_1, k_2]$ where $k_1 = -\|F\| - \sqrt{\|F\|^2 + \frac{\lambda_{min}(Q)}{\lambda_{max}(P)}}$ and $k_2 = -\|F\| + \sqrt{\|F\|^2 + \frac{\lambda_{min}(Q)}{\lambda_{max}(P)}}$.

Using the fact $1 \le k < \sqrt{\|F\|^2 + \frac{\lambda_{min}(Q)}{\lambda_{max}(P)}} - \|F\|$, $\triangle V(e_{i+1}) < 0$, which completes the proof.

Based on the above results, the observer (4) verifies

$$\| z_i - x_i \| \to 0, \text{ when } i \to \infty.$$

4 Controller Design

Here, the requirement for the developed controller to force the system output to follow the reference output model (3) as closely as possible is the following assumption.

Assumption 2 There exist matrices R, G, G_e, M and H given by

$$G = R^T \times \left[R R^T \right]^{-1} C_m \tag{13}$$
$$R = C (F + BK)^{-1} B K \tag{14}$$
$$G_e = (F + BK)^{-1} B K G \tag{15}$$
$$M = B^T [B B^T]^{-1} \tag{16}$$
$$H = M G_e A_m \tag{17}$$

and K is a constant matrix chosen in the way that $(F + BK)$ is Hurwitz invertible matrix. If one of these matrices cannot be found, a different model must be chosen.

In this paper, the proposed control law is

$$u_i = (K - MN) z_i + (H - KG) x_i^m - M L y_i^p. \tag{18}$$

Let's define an auxiliary variable as follows

$$\tilde{x}_i = z_i - G_e x_i^m, \tag{19}$$

where G_e is given by (15), then we have

$$\left\| C z_i - C_m x_i^m \right\| = \left\| C \left(z_i - G_e x_i^m \right) \right\|$$
$$\leq \| C \| \, \| \tilde{x}_i \| .$$

and by passing to the limit, it is deduced that

$$\left\| y_i^p - y_i^m \right\| \leq \| C \| \, \| \tilde{x}_i \| . \tag{20}$$

It is clear from (20) that the convergence of $\tilde{x}_i$ to the origin is sufficient to achieve the tracking goal. Then we have the following result

Theorem 3 *If matrices (13–17) exist, then the control law (18) drives the output function (5) to asymptotically track the output of the reference system (3).*

Proof It follows from (13), (15), (17) and (19) that

$$
\begin{aligned}
\tilde{x}_{i+1} &= z_{i+1} - G_e x_{i+1}^m \\
&= F z_i + L y_i^p + N z_i + B K z_i - B M N z_i \\
&\quad + B(H - KG) x_i^m - B M L y_i^p - G_e A_m x_i^m \\
&= F \tilde{x}_i + F G_e x_i^m + B K G_e x_i^m - B K G_e x_i^m \\
&\quad + B K z_i - B K G x_i^m \\
\tilde{x}_{i+1} &= (F + B K) \tilde{x}_i.
\end{aligned}
\tag{21}
$$

Constructing now the Lyapunov function as

$$
V(x_i) = x_i^T P x_i
\tag{22}
$$

where P is the unique solution of Lyapunov equation

$$
P = (F + BK)^T P(F + BK) + Q
\tag{23}
$$

for a given symmetric positive definite matrix Q. The increment of the Lyapunov function in (22) is given by

$$
\begin{aligned}
\triangle V(\tilde{x}_{i+1}) &= \tilde{x}_{i+1}^T P \tilde{x}_{i+1} - \tilde{x}_i^T P \tilde{x}_i \\
&= \tilde{x}_i^T (F + BK)^T P(F + BK) \tilde{x}_i - \tilde{x}_i^T P \tilde{x}_i \\
&= -\tilde{x}_i^T Q \tilde{x}_i \le 0.
\end{aligned}
$$

This shows that all trajectories of the closed-loop system (21) will converge to the origin. Then it can be obtained from (20) that the tracking error decreases asymptotically towards zero. This completes the proof.

5 Numerical Examples

Algorithm

Step 1: Initialization of the parameters, and move to the next step.

  Step 2: Construction of the matrices G_e, R, G, H and M, and move to the next step.

  Step 3: Simulation of the reference model, and move to the next step.

  Step 4: Construction of the observer and the nominal system in the same time, and move to the next step.

  Step 5: Construction of the output functions based on values of x and z calculated in the previous step, and plot the results.

Table 1 Matrices data

A	B	C
$\begin{pmatrix} 3.1 & 6 \\ 1 & 2.1 \end{pmatrix}$	$\begin{pmatrix} 1 & -2 \\ 9 & -1 \end{pmatrix}$	$\begin{pmatrix} -1 & -2 \end{pmatrix}$

Table 2 Matrices data

A	B	C
$\begin{pmatrix} 1.9 & -2 \\ -2 & 4.9 \end{pmatrix}$	$\begin{pmatrix} 1.2 & -1 \\ 9 & -0.3 \end{pmatrix}$	$\begin{pmatrix} 1 & -2 \end{pmatrix}$

This algorithm has been compiled for different data cited in Tables 1 and 2. In the first step, all parameters are initialized, in order to construct matrices G_e, R, G, H and M in the second step, in step 3, one calculate the reference model's state x^m, based on these values, in step 4, one calculate the observer's state z and the system state x at instant $i = 1$ by utilizing x_0, at the same time, we can therefore calculate z_2 according to the calculated x_1, and so on. After these steps, we must have all the necessary states to make plots and comparisons, then we calculate output functions y and y^p, errors and one plot the results.

To illustrate the utilization of our approach, in this section, we consider two numerical examples. Here, the discrete-time system is given as follows

$$\begin{cases} x_{i+1} = Ax_i + Nx_i + Bu_i \\ y_i \quad = Cx_i \qquad\qquad\quad i \geq 0 \\ x_0 \quad \in \mathbb{R}^2. \end{cases} \tag{24}$$

The state of the system (24) can be estimated by an observer of the form

$$\begin{cases} z_{i+1} = Fz_i + Nz_i + Ly_i + Bu_i \\ \qquad\qquad z_0 \end{cases} \tag{25}$$

and the observer output is given by

$$y_i^p = Cz_i. \tag{26}$$

Example 1 For matrices A, B and C given by Table 1, and N a non linear map defined by

$$N : \mathbb{R}^2 \rightarrow \mathbb{R}^2$$
$$\begin{pmatrix} x \\ y \end{pmatrix} \longmapsto \begin{pmatrix} \frac{\sin(x+y)}{3} + 10 \\ \frac{\cos(x-y)}{3} - 5 \end{pmatrix}$$

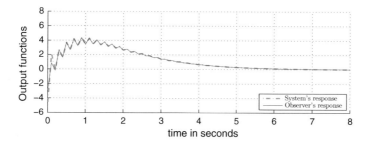

Fig. 1 Comparison of the system's response (24) and the observer's response (26) in the initial configuration

which is $\frac{2}{3}$-contraction, where $F = \begin{pmatrix} 0.1 & 0 \\ 0 & 0.1 \end{pmatrix}$ and $L = \begin{pmatrix} -3 \\ -1 \end{pmatrix}$. It is clear that $F + LC = A$.

By using the above data, Fig. 1 shows that the state of the observer (25) can effectively estimate the state of the system (24). The output function y_i showed in Fig. 1, is the measured response before the use of the proposed control law u_i given by (18).

The control input u_i in (24) is used in order to tracks the output response of the reference system given by

$$\begin{cases} x^m_{i+1} = A_m x^m_i \in \mathbb{R}^2 \\ y^m_i = C_m x^m_i \in \mathbb{R} \end{cases} \tag{27}$$

where

$$A_m = \begin{pmatrix} 0.9 & 1 \\ 0 & 0.9 \end{pmatrix},$$
$$C_m = \begin{pmatrix} 1 & 0.9 \end{pmatrix},$$
$$x^m_0 = \begin{pmatrix} 0 & 1 \end{pmatrix}^T.$$

It's clear that the pair (F, B) is controllable, then we choose K such that

$$K = \begin{pmatrix} 0.0588 & 0.0824 \\ 0.5294 & 0.0412 \end{pmatrix} \text{ and } F + BK = \begin{pmatrix} -0.9 & 0 \\ 0 & 0.8 \end{pmatrix} \tag{28}$$

Matrices (13), (15), (16) and (17) are given, respectively, by

$$G = \begin{pmatrix} -0.2586 & -0.2327 \\ -0.4073 & -0.3665 \end{pmatrix},$$
$$G_e = \begin{pmatrix} -0.2873 & -0.2586 \\ -0.3563 & -0.3207 \end{pmatrix},$$

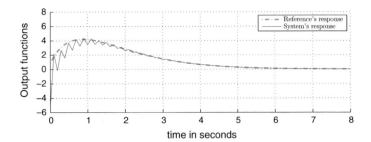

Fig. 2 Comparison of the system's response (24) and the reference response (27)

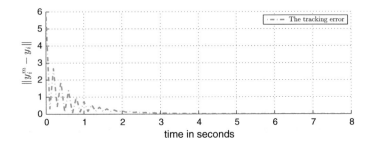

Fig. 3 The tracking performance

$$M = \begin{pmatrix} -0.0588 & 0.1176 \\ -0.5294 & 0.0588 \end{pmatrix},$$

$$H = \begin{pmatrix} -0.0225 & -0.0453 \\ 0.1180 & 0.2374 \end{pmatrix}.$$

Figure 2 shows the effectiveness of our approach, where it can be deduced that with the control law (18), the system's output in (24) can track asymptotically the reference's output, and obviously in Fig. 3, the tracking error decreases asymptotically towards the orgin.

Example 2 For matrices A, B and C given by Table 2, and N a non linear map defined by

$$N : \mathbb{R}^2 \rightarrow \mathbb{R}^2$$
$$\begin{pmatrix} x \\ y \end{pmatrix} \longmapsto \begin{pmatrix} 0 \\ \frac{5sin(x)}{3} \end{pmatrix}$$

where the Lipschitz constant is $\frac{5}{3}$. We consider the matrices F and L by $F = \begin{pmatrix} 0.9 & 0 \\ 0 & 0.9 \end{pmatrix}$, $L = \begin{pmatrix} 1 \\ -2 \end{pmatrix}$, which verify $F + LC = A$.

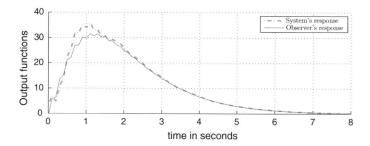

Fig. 4 Comparison of the system's response (24) and the observer's response (26) in the initial configuration

By using the above data, Fig. 4 shows that the state of the observer (25) can effectively estimate the state of the system (24). The output function y_i showed in Fig. 4, is the measured response before the use of the proposed control law u_i given by (18).

The control input u_i in (24) is used in order to tracks the output response of the reference system given by

$$\begin{cases} x_{i+1}^m = A_m x_i^m \in \mathbb{R}^2 \\ y_i^m = C_m x_i^m \in \mathbb{R} \end{cases} \tag{29}$$

where

$$A_m = \begin{pmatrix} 0.9 & 1 & 0.1 \\ 0 & 0.9 & 1 \\ 0 & 0 & 0.7 \end{pmatrix},$$

$$C_m = \begin{pmatrix} 1 & 0.9 & 0.1 \end{pmatrix},$$

$$x_0^m = \begin{pmatrix} 1 & 1 & 2 \end{pmatrix}^T.$$

It's clear that the pair (F, B) is controllable, then we choose K such that

$$K = \begin{pmatrix} 0.0625 & -0.0116 \\ 1.8750 & -0.0139 \end{pmatrix} \text{ and } F + BK = \begin{pmatrix} -0.9 & 0 \\ 0 & 0.8 \end{pmatrix} \tag{30}$$

Matrices (13), (15), (16) and (17) are given, respectively, by

$$G = \begin{pmatrix} 0.4923 & 0.4431 & 0.0492 \\ 0.0615 & 0.0554 & 0.0062 \end{pmatrix},$$

$$G_e = \begin{pmatrix} 0.9846 & 0.8862 & 0.0985 \\ -0.0077 & -0.0069 & -0.0008 \end{pmatrix},$$

$$M = \begin{pmatrix} -0.0347 & 0.1157 \\ -1.0417 & 0.1389 \end{pmatrix},$$

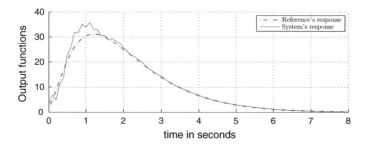

Fig. 5 Comparison of the system's response (24) and the reference response (29)

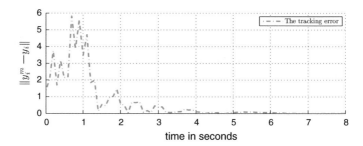

Fig. 6 The tracking performance

$$H = \begin{pmatrix} -0.0316 & -0.0635 & -0.0375 \\ -0.9240 & -1.8583 & -1.0986 \end{pmatrix}.$$

Figure 5 shows the effectiveness of our approach, where it can be deduced that with the control law (18), the system's output in (24) can track asymptotically the reference's output, and also in Fig. 6, we can see that the tracking error decreases asymptotically towards the orgin.

6 Conclusion

In this paper, the problem of observer design and robust tracking for a class of discrete-time nonlinear Lipschitz systems has been considered. Here, the system states are supposed to be inaccessible. Firstly, we set about designing a nonlinear observer. Secondly, based on the observer states, the controller is designed. New sufficient stability conditions of the controlled system augmented by its observer, and the proposed control approach, based on the Lyapunov stability theory, were formulated. We prove that the constructed controller can drive the output function of the system to the desired output generated by a reference model. To illustrate our work an advanced simulation study on numerical examples was performed. It has been shown from the simulation results that the proposed feedback control scheme, based on the designed observer, is efficient and allows good reconstruction of non-measurable state variables and fast-tracking.

References

1. Mobayen, S., Majd, V.J.: Robust tracking control method based on composite nonlinear feedback technique for linear systems with time-varying uncertain parameters and disturbances. Nonlinear Dyn. **70**(1), 171–180 (2012)
2. Wu, H.: Adaptive robust tracking and model following of uncertain dynamical systems with multiple time delays. IEEE Trans. Autom. Control. **49**(4), 611–616 (2004)
3. Ackermann, J., Guldner, J., Sienel, W., Steinhauser, R., Utkin, V.I.: Linear and nonlinear controller design for robust automatic steering. IEEE Trans. Control. Syst. Technol. **3**(1), 132–143 (1995)
4. Çimen, Tayfun: Systematic and effective design of nonlinear feedback controllers via the state-dependent Riccati equation (SDRE) method. Annu. Rev. Control. **34**(1), 32–51 (2010)
5. Ha, I.N., Gilbert, E.L.M.E.R.G.: Robust tracking in nonlinear systems. IEEE Trans. Autom. Control. **32**(9), 763–771 (1987)
6. Rachik, M., Lhous, M.: "An Observer-based control of linear systems with uncertain parameters," Archives of Control Sciences, 26(LXII). No. **4**, 565–576 (2016)
7. Cao, Y.Y., Frank, P.M.: Analysis and synthesis of nonlinear time-delay systems via fuzzy control approach. IEEE Trans. Fuzzy Syst. **8**(2), 200–211 (2000)
8. Findeisen, R., Imsland, L., Allgower, F., Foss, B.A.: State and output feedback nonlinear model predictive control: An overview. Eur. J. Control. **9**(2), 190–206. ISO 690 (2003)
9. Chen, W.H.: Disturbance observer based control for nonlinear systems. IEEE/ASME Trans. Mechatron.S **9**(4), 706–710 (2004)
10. Rodrigues, L., How, J.P.: Observer-based control of piecewise-affine systems. Int. J. Control. **76**(5), 459–477 (2003)
11. Talole, S.E., Kolhe, J.P., Phadke, S.B.: Extended-state-observer-based control of flexible-joint system with experimental validation. IEEE Trans. Ind. Electron. **57**(4), 1411–1419 (2010)
12. Luenberger, D.G.: Observing the state of a linear system. IEEE Trans. Mil.Y Electron. **8**(2), 74–80 (1964)
13. Gauthier, J.P., Hammouri, H., Othman, S.: A simple observer for nonlinear systems applications to bioreactors. IEEE Trans. Autom. Control. **37**, 875–880 (1992)
14. Qian, C., Lin, W.: Output feedback control of a class of nonlinear systems: A nonseparation principle paradigm. IEEE Trans. Autom. Control. **47**, 1710–1715 (2002)
15. Tsinias, J.: A theorem on global stabilization of nonlinear systems by linear feedback. Syst. & Control. Lett. **17**, 357–362 (1991)
16. Tsinias, J.: Backstepping design for time-varying nonlinear systems with unknown parameters. Syst. & Control. Lett. **39**, 219–227 (2000)
17. Aboky, C., Sallet, G., Vivalda, J.C.: Observers for Lipschitz non-linear systems. Int. J. Control. **75**(3), 204–212 (2002)
18. Raghavan, I.R., Hedrick, J.K.: Observer design for a class of nonlinear systems. Int. J. Control. **1**, 171–185 (1994)
19. Thau, F.E.: Observing the state of non-linear dynamic systems. Int. J. Control. **17**(3), 471–479 (1973)
20. Rajamani, R., Cho, Y.M.: Existence and design of observer for nonlinear systems: relation to distance to unobservability. Int. J. Control. **69**(5), 717–731 (1998)
21. Schreier, G., et al.: Observer design for a class of non-linear systems. IFAC Proc. Vol. **30**(18), 483–488 (1997)

On the Output Stabilization for a Class of Infinite Dimensional Bilinear Systems

El Hassan Zerrik and Abderrahman Ait Aadi

Abstract This paper is concerned with the output stabilization for a class of distributed bilinear system evolving in a spatial domain Ω. We give sufficient conditions for strong and weak output stabilization. Also, the question of the output stabilization is discussed using a minimization problem. Examples and simulations are given.

1 Introduction

In this paper, we consider the following bilinear system

$$\begin{cases} \dot{z}(t) = Az(t) + v(t)Bz(t), \\ z(0) = z_0, \end{cases} \tag{1}$$

where $A : D(A) \subset H \to H$ generates a strongly continuous semigroup of contractions $(S(t))_{t \geq 0}$ on a Hilbert space H with a dense domain $D(A) \subset H$, endowed with norm and inner product denoted, respectively, by $\|.\|$ and $\langle ., . \rangle$, $v(.) \in V_{ad}$ (the admissible controls set) is a scalar valued control and B is a bounded linear operator from H to H.

The problem of feedback stabilization of distributed system (1) was studied in many works that lead to various results. In [1], it was shown that the control

$$v(t) = -\langle z(t), Bz(t) \rangle, \tag{2}$$

E. H. Zerrik · A. Ait Aadi (✉)
MACS Laboratory, Department of Mathematics, Moulay Ismail University,
Meknes, Morocco
e-mail: abderrahman.aitaadi@gmail.com

E. H. Zerrik
e-mail: zerrik3@yahoo.fr

© Springer Nature Switzerland AG 2020   177
E. H. Zerrik et al. (eds.), *Recent Advances in Modeling, Analysis and Systems Control:
Theoretical Aspects and Applications*, Studies in Systems, Decision and Control 243,
https://doi.org/10.1007/978-3-030-26149-8_14

weakly stabilizes system (1) provided that B be a weakly sequentially continuous operator such that, for all $\psi \in H$, we have

$$\langle BS(t)\psi, S(t)\psi \rangle = 0, \quad \forall t \geq 0 \Longrightarrow \psi = 0, \tag{3}$$

and if (3) is replaced by the following assumption

$$\int_0^T |\langle BS(s)\psi, S(s)\psi \rangle| ds \geq \gamma \|\psi\|^2, \quad \forall \psi \in H \text{ (for some } \gamma, T > 0), \tag{4}$$

then control (2) strongly stabilizes system (1) (see [2]).

In [3], the authors show that when the resolvent of A is compact, B self-adjoint and monotone, then strong stabilization of system (1) is proved using bounded controls.

Now, let the output state space Y be a Hilbert space with inner product $\langle ., . \rangle_Y$ and the corresponding norm $\|.\|_Y$, and let $C \in \mathcal{L}(H, Y)$ be an output operator.

System (1) is augmented with the output

$$w(t) := Cz(t). \tag{5}$$

The output stabilization means that $w(t) \to 0$ as $t \to +\infty$ using suitable controls. In the case when $Y = H$ and $C = I$, one obtains the classical stabilization of the state.

If Ω be the system evolution domain and $\omega \subset \Omega$, when $C = \chi_\omega$, the restriction operator to a subregion ω of Ω, one is concerned with the behaviour of the state only in a subregion of the system evolution domain. This is what we call regional stabilization.

The notion of regional stabilization has been largely developed since its closeness to real applications, and the existence of systems which are not stabilizable on the whole domain but stabilizable on some subregion ω. Moreover stabilizing a systems on a subregion is cheaper than stabilizing it on the whole domain (see [7–11]). In [5], the author establishes weak and strong stabilization of (5) for a class of semilinear systems using controls that do not take into account the output operator.

In this paper, we study strong and weak stabilization of the output of bilinear systems by controls that depend on the output operator. Moreover, the output stabilization is discussed using a minimization problem. The method is based essentially on the decay of the energy and the semigroup approach. Illustrations by examples and simulations are also given.

This paper is organized as follows : In Sect. 2, we discuss sufficient conditions to achieve strong and weak stabilization of the output (5). In Sect. 3, the stabilization of the output (5) is achieved minimizing a quadratic performance cost. Section 4 is devoted to simulations.

2   Output Stabilization

In this section, we develop sufficient conditions for strong and weak stabilization of the output (5).

Consider system (1) augmented with the output (5).

Definition 1 The output (5) is said to be

1. weakly stabilizable, if there exists a control $v(.) \in V_{ad}$ such that for any initial condition $z_0 \in H$, the corresponding solution $z(t)$ of system (1) is global and satisfies

$$\langle Cz(t), \psi \rangle_Y \to 0, \quad \forall \psi \in Y, \quad \text{as } t \to \infty.$$

2. strongly stabilizable, if there exists a control $v(.) \in V_{ad}$ such that for any initial condition $z_0 \in H$, the corresponding solution $z(t)$ of system (1) is global and verifies

$$\|Cz(t)\|_Y \to 0, \quad \text{as } t \to \infty.$$

3. exponentially stabilizable, if there exists a control $v(.) \in V_{ad}$ such that for any initial condition $z_0 \in H$, the corresponding solution $z(t)$ of system (1) is global and there exist $\alpha, \beta > 0$ such that

$$\|Cz(t)\|_Y \leq \alpha e^{-\beta t}\|z_0\|, \quad \forall t > 0.$$

Remark 1 It is clear that exponential stability of (5) $\Rightarrow$ strong stability of (5) $\Rightarrow$ weak stability of (5).

2.1   Output Strong Stabilization

The following result will be used to prove the strong stabilization.

Theorem 1 *Let A generate a semigroup* $(S(t))_{t \geq 0}$ *of contractions on H and B is a bounded linear operator. If the conditions*

1. $\mathscr{R}e\big(\langle C^*CA\psi, \psi \rangle\big) \leq 0, \ \forall \psi \in D(A),$ *where* C^* *is the adjoint operator of C.*
2. $\mathscr{R}e\big(\langle C^*CB\psi, \psi \rangle \langle B\psi, \psi \rangle\big) \geq 0, \ \forall \psi \in H,$

hold, then control

$$v(t) = -\frac{\langle C^*CBz(t), z(t) \rangle}{1 + |\langle C^*CBz(t), z(t) \rangle|},$$

allows the estimate

$$\left(\int_0^T |\langle C^*CBS(s)z(t), S(s)z(t)\rangle|ds \right)^2$$

$$= O\left(\int_t^{t+T} \frac{|\langle C^*CBz(s), z(s)\rangle|^2}{1 + |\langle C^*CBz(s), z(s)\rangle|}ds \right), \quad as \ \ t \to +\infty. \tag{6}$$

Proof From hypothesis 1 of Theorem 1, we have

$$\frac{1}{2}\frac{d}{dt}\|Cz(t)\|_Y^2 \le \mathscr{R}e\big(v(t)\langle C^*CBz(t), z(t)\rangle\big).$$

In order to make the energy nonincreasing, we consider the control

$$v(t) = -\frac{\langle C^*CBz(t), z(t)\rangle}{1 + |\langle C^*CBz(t), z(t)\rangle|},$$

so that the resulting closed-loop system is

$$\begin{cases} \dot{z}(t) = Az(t) + f(z(t)) \\ z(0) = z_0, \end{cases} \tag{7}$$

where $f(y) = -\dfrac{\langle C^*CBy, y\rangle}{1 + |\langle C^*CBy, y\rangle|}By$, for all $y \in H$.

f is locally Lipschitz, then system (7) has a unique mild solution $z(t)$ (see Theorem 1.4, p. 185 in [6]) defined on a maximal interval $[0, t_{max}]$ by

$$z(t) = S(t)z_0 + \int_0^t S(t-s)f(z(s))ds. \tag{8}$$

Since A generate a semigroup of contractions, we have

$$\frac{d}{dt}\|z(t)\|^2 \le -2\frac{\langle C^*CBz(t), z(t)\rangle\langle Bz(t), z(t)\rangle}{1 + |\langle C^*CBz(t), z(t)\rangle|}.$$

Integrating this inequality, we get

$$\|z(t)\|^2 - \|z(0)\|^2 \le -2\int_0^t \frac{\langle C^*CBz(s), z(s)\rangle\langle Bz(s), z(s)\rangle}{1 + |\langle C^*CBz(s), z(s)\rangle|}ds.$$

It follows that

$$\|z(t)\| \le \|z_0\|. \tag{9}$$

From hypothesis 1 of Theorem 1, we have

$$\frac{d}{dt}\|Cz(t)\|_Y^2 \leq -2\frac{|\langle C^*CBz(t), z(t)\rangle|^2}{1 + |\langle C^*CBz(t), z(t)\rangle|}.$$

We deduce

$$\|Cz(t)\|_Y^2 - \|Cz(0)\|_Y^2 \leq -2\int_0^t \frac{|\langle C^*CBz(s), z(s)\rangle|^2}{1 + |\langle C^*CBz(s), z(s)\rangle|}ds. \tag{10}$$

Using (8) and Schwartz inequality, we get

$$\|z(t) - S(t)z_0\| \leq \|B\|\|z_0\|\left(T\int_0^t \frac{|\langle C^*CBz(s), z(s)\rangle|^2}{1 + |\langle C^*CBz(s), z(s)\rangle|}ds\right)^{\frac{1}{2}}, \quad \forall t \in [0, T]. \tag{11}$$

Since B is bounded and C continuous, we have

$$|\langle C^*CBS(s)z_0, S(s)z_0\rangle| \leq 2K\|B\|\|z(s) - S(s)z_0\|\|z_0\| + |\langle C^*CBz(s), z(s)\rangle|, \tag{12}$$

where K is a positive constant.

Replacing z_0 by $z(t)$ in (11) and (12), we get

$$|\langle C^*CBS(s)z(t), S(s)z(t)\rangle| \leq 2K\|B\|^2\|z_0\|^2\left(T\int_t^{t+T} \frac{|\langle C^*CBz(s), z(s)\rangle|^2}{1 + |\langle C^*CBz(s), z(s)\rangle|}ds\right)^{\frac{1}{2}}$$
$$+ |\langle C^*CBz(t+s), z(t+s)\rangle|, \quad \forall t \geq s \geq 0.$$

Integrating this relation over $[0, T]$ and using Cauchy-Schwartz, we deduce

$$\int_0^T |\langle C^*CBS(s)z(t), S(s)z(t)\rangle|ds \leq \left(2K\|B\|^2T^{\frac{3}{2}} + T\left(1 + K\|B\|\|z_0\|^2\right)\right)$$
$$\times \left(\int_t^{t+T} \frac{|\langle C^*CBz(s), z(s)\rangle|^2}{1 + |\langle C^*CBz(s), z(s)\rangle|}ds\right)^{\frac{1}{2}},$$

which achieves the proof. □

The following result gives sufficient conditions for strong stabilization of the output (5).

Theorem 2 *Let A generate a semigroup $(S(t))_{t\geq 0}$ of contractions on H, B is a bounded linear operator. If the assumptions 1, 2 of Theorem 1 and*

$$\int_0^T |\langle C^*CBS(t)\psi, S(t)\psi\rangle|dt \geq \gamma\|C\psi\|_Y^2, \quad \forall \psi \in H, \ (for \ some \ T, \gamma > 0), \tag{13}$$

holds, then the control

$$v(t) = -\frac{\langle C^*CBz(t), z(t)\rangle}{1 + |\langle C^*CBz(t), z(t)\rangle|}$$

strongly stabilizes the output (5) with decay estimate

$$\|Cz(t)\|_Y = O\left(\frac{1}{\sqrt{t}}\right), \quad as \ t \longrightarrow +\infty. \tag{14}$$

Proof Using (10), we deduce

$$\|Cz(kT)\|_Y^2 - \|Cz((k+1)T)\|_Y^2 \geq 2\int_{kT}^{k(T+1)} \frac{|\langle C^*CBz(t), z(t)\rangle|^2}{1 + |\langle C^*CBz(t), z(t)\rangle|}dt, \ k \geq 0.$$

From (6) and (13), we have

$$\|Cz(kT)\|_Y^2 - \|Cz((k+1)T)\|_Y^2 \geq \beta\|Cz(kT)\|_Y^4, \tag{15}$$

where $\beta = \dfrac{\gamma^2}{2\left(2K\|B\|^2T^{\frac{3}{2}} + T\left(1 + K\|B\|\|z_0\|^2\right)\right)^2}$.

Taking $s_k = \|Cz(kT)\|_Y^2$, the inequality (15) can be written as

$$\beta s_k^2 + s_{k+1} \leq s_k, \quad \forall k \geq 0.$$

Since $s_{k+1} \leq s_k$, we obtain

$$\beta s_{k+1}^2 + s_{k+1} \leq s_k, \quad \forall k \geq 0.$$

Applying $p(s) = \beta s^2$ and $q(s) = s - (I + p)^{-1}(s)$ in Lemma 3.3, p. 531 in [4], we deduce

$$s_k \leq x(k), \quad k \geq 0,$$

where $x(t)$ is the solution of equation $x'(t) + q(x(t)) = 0, \ x(0) = s_0$.

Since $x(k) \geq s_k$ and $x(t)$ decreases give $x(t) \geq 0, \forall t \geq 0$. Furthermore, it is easy to see that $q(s)$ is an increasing function such that

$$0 \leq q(s) \leq p(s), \forall s \geq 0.$$

We obtain $-\beta x(t)^2 \leq x'(t) \leq 0$, which implies that

$$x(t) = O(t^{-1}), \quad as \ t \to +\infty.$$

Finally the inequality $s_k \leq x(k)$, together with the fact that $\|Cz(t)\|_Y$ decreases, we deduce the estimate

$$\|Cz(t)\|_Y = O\left(\frac{1}{\sqrt{t}}\right), \quad \text{as } t \longrightarrow +\infty.$$

$\square$

Example 1 Let us consider a system defined on $\Omega =]0, 1[$ by

$$\begin{cases} \dfrac{\partial z(x,t)}{\partial t} = Az(x,t) + v(t)a(x)z(x,t) & \Omega \times]0, +\infty[\\ z(x,0) = z_0(x) & \Omega \\ z(0,t) = z(1,t) = 0 & t > 0, \end{cases} \quad (16)$$

where $H = L^2(\Omega)$, $Az = -z$, and $a \in L^\infty(]0, 1[)$ such that $a(x) \geq 0$ a.e on $]0, 1[$ and $a(x) \geq c > 0$ on subregion ω of Ω and $v(.) \in L^\infty(0, +\infty)$ the control function. System (16) is augmented with the output

$$w(t) = \chi_\omega z(t), \quad (17)$$

where $\chi_\omega : L^2(\Omega) \longrightarrow L^2(\omega)$, the restriction operator to ω.

The operator A generates a semigroup of contractions on $L^2(\Omega)$ given by $S(t)z_0 = e^{-t}z_0$.

We have

$$\mathcal{R}e\big(\langle Ay, y\rangle\big) = -\|y\|^2 \leq 0, \quad \forall y \in L^2(\Omega).$$

Also

$$\mathcal{R}e\big(\langle \chi_\omega^* \chi_\omega Az, z\rangle\big) = -\|\chi_\omega y\|_{L^2(\omega)}^2 \leq 0, \quad \forall y \in L^2(\Omega),$$

where χ_ω^* is the adjoint operator of χ_ω.

For $z_0 \in L^2(\Omega)$ and $T = 2$, we obtain

$$\int_0^2 \langle \chi_\omega^* \chi_\omega B S(t)z_0, S(t)z_0\rangle dt - \int_0^2 e^{-2t} dt \int_\omega a(x)|z_0|^2 dx$$

$$\geq \beta \|\chi_\omega z_0\|_{L^2(\omega)}^2,$$

with $\beta = c \displaystyle\int_0^2 e^{-2t} dt > 0$.

Applying Theorem 2, we conclude that the control

$$v(t) = -\frac{\displaystyle\int_\omega a(x)|z(x,t)|^2 dx}{1 + \displaystyle\int_\omega a(x)|z(x,t)|^2 dx}$$

strongly stabilizes the output (17) with decay estimate

$$\|\chi_\omega z(t)\|_{L^2(\omega)} = O\left(\frac{1}{\sqrt{t}}\right), \quad \text{as } t \longrightarrow +\infty.$$

2.2 Output Weak Stabilization

The following result provides sufficient conditions for weak stabilization of the output (5).

Theorem 3 *Let A generate a semigroup $(S(t))_{t\geq 0}$ of contractions on H and B is a compact operator. If the conditions*

1. $\mathscr{R}e(\langle C^*CA\psi, \psi \rangle) \leq 0, \ \forall \psi \in D(A),$
2. $\mathscr{R}e(\langle C^*CB\psi, \psi \rangle \langle B\psi, \psi \rangle) \geq 0, \ \forall \psi \in H,$
3. $\langle C^*CBS(t)\psi, S(t)\psi \rangle = 0, \quad \forall t \geq 0 \Longrightarrow C\psi = 0$

hold, then the control

$$v(t) = -\frac{\langle C^*CBz(t), z(t) \rangle}{1 + |\langle C^*CBz(t), z(t) \rangle|} \tag{18}$$

weakly stabilizes the output (5).

Proof Let us consider the nonlinear semigroup $\Gamma(t)z_0 := z(t)$ and let (t_n) be a sequence of real numbers such that $t_n \longrightarrow +\infty$ as $n \longrightarrow +\infty$.

From (9), we have $\Gamma(t_n)z_0$ is bounded in H, then there exists a subsequence $(t_{\phi(n)})$ of (t_n) such that

$$\Gamma(t_{\phi(n)})z_0 \rightharpoonup \psi, \text{ as } n \to \infty.$$

Since B is compact and C continuous, we have

$$\lim_{n\to+\infty} \langle C^*CBS(t)\Gamma(t_{\phi(n)})z_0, S(t)\Gamma(t_{\phi(n)})z_0 \rangle = \langle C^*CBS(t)\psi, S(t)\psi \rangle.$$

For all $n \geq$, we set

$$\Lambda_n(t) := \int_{\phi(n)}^{\phi(n)+t} \frac{|\langle C^*CB\Gamma(s)z_0, \Gamma(s)z_0 \rangle|^2}{1 + |\langle C^*CB\Gamma(s)z_0, \Gamma(s)z_0 \rangle|} ds.$$

It follows that $\forall t \geq 0$, $\Lambda_n(t) \to 0$ as $n \to +\infty$.

Using (6), we get

$$\lim_{n \to +\infty} \int_0^t |\langle C^* C B S(s) \Gamma(t_{\phi(n)}) z_0, S(s) \Gamma(t_{\phi(n)}) z_0 \rangle| ds = 0.$$

Hence, by the dominated convergence Theorem, we have

$$\int_0^t |\langle C^* C B S(s) \psi, S(s) \psi \rangle| ds = 0.$$

We conclude that

$$\langle C^* C B S(s) \psi, S(s) \psi \rangle = 0, \quad \forall s \in [0, t].$$

Using condition 3 of Theorem 3, we deduce that

$$C \Gamma(t_{\phi(n)}) z_0 \rightharpoonup 0, \quad \text{as } n \longrightarrow +\infty. \tag{19}$$

On the other hand, it is clear that (19) holds for each subsequence $(t_{\phi(n)})$ of (t_n) such that $C \Gamma(t_{\phi(n)}) z_0$ weakly converges in Y. This implies that $\forall \varphi \in Y$, we have $\langle C \Gamma(t_n) z_0, \varphi \rangle \to 0$ as $n \longrightarrow +\infty$ and hence

$$C \Gamma(t) z_0 \rightharpoonup 0, \quad \text{as } t \longrightarrow +\infty.$$

$\square$

Example 2 Consider a system defined in $\Omega =]0, +\infty[$, and described by

$$\begin{cases} \dfrac{\partial z(x, t)}{\partial t} = -\dfrac{\partial z(x, t)}{\partial x} + v(t) B z(x, t) & x \in \Omega, \ t > 0 \\ z(x, 0) = z_0(x) & x \in \Omega \\ z(0, t) = z(\infty, t) = 0 & t > 0, \end{cases} \tag{20}$$

where $Az = -\dfrac{\partial z}{\partial x}$ with domain $D(A) = \{z \in H^1(\Omega) \mid z(0) = 0, \ z(x) \to 0 \text{ as } x \to +\infty\}$ and $Bz(.) = \displaystyle\int_0^1 z(x) dx (.)$ is the control operator.

The operator A generates a semigroup of contractions

$$(S(t) z_0)(x) = \begin{cases} z_0(x - t) & \text{if } x \geq t \\ 0 & \text{if } x < t. \end{cases}$$

Let $\omega =]0, 1[$ be a subregion of Ω and system (20) is augmented with the output

$$w(t) = \chi_\omega z(t). \tag{21}$$

We have

$$\langle \chi_\omega^* \chi_\omega A z, z \rangle = -\int_0^1 z'(x) z(x) dx$$
$$= -\frac{z^2(1)}{2} \le 0,$$

so, the assumption 1 of Theorem 3 holds.

The operator B is compact and verifies

$$\langle \chi_\omega^* \chi_\omega B S(t) z_0, S(t) z_0 \rangle = \left(\int_0^{1-t} z_0(x) dx \right)^2, \quad 0 \le t \le 1.$$

Thus

$$\langle \chi_\omega^* \chi_\omega B S(t) z_0, S(t) z_0 \rangle = 0, \quad \forall t \ge 0 \implies z_0(x) = 0, \ a.e \ \text{on} \ \omega.$$

Then, the control

$$v(t) = -\frac{\left(\int_0^1 z(x, t) dx \right)^2}{1 + \left(\int_0^1 z(x, t) dx \right)^2}, \tag{22}$$

weakly stabilizes the output (21).

3   An Output Stabilization Problem

The aim of this section is to characterize a control that achieves strong stabilization of the output (5) minimizing a functional cost.

Consider the problem

$$\begin{cases} \min J(v) = \int_0^{+\infty} |\langle C^* C B z(t), z(t) \rangle|^2 dt + \int_0^{+\infty} |v(t)|^2 dt \\ v \in V_{ad} = \{ v \in L^2(0, \infty) \mid z(t) \text{ is a global solution and } J(v) < +\infty \}, \end{cases} \tag{23}$$

The following result concerns strong stabilization of the output (5) solving problem (23).

Theorem 4 *Let A generate a semigroup of contractions on H, and suppose that*

1. $\mathscr{R}e(\langle C^*CA\psi, \psi\rangle) = 0, \quad \forall \psi \in D(A),$
2. $\mathscr{R}e(\langle C^*CB\psi, \psi\rangle\langle B\psi, \psi\rangle) \geq 0, \quad \forall \psi \in H$ *and*
3. *there exist* $T, \gamma > 0$ *such that*

$$\int_0^T |\langle C^*CBS(t)\psi, S(t)\psi\rangle| dt \geq \gamma \|C\psi\|_Y^2, \quad \forall \psi \in H, \tag{24}$$

then the control

$$v(t) = -\langle C^*CBz(t), z(t)\rangle, \tag{25}$$

is the unique solution of problem (23) that strongly stabilizes the output (5).

Proof It's clear that

$$\frac{d}{dt}\|Cz(t)\|_Y^2 = 2\mathscr{R}e(\langle C^*CAz(t), z(t)\rangle) - 2|\langle C^*CBz(t), z(t)\rangle|^2.$$

From hypothesis 1 of Theorem 4, we have

$$\frac{d}{dt}\|Cz(t)\|_Y^2 = -2|\langle C^*CBz(t), z(t)\rangle|^2. \tag{26}$$

Integrating (26), we get

$$\int_0^t |\langle C^*CBz(s), z(s)\rangle|^2 ds \leq \frac{1}{2}\|Cz_0\|_Y^2.$$

Then $J(v^*) < +\infty$, for all $v^* \in V_{ad}$.
    Using (26), we have

$$\frac{d}{dt}\|Cz(t)\|_Y^2 = |\langle C^*CBz(t), z(t)\rangle + v(t)|^2 - \langle C^*CBz(t), z(t)\rangle^2 - v(t)^2. \tag{27}$$

Integrating (27) for all $t \geq 0$, we get

$$\int_0^t \left(\langle C^*CBz(s), z(s)\rangle^2 + v(s)^2\right) ds + \|Cz(t)\|_Y^2 - \|Cz_0\|_Y^2 = \int_0^t |\langle C^*CBz(s), z(s)\rangle$$
$$+ v(s)|^2 ds. \tag{28}$$

To prove that control (25) is solution of problem (23), let $v \in V_{ad}$, $t \geq 0$ and let us define

$$y(\tau) = z(\tau) - S(\tau - t)z(t), \quad \forall \tau \in [t, t+T].$$

Using the variation of constants formula with $z(t)$ as initial state, gives

$$z(\tau) = S(\tau - t)z(t) + \int_t^\tau v(s)S(\tau - s)Bz(s)ds, \ \forall \tau \in [t, t + T]. \quad (29)$$

Therefore $\forall \tau \in [t, t + T]$, we have

$$y(\tau) = \int_t^\tau v(s)S(\tau - s)Bz(s)ds.$$

Since B is a bounded operator and using Schwartz's inequality, we get

$$\|y(\tau)\| \leq \|B\|\sqrt{T\lambda(t)}, \ \forall \tau \in [t, t + T] \quad (30)$$

where $\lambda(t) := \int_t^{t+T} |v(\tau)|^2 d\tau$.

Using (29) and Gronwall's inequality, there exists $v_{max} > |v|$ such that

$$\|z(\tau)\| \leq \|z(t)\|e^{\|B\|T v_{max}}, \ \forall \tau \in [t, t + T]. \quad (31)$$

Schwartz's inequality and (31), give

$$\int_t^{t+T} \langle C^*CBz(\tau), z(\tau)\rangle d\tau \leq M_1 \|z(t)\|\sqrt{\mu(t)},$$

where $\mu(t) := \int_t^{t+T} |\langle C^*CBz(\tau), z(\tau)\rangle|^2 d\tau$ and $M_1 = \sqrt{T}e^{\|B\|T v_{max}}$.

We have

$$\langle C^*CBS(\tau - t)z(t), S(\tau - t)z(t)\rangle = \langle C^*CB(S(\tau - t)z(t) - z(\tau)), S(\tau - t)z(t)\rangle \\ - \langle C^*CBz(\tau), y(\tau)\rangle + \langle C^*CBz(\tau), z(\tau)\rangle.$$

Using (30) and (31), we deduce

$$|\langle C^*CBS(\tau - t)z(t), S(\tau - t)z(t)\rangle| \leq K(1 + e^{\|B\|T v_{max}})\|z(t)\|\|B\|^2\sqrt{T\lambda(t)} \\ + |\langle C^*CBz(\tau), z(\tau)\rangle|, \ \forall \tau \in [t, t + T]$$

where K is a positive constant.

Integrating this inequality over $[t, t + T]$, yields

$$\int_0^T |\langle C^*CBS(\tau)z(t), S(\tau)z(t)\rangle| d\tau \leq M_2 \|z(t)\|\sqrt{\lambda(t)} + M_1 \|z(t)\|\sqrt{\mu(t)}, \quad (32)$$

where $M_2 = K(1 + e^{\|B\|T v_{max}})\|B\|^2 T\sqrt{T}$.

It follows from (9), (24) and (32) that

$$\gamma \|Cz(t)\|_Y^2 \leq \|z_0\|(M_2\sqrt{\lambda(t)} + M_1\sqrt{\mu(t)}).$$

As λ and μ converge to zero, as $t \to +\infty$, then $\|Cz(t)\|_Y \to 0$, as $t \to +\infty$. Then control (25) strongly stabilizes the output (5).

Let $t \to +\infty$ in (28), we obtain

$$J(v) = \|Cz_0\|_Y^2 + \int_0^{+\infty} |\langle C^*CBz(s), z(s)\rangle + v(s)|^2 ds, \tag{33}$$

which implies that $J(v) \geq \|Cz_0\|_Y^2 = J(v^*)$, so (25) is an optimal control.

Now, let $v_i(t), i = 1, 2$, be two solutions of problem (23).

From (33) we deduce that $v_i(t) = \langle C^*CBz_i(t), z_i(t)\rangle$, where $z_i(t)$ verifies

$$\begin{cases} \dot{z}(t) = Az(t) - \langle C^*CBz(t), z(t)\rangle Bz(t) \\ z(0) = z_0, \end{cases}$$

thus $z_i(t) = z(t), i = 1, 2$ and hence $v_1(t) = v_2(t)$.

4   Simulations

We consider the following algorithm:

Step 1: Initial data: threshold $\varepsilon > 0$, initial condition z_0 and subregion ω.

Step 2: We apply the control $v(t_i) = -\dfrac{\langle \chi_\omega^*\chi_\omega Bz(t_i), z(t_i)\rangle}{1 + |\langle \chi_\omega^*\chi_\omega Bz(t_i), z(t_i)\rangle|}$.

Step 3: We solve the system (1) which gives $z(t_{i+1})$.

Step 4: If $\|\chi_\omega z(t_{i+1})\|_{L^2(\omega)} < \varepsilon$ stop, otherwise.

Step 5: $i = i + 1$ and go to step 2.

On $\Omega =]0, +\infty[$, we consider the system given by

$$\begin{cases} \dfrac{\partial z(x, t)}{\partial t} = -\dfrac{\partial z(x, t)}{\partial x} + v(t) \displaystyle\int_0^1 z(x)dx & \Omega \times]0, +\infty[\\ z(x, 0) = \sin(\pi x) & \Omega \\ z(0, t) = z(\infty, t) = 0 & t > 0. \end{cases} \tag{34}$$

Let $\omega \subset \Omega$ and system (34) is augmented with the output

$$w(t) = \chi_\omega z(t). \tag{35}$$

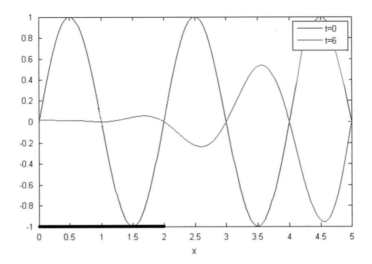

Fig. 1 The stabilization on $\omega =]0, 2[$

Fig. 2 Control function

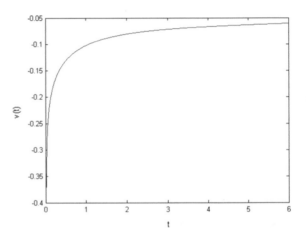

• For $\omega =]0, 2[$, applying control (22), Fig. 1 shows that the state is stabilized on ω with error equals 3.4×10^{-4} and the evolution of control is given by Fig. 2.

• For $\omega =]0, 3[$, Fig. 3 shows that the state is stabilized on ω with error equals 7.8×10^{-4} and the evolution of control is given by Fig. 4.

Fig. 3 The stabilization on
$\omega =]0, 3[$

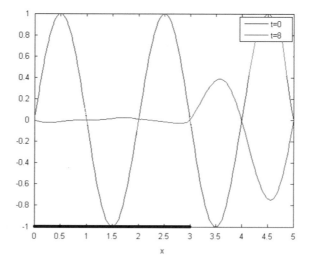

Fig. 4 Control function

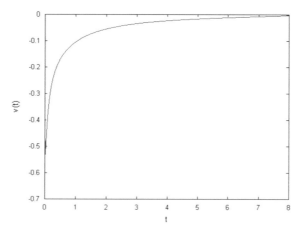

5 Conclusion

This work discuss the question of output stabilization for a class of bilinear systems. Under sufficient conditions, controls depending on the output operator that strongly and weakly stabilizes the output of such systems are given. Also, the output stabilization is discussed using a minimization problem. This work gives an opening to others questions, this is the case of hyperbolic semilinear systems. This will be the purpose of a future research paper.

Acknowledgements The work has been carried out with a grant from Hassan II Academy of Sciences and Technology.

References

1. Ball, J.M., Slemrod, M.: Feedback stabilization of distributed semilinear control systems. J. Appl. Math. Optim. **5**, 169–179 (1979)
2. Berrahmoune, L.: Stabilization and decay estimate for distributed bilinear systems. Syst. Control Lett. **36**, 167–171 (2009)
3. Bounit, H., Hammouri, H.: Feedback stabilization for a class of distributed semilinear control systems. J. Nolinear Anal. Theory Method Appl. **37**(8), 953–969 (1999)
4. Lasiecka, I., Tatau, D.: Uniform boundary stabilisation of semilinear wave equation with nonlinear boundary damping. Differ. Integr. Equ. **6** (1993)
5. Ouzahra, M.: Partial stabilisation of semilinear systems using bounded controls. Int. J. Control. **86**(12), 2253–2262 (2013)
6. Pazy, A.: Semi-Groups of Linear Operators and Applications to Partial Differential Equations. Springer, New York (1983)
7. Zerrik, E., Ait Aadi, A., Larhrissi, R.: On the output feedback stabilization for distributed semilinear systems. Asian J. Control. https://doi.org/10.1002/asjc.2081
8. Zerrik, E., Ait Aadi, A., Larhrissi, R.: Regional stabilization for a class of bilinear systems. IFAC-PapersOnLine, **50**(1), 4540–4545 (2017)
9. Zerrik, E., Ait Aadi, A., Larhrissi, R.: On the stabilization for a class of infinite dimensional bilinear systems with unbounded control operator. J. Nonlinear Dyn. Syst. Theory **18**(4), 418–425 (2018)
10. Zerrik, E., Ouzahra, M.: Regional stabilization for infinite-dimensional systems. Int. J. Control. **76**, 73–81 (2003)
11. Zerrik, E., Ouzahra, M., Ztot, K.: Regional stabilization for infinite bilinear systems. IEE Control Theory Appl. **151**(1), 109–116 (2004)

Bilinear Boundary Control Problem of an Output of Parabolic Systems

El Hassan Zerrik and Abella El Kabouss

Abstract The aim of this paper is to study a bilinear boundary optimal control problem for an output of infinite dimensional parabolic systems. We show the existence of a control minimizing a quadratic functional, that we characterize using an optimality system. Then, we derive sufficient condition for the uniqueness of such a control, and we develop an algorithm that allows its computation. A simulation illustrating the theoretical results is given.

Keywords Parabolic system · Boundary control · Bilinear control · Optimal control

1 Introduction

Bilinear control equations are useful to model a various phenomena in many real applications of physical, chemical, biological, and social systems, as well as nuclear and thermal controls processes (see [6]).

Much attention has been paid to the study of bilinear optimal control of particular systems: in case of internal controls, in [2] authors studied a spatio-temporal optimal control for a Kirchhoff equation using a quadratic cost, in [4] they studied the case of only time dependent optimal control, and in [3] they discussed the case of a spacial optimal control.

Recently, in [11] authors studied a regional bilinear optimal control of a system governed by a fourth-order parabolic operator with bounded and unbounded controls, minimizing a functional cost, then in [10] they studied this problem for a class of infinite bilinear system with unbounded control operator in both cases unbounded

E. Zerrik · A. El Kabouss (✉)
MACS laboratory, Faculty of sciences, Univeristy of Moulay Ismail,
Meknes, Morocco
e-mail: elkabouss.abella@gmail.com

E. Zerrik
e-mail: zerrik3@yahoo.fr

© Springer Nature Switzerland AG 2020   193
E. H. Zerrik et al. (eds.), *Recent Advances in Modeling, Analysis and Systems Control:
Theoretical Aspects and Applications*, Studies in Systems, Decision and Control 243,
https://doi.org/10.1007/978-3-030-26149-8_15

and bounded controls, while in [9] they considered a regional bilinear optimal control problem of a wave equation.

In case of boundary bilinear controls, in [5] author considered the problem of controlling the solution of the heat equation with the convective boundary condition taking the heat transfer coefficient as the bilinear control, they established existence and uniqueness of the optimal control, while in [7] author considered a bilinear boundary optimal control problem for a Kirchhoff plate equation.

In this work we deal with a bilinear boundary control problem for an output infinite parabolic systems, we show existence of an optimal control and give characterization. Also, under a condition the uniqueness is proved.

More precisely, let Ω be an open bounded set of $\mathbb{R}^n$ $(n \geq 2)$, with a regular boundary $\partial\Omega$.

For $T > 0$, we denote by $Q = \Omega \times]0, T[$, $\Gamma = \partial\Omega \times]0, T[$, and we consider a system described by

$$\begin{cases} \dfrac{\partial z}{\partial t}(x, t) + Az(x, t) = 0 & Q, \\ \dfrac{\partial z}{\partial \upsilon_A}(x, t) = u(x, t)Bz(x, t) & \Gamma, \\ z(x, 0) = z_0(x) & \Omega, \end{cases} \tag{1}$$

where A is an elliptic differential operator of the form

$$Az(x) = -\sum_{i,j=1}^{N} \frac{\partial}{\partial x}\left(a_{ij}(x)\frac{\partial}{\partial x}z(x)\right) \quad x \in \Omega.$$

The functions a_{ij} are assumed to belong to $L^\infty(\Omega)$, and to satisfy the condition $a_{ij}(x) = a_{ji}(x)$ for all $i, j \in \{1, \ldots, N\}$ and $x \in \Omega$, moreover, there exists $\gamma_0 > 0$ such that

$$\sum_{i,j=1}^{N} a_{ij}(x)\xi_i\xi_j \geq \gamma_0|\xi|^2, \ \forall \xi \in \mathbb{R}^N.$$

$\dfrac{\partial}{\partial \upsilon_A}$ is the directional derivative of the conormal vector υ_A, and $u \in U = \{u \in L^\infty(\Gamma)/0 \leq u \leq M\}$ (m and M are non-negative constants) is the control where U is the set of admissible controls.

B is a linear continuous operator on $L^2(\Gamma)$.

For $z_0 \in L^2(\Omega)$ and $u \in U$ according to Lemma 5.3 (p. 373 [8]) system (1) has a unique weak solution $z \in L^2(0, T; H^1(\Omega))$ such that $\dfrac{\partial z}{\partial t} \in L^2(0, T; (H^1(\Omega))')$ which satisfies, for $\varphi \in H^1(\Omega)$,

$$\int_0^T \langle \frac{\partial z}{\partial t}, \varphi \rangle dt + \int_0^T < A^{1/2}, zA^{1/2}\varphi >_{L^2(\Omega)} dt = \int_\Gamma uBy\varphi \, dx \, dt.$$

where $\langle ., . \rangle$ denotes the duality between $H^1(\Omega)$ and $(H^1(\Omega))'$.

An optimal control problem my be stated as follows

$$\min_{u \in U} J(u),\qquad(2)$$

where J is the cost functional given by

$$J(u) = \frac{1}{2}\int_0^T \|Cz(x,t) - y_d(x)\|_Y^2 dt + \frac{\beta}{2}\|u\|_{L^2(\Gamma)}^2.\qquad(3)$$

where C is a bounded operator from $L^2(\Omega)$ to a Hilbert space Y, $y_d \in Y$ is a desired output and β is a positive constant.

The paper is organized as follows, in Sect. 2, the existence of an optimal control solution of problem (2) is shown. In Sect. 3, we give a characterization of such a control and we discuss a condition of its uniqueness. In the last section, an example and simulations illustrate the obtained results.

2 Existence of an Boundary Optimal Control

In this section, we show the existence of solution of problem (2). Firstly, we prove a priori estimates which are necessary for the existence of an optimal control.

Lemma 1 *We consider the system:*

$$\begin{cases} \dfrac{\partial z}{\partial t}(x,t) + Az(x,t) = 0 & Q, \\ \dfrac{\partial z}{\partial v_A}(x,t) = u(x,t)Bz(x,t) + f(x,t) & \Gamma, \\ z(x,0) = z_0(x) & \Omega, \end{cases}\qquad(4)$$

where $z_0 \in L^2(\Omega)$ and $f \in L^2(\Gamma)$. Then the weak solution z of system (4) satisfies the following estimates.

$$\|z\|_{L^2(0,T;H^1(\Omega))} \le C_1\|f\|_{L^2(0,T;L^2(\partial\Omega))} + C_2\|z_0\|_{L^2(\Omega)}\qquad(5)$$

$$\left\|\frac{\partial z}{\partial t}\right\|_{L^2(0,T;(H^1(\Omega))')} \le C_3\|f\|_{L^2(0,T;L^2(\partial\Omega))} + C_4\|z\|_{L^2(0,T;H^1(\Omega))}\qquad(6)$$

Proof Multiplying system (4) by z and integrating over $(0, T)$ we have

$$\int_0^T \langle \frac{\partial z}{\partial t}, z \rangle dt + \int_0^T < A^{1/2}z, A^{1/2}z >_{L^2(\Omega)} dt = \int_\Gamma (uBz + f)z\, dxdt.$$

Since B is a bounded operator, we have

$$\frac{1}{2}\int_0^T \frac{d}{dt}\|z(t)\|^2_{L^2(\Omega)}dt + \int_0^T \|A^{1/2}z(t)\|_{L^2(\Omega)}dt \le \int_0^T \|f\|_{L^2(\partial\Omega)}\|z\|_{L^2(\partial\Omega)}dt$$

Hence, for $\alpha > 0$ we have

$$\frac{1}{2}\|z(T)\|^2_{L^2(\Omega)} + \int_0^T \|A^{1/2}z(t)\|_{L^2(\Omega)}dt \le \int_0^T \frac{\alpha^2}{2}\|f\|^2_{L^2(\partial\Omega)} + \frac{(1)}{2\alpha^2}\|z\|^2_{L^2(\partial\Omega)}dt + \frac{1}{2}\|z_0\|^2_{L^2(\Omega)}$$

where C depends on B.

For $n \ge 2$, we have

$$\forall z \in H^1(\Omega), \exists C_0 > 0, \|z\|_{L^2(\partial\Omega)} \le C_0\|z\|_{H^1(\Omega)}$$

(see [1]), we obtain

$$\int_0^T \|z\|_{H^1(\Omega)}dt \le \int_0^T \frac{\alpha^2}{2}\|f\|^2_{H^1(\Omega)} + \frac{C_0}{2\alpha^2}\|z\|^2_{H^1(\Omega)}dt + \frac{1}{2}\|z_0\|^2_{L^2(\Omega)}$$

Choose $\alpha > \sqrt{\dfrac{C_0}{2}}$, we deduce that

$$\|z\|_{L^2(0,T;H^1(\Omega))} \le C_1\|f\|_{L^2(0,T;L^2(\partial\Omega))} + C_2\|z_0\|_{L^2(\Omega)}.$$

For $\varphi \in H^1(\Omega)$, we have

$$\int_0^T |\langle\frac{\partial z}{\partial t}, \varphi\rangle + < A^{1/2}z, A^{1/2}\varphi >_{L^2(\Omega)} | dt = \int_\Gamma |(uBz + f)\varphi| \, dx \, dt.$$

It follows

$$\int_0^T |\langle\frac{\partial z}{\partial t}, \varphi\rangle| dt \le C \int_0^T (\|z\|_{H^1(\Omega)} + \|f\|_{L^2(\partial\Omega)})\|\varphi\|_{H^1(\Omega)}dt$$

Therefore using (5), we obtain

$$\|\frac{\partial z}{\partial t}\|_{L^2(0,T;(H^1(\Omega))')} \le C_3\|f\|_{L^2(0,T;L^2(\partial\Omega))} + C_4\|z\|_{L^2(0,T;H^1(\Omega))}dt$$

Theorem 1 *There exists an optimal control $u \in U$, solution of problem (2).*

Proof The set $\{J(u)|u \in U\}$ is non-empty and is bounded from below by 0.
Let $(u_k)_{k\in\mathbb{N}}$ be a minimizing sequence in U.

$$\lim_{k\to\infty} J(u_k) = \inf_{h\in U} J(h).$$

Then $(J(u_k))_{k\in\mathbb{N}}$ is bounded.

Since $\|u_k\|_{L^2(\Gamma)} \leq \dfrac{2}{\beta} J(u_k)$ then $(u_k)_{k\in\mathbb{N}}$ is bounded.

Then there exists a sup-sequences still denoted $(u_k)_{k\in\mathbb{N}}$ that converges weakly to a limit $u^* \in L^2(\Gamma)$.

Let z_k and z_{u*} solutions of system (1) associated to u_k and u^* respectively.

Using the inequalities (5) and (6) there exists subsequence with the following convergence proprieties.

$$z_k \rightharpoonup z_{u^*} \quad \text{in } L^2(0, T; H^1(\Omega)),$$

$$\frac{\partial z_k}{\partial t} \rightharpoonup \frac{\partial z_{u^*}}{\partial t} \quad \text{in } L^2(0, T; (H^1(\Omega))'),$$

$$u_k \rightharpoonup u^* \quad \text{in } L^2(\Gamma).$$

Using the compact injection of $H^1(\Omega)$ into $H^{1/2+\varepsilon}(\Omega)$ (where $0 < \varepsilon < \dfrac{1}{2}$), we obtain

$$z_k \longrightarrow z_{u^*} \quad \text{in } L^2(0, T; H^{1/2+\varepsilon}(\Omega)).$$

The mapping $\varphi : H^{1/2+\varepsilon}(\Omega) \longrightarrow \varphi_{|\Gamma} \in L^2(\Gamma)$ is continuous then $z_k \longrightarrow z_{u^*}$ in $L^2(\Gamma)$ Since B is a linear operator then $u_k B z_k \longrightarrow u^* B z_{u^*}$ in $L^2(\Gamma)$ as $k \longrightarrow \infty$, we deduce that z_{u^*} solves system (1) with control u^*.

Since U is convex, $u^* \in U$.

J is lower semi-continuous with respect to weak convergence, we obtain

$$J(u^*) \leq \lim_{k \to \infty} \inf J(u_k),$$

leading to $J(u^*) = \inf_{u \in U} J(u_k)$.

3   Characterization of an Optimal Control

In this section we derive a characterization of an optimal control solution of problem (2).

Firstly, we examine the differentiability of the mapping $u \longrightarrow z_u$ with respect to u, given by the following lemma.

Lemma 2 *The mapping*

$$u \in U \to z_u \in L^2(0, T; H^1(\Omega))$$

is differentiable in the sense:

$$\frac{z_{u+\varepsilon h} - z_u}{\varepsilon} \rightharpoonup \psi \quad \text{weakly in } L^2(0, T; H^1(\Omega)),$$

as $\varepsilon \to 0$, for any $u, u + \varepsilon h \in U$, where ψ is solution of

$$\begin{cases} \dfrac{\partial \psi}{\partial t}(x, t) + A\psi(x, t) = 0 & Q, \\ \dfrac{\partial \psi}{\partial v_A}(x, t) = u(x, t)B\psi(x, t) + h(x, t)Bz(x, t) & \Gamma, \\ \psi(x, 0) = 0 & \Omega. \end{cases} \tag{7}$$

Proof Let consider $z_{u+\varepsilon h}$ solution of system (1) with control $u + \varepsilon h$, then $\varphi = \dfrac{z_{u+\varepsilon h} - z_u}{\varepsilon}$ is solution of the following system

$$\begin{cases} \dfrac{\partial \varphi}{\partial t}(x, t) + A\varphi(x, t) = 0 & Q, \\ \dfrac{\partial \varphi}{\partial v_A}(x, t) = u(x, t)B\varphi(x, t) + h(x, t)Bz(x, t) & \Gamma, \\ \varphi(x, 0) = 0 & \Omega, \end{cases}$$

Using estimates (5) and (6) we obtain

$$\|\frac{z_{u+\varepsilon h} - z_u}{\varepsilon}\|_{L^2(0,T;H^1(\Omega))} \le C_1 \|hBz\|_{L^2(0,T;L^2(\partial\Omega))}$$
$$\le C_2 \|z_0\|_{L^2(\Omega)}$$

and

$$\|\frac{\partial}{\partial t}(\frac{z_{u+\varepsilon h} - z_u}{\varepsilon})\|_{L^2(0,T;(H^1(\Omega))')} \le \|hBz\|_{L^2(0,T;L^2(\partial\Omega))}$$
$$\le C_3 \|z_0\|_{L^2(\Omega)}$$

where C_2 and C_3 are independent of ε.

Thus as $\varepsilon \longrightarrow 0$ we have

$$\frac{z_{u+\varepsilon h} - z_u}{\varepsilon} \rightharpoonup \psi \quad \text{weakly in } L^2(0, T; H^1(\Omega)),$$

Similar to the proof of Theorem (1), we deduce that ψ is the weak solution of system (7).

Proposition 1 *Let $u^* \in U$ be an optimal control and let $h \in L^2(\Gamma)$ such that $u^* + \varepsilon h \in U$, for $\varepsilon > 0$, z_{u^*} is the solution of system (1) corresponding to u^*, and p solution of the following system*

$$\begin{cases} \dfrac{\partial p}{\partial t}(x,t) + Ap(x,t) = C^*y_d(x) - C^*Cz_{u^*}(x,t) & Q \\[2mm] \dfrac{\partial p}{\partial v_A}(x,t) = B^*u^*(x,t)p(x,t) & \Gamma, \\[2mm] p(x,T) = 0 & \Omega. \end{cases} \qquad (8)$$

Then an optimal control is given by

$$u^*(x,t) = \max\left(m, \min\left(-\frac{1}{\beta}Bz_{|\Gamma}(x,t)p_{|\Gamma}(x,t), M\right)\right)$$

Proof Let u^* be an optimal control and z_{u^*} the corresponding solution to u^*, $h \in L^2(\Gamma)$ such that $u^* + \varepsilon h \in U$ for $\varepsilon > 0$. We calculate the directional derivative of the cost functional $J(u)$ with respect to u in the direction of h

$$dJ(u).h = \lim_{\varepsilon \to 0^+} \frac{J(u^* + \varepsilon h) - J(u^*)}{\varepsilon}$$

$$= \int_0^T \frac{1}{2\varepsilon}(\|Cz_\varepsilon - y_d\|_Y^2 - \|Cz_{u^*} - y_d\|_Y^2)dt + \frac{\beta}{2}\int_\Gamma (2hu^* + \varepsilon h^2)\,dxdt$$

$$= \lim_{\varepsilon \to 0^+} \int_0^T \frac{1}{2\varepsilon} < Cz_\varepsilon - y_d, Cz_\varepsilon - z_{u^*} >_Y dt + \frac{\beta}{2}\int_\Gamma (2hu^* + \varepsilon h^2)\,dxdt$$

$$= \lim_{\varepsilon \to 0^+} \int_Q \left(\frac{z_\varepsilon - z_{u^*}}{\varepsilon}\right)\left(\frac{C^*Cz_\varepsilon + C^*Cz_{u^*} - 2C^*y_d}{2}\right)dQ + \frac{\beta}{2}\int_\Gamma (2hu^* + \varepsilon h^2)\,dxdt$$

$$= \int_Q \psi(C^*Cz_{u^*} - C^*y_d)dQ + \beta\int_\Gamma u^*h\,dxdt\,.$$

where ψ is solution of system (7) corresponding to u^*.
Using (8) we have

$$dJ(u).h = \left(-\int_0^T < \psi, \frac{\partial p}{\partial t} > - < \psi, Ap > \right)dt + \int_\Gamma u^*hdxdt.$$

$$= \int_0^T \left(< \frac{\partial \psi}{\partial t}, p > + < A^{1/2}\psi, A^{1/2}p > - < hB\psi, p >_{L^2(\partial\Omega)}\right)dt + \int_\Gamma u^*hdxdt.$$

Using system (7) we obtain

$$dJ(u).h = \int_\Gamma hBzp\,d\Gamma + \beta\int_\Gamma hu^*\,d\Gamma.$$

Since J achieves its minimum at u^*, we have

$$0 \leq \int_\Gamma hBzp\,d\Gamma + \beta\int_\Gamma hu^*\,d\Gamma.$$

Taking $h = \max\left(m, \min\left(-\frac{1}{\beta}Bz_{|_\Gamma}(x,t)p_{|_\Gamma}(x,t), M\right)\right) - u^*$, we show that $h(u^* + \frac{1}{\beta}Bz_{|_\Gamma}p_{|_\Gamma})$ is negative and then

$$\left(\max\left(m, \min\left(-\frac{1}{\beta}Bz_{|_\Gamma}(x,t)p_{|_\Gamma}(x,t), M\right)\right) - u^*\right)\left(u^* + \frac{1}{\beta}Bz_{|_\Gamma}p_{|_\Gamma}\right) = 0.$$

If $M \leq -\frac{1}{\beta}Bz_{|_\Gamma}p_{|_\Gamma}$ we have $(M - u^*)\left(u^* + \frac{1}{\beta}Bz_{|_\Gamma}p_{|_\Gamma}\right) = 0$, thus $u^* = M$.

If $m \leq -\frac{1}{\beta}Bz_{|_\Gamma}p_{|_\Gamma} \leq M$ we have $\left(-\frac{1}{\beta}Bz_{|_\Gamma}p_{|_\Gamma} - u^*\right)\left(u^* + \frac{1}{\beta}Bz_{|_\Gamma}p_{|_\Gamma}\right) = 0$.

Therefore $u^* = -\frac{1}{\beta}Bz_{|_\Gamma}p_{|_\Gamma}$.

Now, if $m \geq -\frac{1}{\beta}Bz_{|_\Gamma}p_{|_\Gamma}$, we have $(m - u^*)\left(u^* + \frac{1}{\beta}Bz_{|_\Gamma}p_{|_\Gamma}\right) = 0$ and then $u^* = m$.

We conclude that

$$u^*(x,t) = \max\left(m, \min\left(-\frac{1}{\beta}Bz_{|_\Gamma}(x,t)p_{|_\Gamma}(x,t), M\right)\right).$$

Corollary 1 *If we consider $U = \{u \in L^\infty(0,T), m \leq u \leq M\}$, an optimal control is given by*

$$u^*(t) = \max\left(m, \min\left(-\frac{1}{\beta}\int_{\partial\Omega} Bz_{|_\Gamma}(x,t)p_{|_\Gamma}(x,t)dx, M\right)\right). \tag{9}$$

Proposition 2 *Assume that z and p are bounded on $\overline{\Omega}$, for β large enough, the solution of problem (2) is unique.*

Proof For $i = 1, 2$, let z_i and p_i be solutions of system (1) and (7) respectively, then using (5) and (6) we have

$$\|z_i\|_{L^2(0,T;H^1(\Omega))} \leq C_2\|z_0\|_{L^2(\Omega)}$$

and

$$\|p_i\|_{L^2(0,T;H^1(\Omega))} \leq C_1\|z_i - y_d\|_{L^2(\Omega)} + C_2\|z_0\|_{L^2(\Omega)}.$$

Since $z_1 - z_2$ is the solution of the following system

$$\begin{cases} \dfrac{\partial(z_1 - z_2)}{\partial t}(x,t) + A(z_1 - z_2)(x,t) = 0 & Q \\ \dfrac{\partial(z_1 - z_2)}{\partial\upsilon_A}(x,t) = u_1(x,t)B(z_1 - z_2)(x,t) + (u_1 - u_2)(x,t)Bz_2(x;t) & \Gamma, \\ (z_1 - z_2)(x,0) = 0 & \Omega, \end{cases} \tag{10}$$

then using (5), we have

$$\|z_1 - z_2\|_{L^2(0,T;H^1(\Omega))} \leq C\|z_0\|_{L^2(\Omega)}\|u_1 - u_2\|_{L^\infty(\Gamma)}$$

Similarly for $(p_1 - p_2)$ we have

$$\begin{aligned}
\|p_1 - p_2\|_{L^2(0,T;H^1(\Omega))} &\leq \|B^*(u_1 - u_2)p_2\|_{L^2(\Gamma)} \\
&\leq C\|u_1 - u_2\|_{L^\infty(\Gamma)}\|p_2\|_{L^2(0,T;H^1(\Omega))} \\
&\leq C\|z_2 - y_d\|_{L^2(\Omega)}\|u_1 - u_2\|_{L^\infty(\Gamma)}.
\end{aligned}$$

Then

$$\|u_1 - u_2\|_{L^\infty(\Gamma)} \leq \frac{1}{\beta}(\|p_1 B(z_1 - z_2)\|_{L^\infty(\Gamma)} + \frac{1}{\beta}\|Bz_2(p_1 - p_2)\|_{L^\infty(\Gamma)}),$$

which gives

$$\begin{aligned}
\|u_1 - u_2\|_{L^\infty(\Gamma)} &\leq \frac{1}{\beta}(\|p_1\|_{L^\infty(\Gamma)}\|B(z_1 - z_2)\|_{L^2(\Gamma)} + \|Bz_2\|_{L^2(\Gamma)}\|(p_1 - p_2)\|_{L^\infty(\Gamma)}) \\
&\leq \frac{C}{\beta}(\|(z_1 - z_2)\|_{L^2(0,T;H^1(\Omega))} + \|(p_1 - p_2)\|_{L^2(0,T;H^1(\Omega))}) \\
&\leq \frac{C}{\beta}(\|(z_1 - z_2)\|_{L^2(0,T;H^1(\Omega))} + \|z_2 - y_d\|_{L^2(\Omega)}\|u_1 - u_2\|_{L^\infty(\Gamma)}).
\end{aligned}$$

It follows

$$\|u_1 - u_2\| \leq \frac{C}{\beta}(\|(u_1 - u_2)\|_{L^\infty(\Gamma)}(2\|z_0\|_{L^2(\Omega)} + \|y_d\|_{L^2(\Omega)})). \tag{11}$$

So for β large enough such that $(2C\|z_0\|_{L^2(\Omega)} + C\|y_d\|_{L^2(\Omega)}) < \beta$.
   Then the solution of problem (2) is unique.

4 Algorithm and Simulations

An optimal control solution of problem (2) can be calculated by the following for-mula:

$$u_{k+1}(x,t) = \max\left(m, \min\left(-\frac{1}{\beta}Bz_{k|\Gamma}(x,t)p_{k|\Gamma}(x,t), M\right)\right) \tag{12}$$

where z_k is the solution of system (1) associated with u_k and p_k is the one of the adjoint equation (8). This allows to consider the following algorithm:

Algorithm

Step 1: initials data.

* An initial state z_0, a time T, a control $u_1 = 0$, a desired output y_d.
* A threshold accuracy $\varepsilon > 0$, and $k = 1$.

Step 2:
* Solving equation (1) gives z_k.
* Solving equation (8) gives p_k.
* Calculate u_{k+1} by formula (12).

Step 3 : while $\|u_{k+1} - u_k\|_{L^2(\Gamma)} > \varepsilon$, k = k + 1 go to step 2.

Example 1 On $\Omega =]0, 1[\times]0, 1[$, we consider the following system

$$\begin{cases} \dfrac{\partial z}{\partial t}(x, y, t) = 0.01\Delta z(x, y, t) & Q \\ \dfrac{\partial z}{\partial \upsilon}(x, y, t) = -u(t)\mathbb{1}_{\Gamma_1}z(x, y, t) & \Gamma, \\ z(x, y, 0) = z_0(x, y) & \Omega, \end{cases} \tag{13}$$

where $z_0(x, y) = x(y - x)$, $u \in U = \{u \in L^2(0, T)/0 \le u \le 0.04\}$, and $\mathbb{1}_{\Gamma_1}$ denotes the characteristic function to $\Gamma_1 = \{1\}\times]0, 1[$.

Let $\omega =]0, 1[\times]0, 0, 25[$, $C = \chi_\omega$ indicates the restriction function to $\omega \subset \Omega$ with $Y = L^2(\omega)$.

We consider a desired output $y_d(x) = 0 \in L^2(\omega)$. Applying control (9), Fig. 1 shows that the final state on ω is close to the desired one on ω with error $\|\chi_\omega z(x, y, T) - y_d(x, y)\|^2_{L^2(\omega)} = 8.26 \cdot 10^{-5}$, and a cost $J(u^*) = 1.22 \cdot 10^{-3}$.

An optimal control is presented by Fig. 2.

Fig. 1 Final state

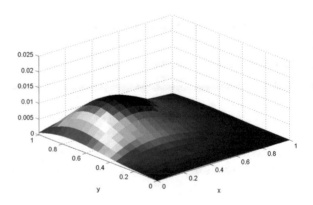

Fig. 2 The evolution of an optimal control

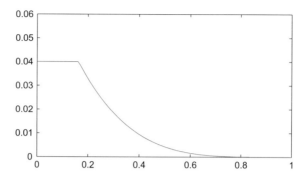

5 Conclusion

In this paper, we study a boundary bilinear optimal control problem for an output of infinite dimensional parabolic systems. This approach allows to study many open questions, for instance, a boundary bilinear optimal control of an output of a wave equation. This is under consideration.

Acknowledgements The work has been carried out with a grant from Hassan II Academy of Sciences and Technology.

References

1. Adams, R.A., Fournier, J.J.: Sobolev Spaces, vol. 140. Academic (2003)
2. Bradley, M., Lenhart, S.: Bilinear optimal control of a Kirchhoff plate. Syst. Control Lett. **22**(1), 27–38 (1994)
3. Bradley M.E., Lenhart, S.: Bilinear spatial control of the velocity term in a Kirchhoff plate equation. Electron. J. Differ. Equ. (EJDE)[electronic only], 2001:Paper–No (2001)
4. Bradley, M.E., Lenhart, S., Yong, J.: Bilinear optimal control of the velocity term in a Kirchhoff plate equation. J. Math. Anal. Appl. **238**(2), 451–467 (1999)
5. Lenhart, S., Wilson, D.: Optimal control of a heat transfer problem with convective boundary condition. J. Optim. Theory Appl. **79**(3), 581–597 (1993)
6. Mohler R.R.: Bilinear Control Processes: with Applications to Engineering, Ecology and Medicine. Academic, Inc. (1973)
7. Park, J.Y.: Bilinear boundary optimal control of the velocity terms in a Kirchhoff plate equation. Trends Math. **9**, 41–44 (2006)
8. Tröltzsch, F.: Optimal Control of Partial Differential Equations: Theory, Methods, and Applications, vol. 112. American Mathematical Society (2010)
9. Zerrik, E., El Kabouss, A.: Regional optimal control of a bilinear wave equation. Int. J. Control, **92**(4),940–949 (2019)
10. Zerrik, E., El Kabouss, A.: Regional optimal control problem of a class of infinite-dimensional bi-linear systems. Int. J. Control **90**(7), 1495–1504 (2017)
11. Zerrik, E., El Kabouss, A.: Regional optimal control of a class of bilinear systems. IMA J. Math. Control. Inf. **34**(4), 1157–1175 (2016)

Optimal Control of a Parabolic Trough Receiver Distributed Model

El Hassan Zerrik and Nihale El Boukhari

Abstract This paper examines an optimal control problem for a parabolic trough receiver. We consider a simplified bilinear distributed model, where the control stands for the velocity of the heat transfer fluid. Using the tools of semigroup theory, we characterize the optimal control, that steers the fluid temperature close to the required level. Then we give an algorithm for the numerical implementation of the optimal control. The obtained results are illustrated through simulations.

1 Introduction

Solar thermal energy is the technology of generating electricity by concentrating solar radiation. It is an environment-friendly alternative to fossil energy, and offers a sustainable solution to the increasingly global demand for energy and the fluctuating prices of fossil fuels.

This paper deals with the optimization of a solar power plant performance. A simplified schematic of a solar plant is shown in Fig. 1. The plant is equipped with parabolic solar troughs, where parabolic mirrors (or reflectors) concentrate sunlight onto a heat collection element (HCE), located in the focal line of the reflectors. The HCE is depicted in Fig. 2, and consists of:

- A receiver tube: usually a stainless steel tube, about 70 mm in diameter, with a special coating.
- A glass envelope: a transparent cylindrical glass body with an anti-reflection coating. It covers the receiver tube, with vacuum separating both tubes to reduce heat loss.
- Bellows at both ends to ensure glass-to-metal seal and prevent heat loss.

E. H. Zerrik · N. El Boukhari (✉)
MACS Team, Department of Mathematics, Faculty of Sciences, Moulay Ismail University, Meknes, Morocco
e-mail: elboukhari.nihale@gmail.com

E. H. Zerrik
e-mail: zerrik3@yahoo.fr

© Springer Nature Switzerland AG 2020
E. H. Zerrik et al. (eds.), *Recent Advances in Modeling, Analysis and Systems Control: Theoretical Aspects and Applications*, Studies in Systems, Decision and Control 243, https://doi.org/10.1007/978-3-030-26149-8_16

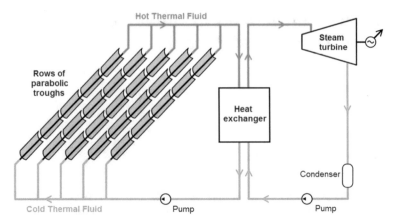

Fig. 1 Schematic of a solar power plant

Fig. 2 Heat collection
element

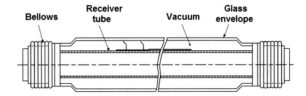

A heat transfer fluid, which is usually a synthetic oil, flows through the receiver tube to absorb solar radiation. Once the heat transfer fluid is heated, it is transferred to the heat exchanger to generate high-temperature steam, which is used to generate electricity via a steam turbine (For further details see [3, 8, 9, 15, 18]). For a maximal efficiency of the heat exchanger, the temperature of the heat transfer fluid has to be close to a required level $T_d \simeq 400\,°C$. The fluid temperature depends on the intensity of solar radiation, the geometry and the optical properties of the reflectors, the thermal properties of the HCE, and on the fluid velocity. In order to reach the required temperature, it is imperative to determine the optimal fluid velocity. This can be achieved by formulating and solving a minimization problem.

We aim to solve an optimal control problem, where the control stands for the fluid velocity. To this end, the original model describing the heat balance within the HCE is simplified into a distributed bilinear system, taking the form $\dot{y}(t) = Ay(t) + u(t)By(t) + b$. A rich literature is devoted to the optimal control of such systems (see for instance [1, 2, 5, 13, 19]). Nevertheless, in previous works, the operator B is usually assumed to be relatively bounded with respect to A. By contrast, the solar trough receiver studied here is modeled by a bilinear system whose operator A is bounded, while the control operator B is unbounded and is the generator of a strongly continuous semigroup. Thereby, the optimal control problem will be studied using a new method, based on the tools of semigroup theory.

This is the purpose of the present paper, which is organized as follows. In Sect. 2 the original model of the solar receiver is simplified into a distributed bilinear system, modeling the evolution of the fluid temperature. Section 3 is devoted to the characterization of the optimal control. Finally, simulations of the simplified model are provided in Sect. 4.

2   The Simplified Model

We consider a heat collection element, as shown in Fig. 2. The heat transfer fluid flows through a receiver tube of length L, which is enclosed in an evacuated glass envelope. We assume that:

- The fluid is incompressible and the flow is laminar, then the fluid velocity is space-invariant, and depends only on time.
- The fluid velocity and temperature are uniformly distributed over the tube section, hence the receiver tube can be modeled by a one-dimensional segment $[0, L]$.

Upon the above assumptions, it has been shown in [11] that the heat balance between the receiver tube and the heat transfer fluid can be modeled by the following system

$$\begin{cases} \dfrac{\partial T_f}{\partial t} = a_f[T_m - T_f] - u(t)\dfrac{\partial T_f}{\partial x} \\ \dfrac{\partial T_m}{\partial t} = -a_m[T_m - T_f] - b_m[T_m - T_a(t)] + q_{Sol}(t) \end{cases} \tag{1}$$

with the boundary condition

$$T_f(0, t) = T_0$$

and the initial conditions

$$T_f(x, 0) = T_0, \quad T_m(x, 0) = T_{m0}(x)$$

where T_f (°C) is the temperature of the heat transfer fluid, T_m (°C) is the temperature of the receiver tube, and T_a (°C) is the ambient temperature. a_f, a_m, and b_m are nonnegative coefficients, depending on the physical properties of the receiver tube and the fluid. q_{Sol} (°C s^{-1}) is the temperature variation gained from solar radiation, and transmitted to the receiver tube. $u(t)$ (m s^{-1}) is the velocity of the heat transfer fluid.

Moreover, we assume that, for a relatively large q_{Sol}, the receiver tube temperature reaches rapidly a given value T_{max}, due to the quick metal heat absorption. Then the temperature of the receiver tube $T_m(., .)$ is assumed to be time-invariant and close to T_{max}. Therefore, the parabolic trough model (1) is simplified into a system of one partial differential equation, given by:

Fig. 3 Temperature of the
receiver tube

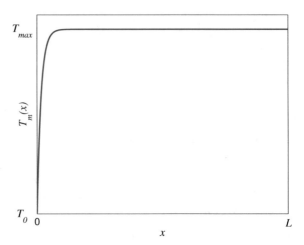

$$
\begin{cases}
\dfrac{\partial T_f}{\partial t}(x,t) = -a_f T_f(x,t) + a_f T_m(x) - u(t)\dfrac{\partial T_f}{\partial x}(x,t) \\
T_f(0,t) = T_0 \\
T_f(x,0) = T_0
\end{cases}
\tag{2}
$$

where the receiver tube temperature T_m is approximated by a smooth function over $[0, L]$, satisfying $T_m(0) = T_0$, as depicted in Fig. 3.

We consider the state space $H = L^2(0, L)$, and we define the state $y(t) \in H$ by

$$
y(t) = T_f(., t) - T_0
$$

Denote $B = -\dfrac{\partial}{\partial x}$, and $b = a_f(T_m(.) - T_0) \in H$, then Eq. (2) takes the form of the following bilinear system:

$$
\begin{cases}
\dot{y}(t) = u(t)By(t) + a_f y(t) + b \\
y(0) = 0
\end{cases}
\tag{3}
$$

The domain of B is $\mathscr{D}(B) = \{z \in H^1(0, L) : z(0) = 0\}$. Then $y(0), b \in \mathscr{D}(B)$.

In addition, B is the infinitesimal generator of the following C_0 semigroup of bounded linear operators:

$$
[S(t)y](x) = \begin{cases}
y(x - t) & \text{if } x - t \geq 0 \\
0 & \text{if } x - t < 0
\end{cases}
$$

In order for the fluid temperature T_f to get close to $T_d = 400\,°\mathrm{C}$, at a given time T, we will search the optimal control $u(.)$ that steers $y(T)$ close to $y_d = T_d - T_0$. This is the purpose of the next section.

3   Optimal Control Problem

We consider the following system

$$\begin{cases} \dot{y}(t) = u(t)By(t) + ay(t) + b(t) \\ y(0) = y_0 \end{cases} \tag{4}$$

such that:

- $B : \mathscr{D}(B) \subset H \to H$ is the infinitesimal generator of a C_0 semigroup $(S(t))_{t \geq 0}$.
- H is a separable Hilbert space, endowed with the inner product $\langle ., . \rangle$ and the associated norm $\|.\|$.
- $u \in L^\infty(0, T)$ is the control, such that $u(t) \geq 0$ almost everywhere (a.e.) on $[0, T]$.
- $a \in \mathbb{R}$, $b \in L^2(0, T; \mathscr{D}(B))$, and $y_0 \in \mathscr{D}(B)$.

Our objective is to find a control that steers the state $y(T)$ close to a desired state $y_d \in H$. This can be achieved by minimizing the following cost functional:

$$J(u) = \frac{\gamma}{2} \|y(T) - y_d\|^2 + \frac{r}{2} \int_0^T u(s)^2 ds \tag{5}$$

over the set of admissible controls, given by:

$$U_{ad} = \{u \in L^\infty(0, T) \ : \ u_{min} \leq u(t) \leq u_{max} \ \text{a.e.} \} \tag{6}$$

where $\gamma \geq 0$, $r > 0$, and $0 \leq u_{min} < u_{max}$.
   Thereby, the optimal control problem is stated as:

$$\begin{cases} \min J(u) \\ u \in U_{ad} \end{cases} \tag{7}$$

We first give the expression of the unique mild solution of system (4).

Lemma 1 *For any $u \in L^\infty(0, T)$ such that $u(t) \geq 0$ a.e., system (4) has a unique mild solution, written as*

$$y(t) = U(t, 0)y_0 + \int_0^t U(t, s) \left[ay(s) + b(s)\right] ds \tag{8}$$

where $U(t, s) = S \left(\int_s^t u(\tau)d\tau \right)$.
   In addition, since $y_0 \in \mathscr{D}(B)$, then $y(t) \in \mathscr{D}(B)$, $\forall t \geq 0$.

Proof Let $u \in L^\infty(0, T)$ such that $u(t) \geq 0$ a.e. on $[0, T]$. Then $(U(t, s))_{t \geq s}$ is a strongly continuous evolution family, generated by the family of operators $B(t) = u(t)B$ (see Definition 9.2, Chap. VI in [7]). In addition, there exist constants

$M \geq 1$ and $\omega \in \mathbb{R}$ such that $\|U(t, s)\| \leq Me^{|\omega| \|u\|_\infty (t-s)}$, then $U(t, s)$ is exponentially bounded. By virtue of corollary 9.20, chapter VI in [7], it follows that the family of operators $(u(t)B + aI)$ generates an exponentially bounded evolution family $(\Gamma(t, s))_{t \geq s}$, given by:

$$\Gamma(t, s)y = U(t, s)y + \int_s^t U(t, \tau)a\Gamma(\tau, s)y\,d\tau$$

Then system (4) has a unique mild solution, written as:

$$y(t) = \Gamma(t, 0)y_0 + \int_0^t \Gamma(t, s)b(s)ds$$

Replacing $\Gamma(t, 0)$ and $\Gamma(t, s)$ by their respective expressions, and applying Fubini's theorem yield

$$y(t) = U(t, 0)y_0 + \int_0^t U(t, s) [ay(s) + b(s)]\,ds$$

Finally, since $y_0 \in \mathscr{D}(B)$ then, by Proposition 9.3, chapter VI in [7], $y(t) \in \mathscr{D}(B)$, $\forall t \geq 0$. $\qquad\square$

Hereafter we prove some technical results, that will be useful to characterize the optimal control.

Proposition 1 *Let* $u, h \in L^\infty(0, T)$ *such that* $u(t), h(t) \geq 0$ *a.e., then*

$$\lim_{\varepsilon \to 0^+} \frac{y_{u+\varepsilon h}(t) - y_u(t)}{\varepsilon} = z_h(t) \qquad (9)$$

where z_h *is the mild solution of the following equation:*

$$\begin{cases} \dot{z}_h(t) = u(t)Bz_h(t) + az_h(t) + h(t)By_u(t) \\ z_h(0) = 0 \end{cases} \qquad (10)$$

and is written as:

$$z_h(t) = \int_0^t S\left(\int_s^t u(\tau)d\tau\right) [az_h(s) + h(s)By_u(s)]\,ds \qquad (11)$$

Proof Let $u, h \in L^\infty(0, T)$, such that $u(t), h(t) \geq 0$ a.e. on $[0, T]$. Let $\varepsilon > 0$. Without loss of generality, we can assume that $\varepsilon \|h\| \leq 1$.

Let y_u and $y_{u+\varepsilon h}$ be the respective mild solutions of system (4), relatively to u and $u + \varepsilon h$, and let z_h be the mild solution of Equation (10). Denote $U(t, s) = S\left(\int_s^t u(\tau)d\tau\right)$. By replacing $y_u(s)$ in Eq. (11) by its expression, we obtain

$$z_h(t) = \int_0^t U(t,s)az_h(s)ds + \int_0^t h(s)BU(t,0)y_0 ds$$
$$+ \int_0^t \int_0^s h(s)BU(t,\theta)[ay_u(\theta)+b(\theta)]d\theta ds$$

Applying Fubini's theorem to the above expression yields

$$z_h(t) = \int_0^t U(t,s)az_h(s)ds + \left(\int_0^t h(\tau)d\tau\right) BU(t,0)y_0$$
$$+ \int_0^t \left(\int_s^t h(\theta)d\theta\right) BU(t,s)[ay_u(s)+b(s)]ds \tag{12}$$

For $0 \le s \le t \le T$, we denote

$$\begin{cases} Y_0(t) = U(t,0)y_0, & Y(s) = U(t,s)ay_u(s) \\ Z(s) = U(t,s)b(s), & H_\varepsilon(t,s) = S\left(\varepsilon\int_s^t h(\tau)d\tau\right) \end{cases}$$

Then

$$y_u(t) = Y_0(t) + \int_0^t [Y(s)+Z(s)]ds$$

$$y_{u+\varepsilon h}(t) = H_\varepsilon(t,0)Y_0(t) + \int_0^t H_\varepsilon(t,s)U(t,s)ay_{u+\varepsilon h}(s)ds + \int_0^t H_\varepsilon(t,s)Z(s)ds$$

and, by expression (12), we have

$$z_h(t) = \int_0^t U(t,s)az_h(s)ds + \left(\int_0^t h(\tau)d\tau\right)BY_0(t) + \int_0^t \left(\int_s^t h(\tau)d\tau\right)B[Y(s)+Z(s)]ds$$

which yields

$$\frac{y_{u+\varepsilon h}(t) - y_u(t)}{\varepsilon} - z_h(t) = \frac{H_\varepsilon(t,0)Y_0(t) - Y_0(t)}{\varepsilon} - \int_0^t h(\tau)d\tau\, BY_0(t)$$
$$+ \int_0^t \left[\frac{H_\varepsilon(t,s)Y(s) - Y(s)}{\varepsilon} - \int_s^t h(\tau)d\tau\, BY(s)\right]ds$$
$$+ \int_0^t \left[\frac{H_\varepsilon(t,s)Z(s) - Z(s)}{\varepsilon} - \int_s^t h(\tau)d\tau\, BZ(s)\right]ds$$
$$+ \int_0^t H_\varepsilon(t,s)U(t,s)a\left[\frac{y_{u+\varepsilon h}(s) - y_u(s)}{\varepsilon} - z_h(s)\right]ds$$
$$+ \int_0^t [H_\varepsilon(t,s)az_h(s) - az_h(s)]ds$$

There exist $M \geq 1$ and $\omega \in \mathbb{R}$ such that $\|S(t)\| \leq M e^{\omega t}$, $\forall t \geq 0$. Then applying Gronwall's lemma to the above equality yields

$$\left\| \frac{y_{u+\varepsilon h}(t) - y_u(t)}{\varepsilon} - z_h(t) \right\| \leq \widehat{M} \left\| \frac{H_\varepsilon(t, 0) Y_0(t) - Y_0(t)}{\varepsilon} - \int_0^t h(\tau) d\tau \, B Y_0(t) \right\|$$

$$+ \widehat{M} \left\| \int_0^t \left[\frac{H_\varepsilon(t, s) Y(s) - Y(s)}{\varepsilon} - \int_s^t h(\tau) d\tau \, B Y(s) \right] ds \right\|$$

$$+ \widehat{M} \left\| \int_0^t \left[\frac{H_\varepsilon(t, s) Z(s) - Z(s)}{\varepsilon} - \int_s^t h(\tau) d\tau \, B Z(s) \right] ds \right\|$$

$$+ \widehat{M} \left\| \int_0^t [H_\varepsilon(t, s) a z_h(s) - a z_h(s)] ds \right\|$$

where $\widehat{M} = M e^{|\omega a| T (\|u\| + 1)}$.

By passing to the limit, we have

$$\lim_{\varepsilon \to 0} \| H_\varepsilon(t, s) a z_h(s) - a z_h(s) \| = 0 \quad \text{a.e. on } [0, t]$$

Since $Y_0(t)$, $Y(s)$ and $Z(s)$ belong to $\mathscr{D}(B)$, then, by using the semigroup property $\lim\limits_{\tau \to 0^+} \dfrac{S(\tau) z - z}{\tau} = Bz$, $\forall z \in \mathscr{D}(B)$, we obtain

$$\lim_{\varepsilon \to 0^+} \left\| \frac{H_\varepsilon(t, 0) Y_0(t) - Y_0(t)}{\varepsilon} - \int_0^t h(\tau) d\tau \, B Y_0(t) \right\| = 0$$

$$\lim_{\varepsilon \to 0^+} \left\| \frac{H_\varepsilon(t, s) Y(s) - Y(s)}{\varepsilon} - \int_s^t h(\tau) d\tau \, B Y(s) \right\| = 0 \quad \text{a.e. on } [0, t]$$

$$\lim_{\varepsilon \to 0^+} \left\| \frac{H_\varepsilon(t, s) Z(s) - Z(s)}{\varepsilon} - \int_s^t h(\tau) d\tau \, B Z(s) \right\| = 0 \quad \text{a.e. on } [0, t]$$

By applying the dominated convergence theorem, the above equalities yield

$$\lim_{\varepsilon \to 0^+} \left\| \int_0^t [H_\varepsilon(t, s) a z_h(s) - a z_h(s)] ds \right\| = 0$$

$$\lim_{\varepsilon \to 0^+} \left\| \int_0^t \left[\frac{H_\varepsilon(t, s) Y(s) - Y(s)}{\varepsilon} - \int_s^t h(\tau) d\tau \, B Y(s) \right] ds \right\| = 0$$

$$\lim_{\varepsilon \to 0^+} \left\| \int_0^t \left[\frac{H_\varepsilon(t, s) Z(s) - Z(s)}{\varepsilon} - \int_s^t h(\tau) d\tau \, B Z(s) \right] ds \right\| = 0$$

Therefore

$$\lim_{\varepsilon \to 0^+} \left\| \frac{y_{u+\varepsilon h}(t) - y_u(t)}{\varepsilon} - z_h(t) \right\| = 0 \qquad \square$$

Corollary 1 *Let $u \in L^\infty(0, T)$ such that $u(t) \geq 0$ a.e.*
For any $h \in L^\infty(0, T)$ satisfying $h(t) \leq 0$ and $h(t) + u(t) \geq 0$ a.e., we have

$$\lim_{\varepsilon \to 0^+} \frac{y_{u+\varepsilon h}(t) - y_u(t)}{\varepsilon} = z_h(t) \tag{13}$$

where z_h is the mild solution of (10), and satisfies expression (11).

Proof Let $h \in L^\infty(0, T)$ such that $h(t) \leq 0$ and $u(t) + h(t) \geq 0$ a.e. Let $0 < \varepsilon \leq 1$, and denote $v = u + \varepsilon h$. Then

$$\frac{y_{u+\varepsilon h}(t) - y_u(t)}{\varepsilon} = -\frac{y_{v+\varepsilon(-h)}(t) - y_v(t)}{\varepsilon}$$

By Proposition 1, we have

$$\frac{y_{v+\varepsilon(-h)}(t) - y_v(t)}{\varepsilon} = z_{(-h)}^\varepsilon(t) + \theta(\varepsilon)$$

where $\lim_{\varepsilon \to 0} \theta(\varepsilon) = 0$, and $z_{(-h)}^\varepsilon$ is the solution of Equation (10), with v and $-h$ instead of u and h, respectively. A standard calculus leads to

$$\lim_{\varepsilon \to 0^+} z_{(-h)}^\varepsilon(t) = -z_h(t)$$

Hence

$$\lim_{\varepsilon \to 0^+} \frac{y_{v+\varepsilon(-h)}(t) - y_v(t)}{\varepsilon} = -z_h(t)$$

which yields the limit (13). $\qquad\qquad\qquad\qquad\qquad\qquad\qquad\qquad\qquad\quad$ □

Proposition 2 *Let $u \in L^\infty(0, T)$ such that $u(t) \geq 0$ a.e.*
For any $h \in L^\infty(0, T)$ of constant sign, satisfying $u(t) + h(t) \geq 0$ a.e., we have

$$\lim_{\varepsilon \to 0^+} \frac{J(u + \varepsilon h) - J(u)}{\varepsilon} = \int_0^T J_u'(t) h(t) dt$$

where

$$J_u'(t) = \langle \varphi(t), By(t) \rangle_H + ru(t) \tag{14}$$

such that y is the mild solution of system (4), associated to u, and φ is the mild solution of the following adjoint equation

$$\begin{cases} \dot{\varphi}(t) = -u(t)B^*\varphi(t) - a\varphi(t) \\ \varphi(T) = \gamma[y(T) - y_d] \end{cases} \tag{15}$$

Here B^* denotes the adjoint operator of B.

Proof Let $h \in L^{\infty}(0, T)$ have a constant sign, and such that $u(t) + h(t) \geq 0$ a.e. Using the limits (9) and (13), we have

$$\lim_{\varepsilon \to 0^+} \frac{J(u + \varepsilon h) - J(u)}{\varepsilon} = \langle \gamma[y_u(T) - y_d], z_h(T) \rangle + \int_0^T ru(t)h(t)dt$$

For $n \in \rho(B)$, the resolvent set of B, let $B_n = nB(nI - B)^{-1}$ be the Yosida approximation of B. Let z_n and φ_n be the respective mild solutions of the following equations

$$\begin{cases} \dot{z}_n(t) = u(t)B_n z_n(t) + az_n(t) + h(t)By_u(t) \\ z_h(0) = 0 \end{cases}$$

$$\begin{cases} \dot{\varphi}_n(t) = -u(t)B_n^* \varphi_n(t) - a\varphi_n(t) \\ \varphi_n(T) = \gamma(y_u(T) - y_d) \end{cases}$$

Since B_n is bounded, and $By_u(t) \in H$ a.e. on $[0, T]$, then $\dot{z}_n, \dot{\varphi}_n \in L^2(0, T; H)$. Therefore

$$\begin{aligned} \int_0^T \langle \dot{\varphi}_n(t), z_n(t) \rangle dt &= -\int_0^T \langle u(t)B_n^* \varphi_n(t) + a\varphi_n(t), z_n(t) \rangle dt \\ &= -\int_0^T \langle \varphi_n(t), u(t)B_n z_n(t) + az_n(t) \rangle dt \\ &= -\int_0^T \langle \varphi_n(t), \dot{z}_n(t) - h(t)By_u(t) \rangle dt \\ &= -\int_0^T \langle \varphi_n(t), \dot{z}_n(t) \rangle dt + \int_0^T \langle \varphi_n(t), By_u(t) \rangle h(t)dt \end{aligned}$$

Since $\dot{z}_n, \dot{\varphi}_n \in L^2(0, T; H)$, then

$$\begin{aligned} \int_0^T \langle \dot{\varphi}_n(t), z_n(t) \rangle dt + \int_0^T \langle \varphi_n(t), \dot{z}_n(t) \rangle dt &= \langle \varphi_n(T), z_n(T) \rangle - \langle \varphi_n(0), z_n(0) \rangle \\ &= \langle \gamma[y_u(T) - y_d], z_n(T) \rangle \end{aligned}$$

It follows that

$$\langle \gamma(y_u(T) - y_d), z_n(T) \rangle = \int_0^T \langle \varphi_n(t), By_u(t) \rangle h(t)dt$$

By virtue of Proposition 5.4, Chap. 2 of [12], we have

$$\lim_{n \to +\infty} \|z_n - z_h\|_{C([0,T];H)} = \lim_{n \to +\infty} \|\varphi_n - \varphi\|_{C([0,T];H)} = 0$$

where z_h and φ are the respective mild solutions of (10) and (15). Then the above equality yields

$$\langle \gamma[y_u(T) - y_d], z_h(T)\rangle = \int_0^T \langle \varphi(t), By_u(t)\rangle h(t)dt$$

It follows that $\lim\limits_{\varepsilon \to 0^+} \dfrac{J(u + \varepsilon h) - J(u)}{\varepsilon} = \displaystyle\int_0^T [\langle \varphi(t), By_u(t)\rangle + ru(t)]h(t)dt.$ □

The next proposition characterizes the optimal control, solution of problem (7).

Proposition 3 *Let u^* be an optimal control, then u^* is given by:*

$$u^*(t) = \max\left(u_{min}; \ \min\left(u_{max}; \ -\frac{1}{r}\langle \varphi(t), By(t)\rangle\right)\right) \tag{16}$$

where y and φ are respectively the mild solutions of (4) and (15), associated to u^*.

Proof Let u^* be an optimal control, and let $h \in L^\infty(0, T)$ have a constant sign, and such that $u^* + h \in U_{ad}$, where U_{ad} is given by (6). The convexity of U_{ad} yields $u^* + \varepsilon h \in U_{ad}, \forall 0 \leq \varepsilon \leq 1$. Hence

$$J(u^* + \varepsilon h) \geq J(u^*), \quad \forall 0 \leq \varepsilon \leq 1$$

Using Proposition 2, we obtain

$$\int_0^T J'_{u^*}(t)h(t)dt = \lim_{\varepsilon \to 0^+} \frac{J(u^* + \varepsilon h) - J(u^*)}{\varepsilon} \geq 0 \tag{17}$$

- If $u_{min} < u^*(t) < u_{max}$ over a nonempty open set $I \subset [0, T]$:
  Let $h \in \mathscr{C}_c^\infty(I)$ (the space of smooth functions, compactly supported in I) of constant sign. If $\|h\|_{L^\infty(I)}$ is sufficiently small, then $u^* + h, u^* - h \in U_{ad}$. It follows from inequality (17), applied to h and $-h$, that

$$\int_I J'_{u^*}(t)h(t)dt = 0, \quad \forall h \in \mathscr{C}_c^\infty(I) \text{ of constant sign}$$

  which yields $J'_{u^*}(t) = 0$ a.e. on I. Then by expression (14) we have

$$u^*(t) = -\frac{1}{r}\langle \varphi(t), By(t)\rangle \text{ a.e. on } I$$

- If $u^*(t) = u_{min}$ over a nonempty open set $I \subset [0, T]$:
  Let $h \in \mathscr{C}_c^\infty(I)$ such that $h(t) \geq 0, \forall t \in I$. Then for $\|h\|_{L^\infty(I)}$ sufficiently small we have $u^* + h \in U_{ad}$. Hence inequality (17) yields $\displaystyle\int_I J'_{u^*}(t)h(t)dt \geq 0$.

It follows that $J'_{u^*}(t) \geq 0$ a.e. on I. Thus:

$$-\frac{1}{r}\langle \varphi(t), By(t) \rangle \leq u^*(t) = u_{min} \text{ a.e. on } I$$

- Similarly to the above case, if $u^*(t) = u_{max}$ over a nonempty open set $I \subset [0, T]$, then

$$-\frac{1}{r}\langle \varphi(t), By(t) \rangle \geq u_{max} \text{ a.e. on } I$$

Therefore, $u^*(t) = \max\left(u_{min}; \min\left(u_{max}; -\frac{1}{r}\langle \varphi(t), By(t) \rangle\right)\right)$ a.e. on $[0, T]$. $\square$

Finally, we propose the below algorithm, in order to implement numerically the optimal control, given by (16).

Algorithm

- Step 1: Choose an initial control $u_0 \in U_{ad}$, a precision $\varepsilon > 0$, and a step length λ. Initialize with $k = 0$.
- Step 2: Compute y_k, solution of (4), and φ_k, solution of (15), relatively to u_k.
- Step 3: Compute

$$J'_{u_k}(t) = \langle \varphi_k(t), By_k(t) \rangle + ru_k(t)$$

$$u_{k+1}(t) = \max\left(u_{min}; \min(u_{max}; u_k(t) - \lambda J'_{u_k}(t))\right)$$

- Step 4: If $\|u_{k+1} - u_k\| > \varepsilon$, $k = k + 1$, go to step 2. Otherwise $u^* = u_k$.

4 Simulations

We consider the HCE simplified model (2), with the following data:

$$L = 620 \text{m}, \ a_f = 0.002, \ T_0 = 150\,^\circ\text{C}$$

$T_m(x)$ is depicted in Fig. 3, with $T_{max} = 460\,^\circ\text{C}$.

Our purpose is to find the optimal fluid velocity u^* that steers the fluid temperature T_f as close as possible to $T_d = 400\,^\circ\text{C}$, at time $T = 900$ s. To this end, we consider system (3) and the optimal control problem (7), with the following cost functional:

$$J(u) = \frac{1}{2}\int_0^L [T_f(x, T) - T_d]^2 dx + \frac{10}{2}\int_0^T u(s)^2 ds$$

The set of admissible controls is

$$U_{ad} = \{u \in L^\infty(0, T) \mid 0 \leq u(t) \leq 0.5 \text{ a.e.}\}$$

Fig. 4 Optimal fluid velocity (m s^{-1})

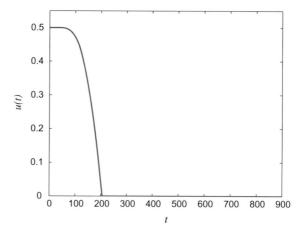

Fig. 5 Fluid temperature (°C) at time $T = 900$ s

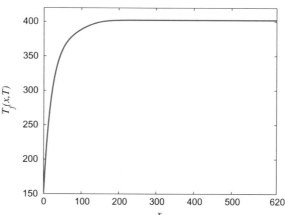

Using Proposition 3, the optimal control is given by

$$u^*(t) = \max\left(0;\ \min\left(0.5;\ 0.1\int_0^L \varphi(x,t)\frac{\partial T_f}{\partial x}(x,t)dx\right)\right)$$

where φ is the solution of the adjoint Eq. (15), with $\gamma = 1$ and $B^* = \dfrac{\partial}{\partial x}$, such that $\mathscr{D}(B^*) = \{z \in H^1(0, L) : z(L) = 0\}$. Hence φ is solution of:

$$\begin{cases} \dfrac{\partial \varphi}{\partial t}(x,t) = -u^*(t)\dfrac{\partial \varphi}{\partial x}(x,t) - a_f\varphi(x,t) \\ \varphi(L,t) = 0 \\ \varphi(x,T) = T_f(x,T) - T_d \end{cases}$$

Simulations are carried out using the above algorithm, with $u_0 = 0.5$, $\lambda = 10^{-6}$, and $\varepsilon = 10^{-5}$.

Figure 4 depicts the optimal fluid velocity (m s^{-1}), while the corresponding fluid temperature (°C) at the final time $T = 900$ s is plotted in Fig. 5. At $T = 900$ s, the average fluid temperature is about $\widehat{T_f} \simeq 389.57\,°C$.

Figure 4 shows that the fluid flows with a maximal velocity during the first 100 seconds. Then the velocity drops down until vanishing, so that the fluid absorbs the necessary amount of solar radiation to reach the desired temperature.

As shown by the above simulations, the optimal fluid velocity, designed via our approach, drives the fluid temperature close to the desired level. This is crucial to the performance of the solar plant.

References

1. Addou, A., Benbrik, A.: Existence and uniqueness of optimal control for a distributed parameter bilinear system. J. Dyn. Control Syst. **8**(2), 141–152 (2002)
2. Bradley, M.E., Lenhart, S., Yong, J.: Bilinear optimal control of the velocity term in a Kirchhoff plate equation. J. Math. Anal. Appl. **238**(2), 451–467 (1999)
3. Camacho, E.F., Berenguel, M., Rubio, F.R.: Advanced Control of Solar Plants. Springer, London (1997)
4. Carotenuto, L., La Cava, M., Raiconi, G.: Regular design for the bilinear distributed parameter of a solar power plant. Int. J. Syst. Sci. **16**(7), 885–900 (1985)
5. Clérin, J.-M.: Problèmes de contrôle optimal du type bilinéaire gouvernés par des équations aux dérivées partielles d'évolution, Ph.D. thesis, University of Avignon (2009). http://tel.archives-ouvertes.fr
6. El Boukhari, N., Zerrik, E.: Optimal control of a parabolic solar collector. ARIMA J. **30**, 16–30 (2019)
7. Engel, J.C., Nagel, R.: One-Parameter Semigroups for Linear Evolution Equations, Graduate Texts in Mathematics. Springer, New York (2000)
8. Fernández-García, A., Zarza, E., Valenzuela, L., Pérez, M.: Parabolic-trough solar collectors and their applications. Renew. Sustain. Energy Rev. **14**, 1695–1721 (2010)
9. Forristall, R.: Heat transfer analysis and modeling of a parabolic trough solar receiver implemented in engineering Equation Solver. Technical report, National Renewable Energy Laboratory (NREL), Colorado (2003)
10. Johansen, T.A., Storaa, C.: Energy-based control of a distributed solar collector field. Automatica **38**(7), 1191–1199 (2002)
11. Klein, A.A., Duffie, J.A., Beckman, W.A.: Transient considerations of flat-plate solar collectors. J. Eng. Power **96**(2), 109–113 (1974)
12. Li, X., Yong, J.: Optimal Control Theory for Infinite Dimensional Systems. Birkhäuser, Boston (1995)
13. Liang, M.: Bilinear optimal control for a wave equation. Math. Mod. Meth. Appl. Sci. **9**(1), 45–68 (1999)
14. Orbach, A., Rorres, C., Fischl, R.: Optimal control of a solar collector loop using a distributed-lumped model. Automatica **27**(3), 535–539 (1981)
15. Patnode, A.M.: Simulation and Performance Evaluation of Parabolic Trough Solar Power Plants, MSc Thesis, University of Wisconsin-Madison (2006)
16. Rorres, C., Orbach, A., Fischl, R.: Optimal and suboptimal control policies for a solar collector system. IEEE Trans. Autom. Control **25**(6), 1085–1091 (1980)

17. Silva, R.N., Lemos, J.M., Rato, L.M.: Variable sampling adaptive control of a distributed collector solar field. IEEE Trans. Control Syst. Technol. **11**(5), 765–772 (2003)
18. Stuetzle, T.A.: Automatic Control of the 30MWe SEGS VI Parabolic Trough Plant, MSc Thesis, University of Wisconsin-Madison (2002)
19. Zerrik, E., El Boukhari, N.: Constrained optimal control for a class of semilinear infinite dimensional systems. J. Dyn. Control Syst. **24**(1), 65–81 (2018)

Strong and Exponential Stabilization for a Class of Second Order Semilinear Systems

El Hassan Zerrik and Lahcen Ezzaki

Abstract In this paper, we deal with stabilization for a class of second order systems. Sufficient conditions for exponential and strong stabilization are given. The obtained results are illustrated by many examples and simulations.

1 Introduction

Semilinear second order systems are special kinds of nonlinear systems capable of representing a variety of important physical processes. The stability of such systems has been considered in many works : Haraux [4], studied the exponential stability of a linear second order system that verify observability inequality. Bardos et al. [1] show that, when the system evolution domain Ω and the damping term are of class C^∞, the exponential stability is proved if and only if the "geometric control condition" is satisfied. In [2], the authors proved the exponential decay of solution of one dimensional wave equation under broad hypotheses on the damping term. In [7], the authors studied the exponential decay of energy of wave equation with potential and indefinite damping.

The strong stabilization of a bilinear second order system has been studied by Couchouron [3] with a diagonal control operator.

Also, Tebou [12] extended the result given in [4] concerning linear systems to semilinear case, so he established an equivalence between the stabilization of a semilinear wave equation and the observability of the corresponding undamped semilinear wave equation. In [10], the authors considered the wave equation with a nonlinear internal damping, and proved that the energy of solutions decays exponentially to zero. Also in [5], the author studied the weak stabilization of the wave equation perturbed by a non monotone damping term.

E. H. Zerrik · L. Ezzaki (✉)
Faculty of Sciences, Moulay Ismail University, Meknes, Morocco
e-mail: ezzaki.lahcen@yahoo.fr

E. H. Zerrik
e-mail: zerrik3@yahoo.fr

© Springer Nature Switzerland AG 2020
E. H. Zerrik et al. (eds.), *Recent Advances in Modeling, Analysis and Systems Control: Theoretical Aspects and Applications*, Studies in Systems, Decision and Control 243,
https://doi.org/10.1007/978-3-030-26149-8_17

In this work we study strong and exponential stabilization of the following system

$$\begin{cases} y_{tt}(t) = Ay(t) + u(t)By_t(t), \\ y(0) = y_0, \ \ y_t(0) = y_1, \end{cases} \tag{1}$$

where A is an unbounded dissipative operator on the Hilbert space H endowed with the inner product $\langle ., . \rangle$ and the corresponding norm $\| \ \|_H$, $B : H \to H$ is a possibly nonlinear and locally Lipschitz operator, u denotes the control function, and $X = V \times H$ is the state space with $V = D((-A)^{\frac{1}{2}})$ endowed with the norm $\|v\|_V$.

In [6], Ghisi et al. studied the exponential stabilization of a special case of system (1) where the identity is the control operator, the used method requires that the control operator must commute with A. In this paper, we study the exponential and strong stabilization of bilinear and semilinear second order systems, where the control operator does not necessarily commute with A.

This paper is organized as follows : in the second section, we give sufficient conditions for exponential and strong stabilization of bilinear systems. Third section is devoted to exponential and strong stabilization of semilinear systems. Finally in last section the obtained results are successfully illustrated by numerical simulations.

2   Stabilization of Bilinear Systems

In this section we give controls that ensure strong and exponential stabilization of system (1) when B is a linear operator.

The system (1) can be written in the form

$$\begin{cases} \dfrac{d}{dt} \tilde{y}(t) = \tilde{A}\tilde{y}(t) + \tilde{B}\tilde{y}(t) \\ \tilde{y}(0) = \tilde{y}_0 \end{cases} \tag{2}$$

where $\tilde{y}(t) = (y, y_t)$ and $\tilde{A}$ given by

$$\tilde{A}\tilde{y}(t) = \begin{pmatrix} 0 & I \\ A & 0 \end{pmatrix} \tilde{y}(t)$$

and the operator $\tilde{B}$ is given by

$$\tilde{B}\tilde{y}(t) = \begin{pmatrix} 0 \\ u(t)By_t(t) \end{pmatrix}.$$

the system (2) has a unique weak solution in $C([0, T], X)$ (see [11]) and given by the variation of constants formula

$$\tilde{y}(t) = S(t)\tilde{y}_0 + \int_0^t S(t-s)\tilde{B}\tilde{y}(s)ds. \tag{3}$$

where $S(t)$ is a semigroup generated by the operator $\tilde{A}$.

We consider the following uncontrolled system

$$\begin{cases} z_{tt}(t) = Az(t), \\ z(0) = z_0, z_t(0) = z_1, \end{cases} \tag{4}$$

2.1 Strong Stabilization

In this section we give sufficient conditions for strong stabilization of system (1).

Theorem 1 *Suppose that the solution z of system (4) satisfies*

$$\|z_0\|_V^2 + \|z_1\|_H^2 \le \alpha \int_0^T |\langle Bz_t(t), z_t(t)\rangle|dt, \ (for \ some \ T, \alpha > 0) \tag{5}$$

then the control

$$u(t) = -\langle By_t(t), y_t(t)\rangle \tag{6}$$

strongly stabilizes system (1).

Proof From the relation

$$\langle Bz_t, z_t\rangle = \langle B(z_t - y_t), z_t\rangle - \langle By_t, y_t - z_t\rangle + \langle By_t, y_t\rangle,$$

and as B is bounded, we have

$$|\langle Bz_t, z_t\rangle| \le \|B\|_H \|z_t - y_t\|_H \|z_t\|_H + \|B\|_H \|y_t - z_t\|_H \|y_t\|_H + |\langle By_t, y_t\rangle| \tag{7}$$

Differentiating the energy $E_y(t) = \dfrac{1}{2}\{\|y_t\|_H^2 + \|y\|_V^2\}$ and since A is dissipative, we get

$$\frac{d}{dt}E_y(t) \le -\langle By_t, y_t\rangle^2$$

Then

$$E_y(t) \le -\int_0^t \langle By_s(s), y_s(s)\rangle^2 ds + E_y(0)$$

Which implies that

$$E_y(t) \le E_y(0) \tag{8}$$

Moreover, we have

$$\|y_t\|_H^2 \le \|(y, y_t)\|_X^2 \le 2E_y(t) \quad and \quad \|z_t - y_t\|_H^2 \le \|(\psi - \phi)\|_X^2 \qquad (9)$$

where $\psi = (z, z_t)$ and $\phi = (y, y_t)$.

Using (7), (8) and (9), we obtain

$$|\langle Bz_t, z_t\rangle| \le 2\|B\|_H\|\psi - \phi\|_X\sqrt{E_y(0)} + |\langle By_t, y_t\rangle| \qquad (10)$$

From (3), we have

$$\|\psi - \phi\|_X \le \sqrt{E_y(0)}\Big(\int_0^t |\langle By_s(s), y_s(s)\rangle|^2 ds\Big)$$

It follows from Schwartz's inequality, that

$$\|\psi - \phi\|_H \le \sqrt{TE_y(0)}\Big(\int_0^T |\langle By_s(s), y_s(s)\rangle|^2 ds\Big)^{\frac{1}{2}} \qquad (11)$$

Integrating (10) over the interval $[0, T]$ and with (11), we obtain

$$\int_0^T |\langle Bz_t, z_t\rangle| dt \le (2E_y(0) + 1)\|B\|_H\sqrt{T}\Big(\int_0^T |\langle By_s(s), y_s(s)\rangle|^2 ds\Big)^{\frac{1}{2}}$$

Condition (5) gives

$$\|z_0\|_V^2 + \|z_1\|_H^2 \le \alpha(2E_y(0) + 1)\|B\|_H\sqrt{T}\Big(\int_0^T |\langle By_s(s), y_s(s)\rangle|^2 ds\Big)^{\frac{1}{2}}$$

Taking $(z_0, z_1) = (y_0, y_1)$, we have

$$\|y_0\|_V^2 + \|y_1\|_H^2 \le \alpha(2E_y(0) + 1)\|B\|_H\sqrt{T}\Big(\int_0^T |\langle By_s(s), y_s(s)\rangle|^2 ds\Big)^{\frac{1}{2}}$$

Now replacing (y_0, y_1) by (y, y_t), we get

$$\|y(t)\|_V^2 + \|y_t(t)\|_H^2 \le \beta\Big(\int_t^{t+T} |\langle By_s(s), y_s(s)\rangle|^2 ds\Big)^{\frac{1}{2}}$$

Then $\|(y, y_t)\|_X \longrightarrow 0$, as $t \longrightarrow +\infty$, which completes the proof.

Example 1 We consider on $\Omega =]0, 1[$, the following system

$$\begin{cases} y_{tt}(x, t) + y_{xx}(x, t) = u(t)y_t(x, t), & \Omega \times]0, +\infty[\\ y(x, 0) = y_0(x), \ y_t(x, 0) = y_1(x), & \Omega \\ y(0, t) = y(1, t) = 0, &]0, +\infty[\end{cases} \qquad (12)$$

where $u(t) \in L^\infty(0, +\infty)$ is a real valued control.

System (12) has a unique weak solution on the state space $X = H_0^1(\Omega) \times L^2(\Omega)$ [8].

The solution of the uncontrolled system

$$\begin{cases} z_{tt}(x,t) + z_{xx}(x,t) = 0, & \Omega \times]0, +\infty[\\ z(x,0) = z_0(x), \, z_t(x,0) = z_1(x), & \Omega \\ z(0,t) = z(1,t) = 0, &]0, +\infty[\end{cases}$$

is given by

$$z(x,t) = \sum_{k=1}^{\infty} (a_k \cos(k\pi t) + b_k \sin(k\pi t)) \sin(k\pi x).$$

The assumption (5) is verified, indeed

$$\begin{aligned} \int_0^2 \langle Bz_t(t), z_t(t) \rangle dt &= \int_0^2 \int_0^1 |z_t(x,t)|^2 \, dx dt \\ &= \sum_{k=1}^{\infty} (k\pi)^2 (a_k^2 + b_k^2) \\ &= \|z_t(0)\|_{L^2(\Omega)}^2 + \|z_x(0)\|_{L^2(\Omega)}^2 \end{aligned}$$

It follows that the control

$$u(t) = -\int_0^1 |y_t(x,t)|^2 dx \tag{13}$$

strongly stabilizes system (12).

2.2 Exponential Stabilization

The following result provides sufficient conditions for exponential stabilization of system (1).

Theorem 2 *Suppose that the solution z of system (4) satisfies the condition (5), then the control*

$$u(t) = \begin{cases} -\dfrac{\langle By_t(t), y_t(t) \rangle}{\|(y, y_t)\|_X^2}, & (y, y_t) \neq (0, 0) \\ 0, & (y, y_t) = (0, 0) \end{cases} \tag{14}$$

stabilizes exponentially system (1), in other word there exist $M > 0$ and $\lambda > 0$ such that for every $(y_0, y_1) \in X$

$$\|y(t)\|_V^2 + \|y_t(t)\|_H^2 \leq Me^{-\lambda t}(\|y_0\|_V^2 + \|y_1\|_H^2).$$

Proof Let us consider the solution z of system (4) with $z(0) = y_0$, $z_t(0) = y_1$ and the natural energy associated with (1)

$$E_y(t) = \frac{1}{2}\{\|y(t)\|_V^2 + \|y_t(t)\|_H^2\}$$

Since A is dissipative, we have

$$\frac{d}{dt}E_y(t) \leq -\frac{\langle By_t, y_t \rangle^2}{\|y(t)\|_V^2 + \|y_t(t)\|_H^2} \tag{15}$$

Integrating this inequality, we get

$$E_y(t) \leq -\int_0^t \frac{\langle By_s(s), y_s(s) \rangle^2}{\|y(s)\|_V^2 + \|y_s(s)\|_H^2}ds + E_y(0)$$

Then

$$E_y(t) \leq E_y(0) \tag{16}$$

Furthermore, we have

$$\|y_t\|_H^2 \leq \|(y, y_t)\|_X^2 \leq 2E_y(t) \quad and \quad \|z_t - y_t\|_H^2 \leq \|(\psi - \phi)\|_X^2 \tag{17}$$

where $\psi = (z, z_t)$ and $\phi = (y, y_t)$.

Using (7) and combining (16) and (17), we obtain

$$|\langle Bz_t, z_t \rangle| \leq 2\|B\|_H\|\psi - \phi\|_X\sqrt{2E_y(0)} + |\langle By_t, y_t \rangle| \tag{18}$$

Furthermore, from (3), we have

$$\|\psi - \phi\|_X \leq \|B\|_H\left\{\int_0^t \frac{|\langle By_s(s), y_s(s) \rangle|}{2E_y(s)}\sqrt{2E_s(s)}ds\right\}$$

Schwartz's inequality, gives

$$\|\psi - \phi\|_X \leq \|B\|_H\sqrt{T}\left\{\int_0^T \frac{|\langle By_s(s), y_s(s) \rangle|^2}{2E_y(s)}ds\right\}^{\frac{1}{2}} \tag{19}$$

Using (16), we have

$$|\langle By_t, y_t \rangle| \leq \frac{|\langle By_t, y_t \rangle|}{2E_y(t)}\sqrt{2E_y(t)}\sqrt{2E_y(0)} \tag{20}$$

Integrating (18) over the interval $[0, T]$ and taking into account (19) and (20), we obtain

$$\int_0^T |\langle Bz_t, z_t \rangle| dt \le (2\|B\|_H + 1)\|B\|_H \sqrt{T2E_y(0)} \left\{ \int_0^T \frac{|\langle By_s(s), y_s(s) \rangle|^2}{2E_y(s)} ds \right\}^{\frac{1}{2}}$$

It follows from (5) that

$$\|z_0\|_V^2 + \|z_1\|_H^2 \le \alpha(2\|B\|_H + 1)\|B\|_H \sqrt{2TE_y(0)} \left\{ \int_0^T \frac{|\langle By_s(s), y_s(s) \rangle|^2}{2E_y(s)} ds \right\}^{\frac{1}{2}}$$

Taking $(z_0, z_1) = (y_0, y_1)$, we get

$$\|y_0\|_V^2 + \|y_1\|_H^2 \le (2\|B\|_H + 1)\|B\|_H \sqrt{2TE_y(0)} \left\{ \int_0^T \frac{|\langle By_s(s), y_s(s) \rangle|^2}{2E_y(s)} ds \right\}^{\frac{1}{2}}$$

Now replacing (y_0, y_1) by (y, y_t), we get

$$\|y(t)\|_V^2 + \|y(t)\|_H^2 \le (2\|B\|_H + 1)\|B\|_H \sqrt{2TE_y(t)} \left\{ \int_t^{t+T} \frac{|\langle By_s(s), y_s(s) \rangle|^2}{2E_y(s)} ds \right\}^{\frac{1}{2}}$$

Then

$$E_y(t) \le \gamma \int_t^{t+T} \frac{|\langle By_s(s), y_s(s) \rangle|^2}{2E_y(s)} ds \qquad (21)$$

where $\gamma = (2\|B\|_H + 1)\|B\|_H \sqrt{2T}$.

Integrating inequality (15) from nT to $(n+1)T$, and since $E_y(t)$ decreases, we obtain

$$E_y(nT) - E_y((n+1)T) \ge \int_{nT}^{(n+1)T} \frac{|\langle By_s(s), y_s(s) \rangle|^2}{2E_y(s)} ds$$

From (21), we get

$$E_y((n+1)T) \le r E_y(nT)$$

where $r = \frac{1}{\gamma}$.

By recurrence, we show that

$$E_y((n+1)T) \le r^n E_y(0)$$

Since $E_y(t)$ decreases and taking n the integer part of $\frac{t}{T}$, we deduce that

$$E_y(t) \le M e^{-\lambda t} E_y(0)$$

where $M = (2\|B\|_H + 1)\|B\|_H \sqrt{T}$ and $\lambda = \dfrac{\ln((2\|B\|_H + 1)\|B\|_H \sqrt{T})}{T}$.

In other word system (1) is exponentially stabilizable by the control (14).

Example 2 On $\Omega =]0, 1[$, we consider the following system

$$\begin{cases} y_{tt}(x,t) + y_{xxxx}(x,t) = u(t)y_t(x,t), & \Omega \times]0, +\infty[\\ y(x,0) = y_0(x), \ y_t(x,0) = y_1(x), & \Omega \\ y(0,t) = y(1,t) = y_{xx}(0,t) = y_{xx}(1,t) = 0, &]0, +\infty[\end{cases} \quad (22)$$

where $u(t) \in L^\infty(0, +\infty)$ is a real valued control.

System (22) has a unique weak solution on the state space $X = (H^2(\Omega) \cap H_0^1(\Omega)) \times L^2(\Omega)$ [8].

Moreover, the solution of the following system

$$\begin{cases} z_{tt}(x,t) + z_{xxxx}(x,t) = 0, & \Omega \times]0, +\infty[\\ z(x,0) = z_0(x), z_t(x,0) = z_1(x), & \Omega \\ z(0,t) = z(1,t) = z_{xx}(0,t) = z_{xx}(1,t) = 0, &]0, +\infty[\end{cases}$$

takes the form

$$z(x,t) = \sum_{m=1}^{\infty} (a_m \cos((m\pi)^2 t) + b_m \sin((m\pi)^2 t)) \sin(m\pi x).$$

The condition (5) is verified, indeed

$$\int_0^2 \langle Bz_t(t), z_t(t) \rangle dt = \int_0^2 \int_0^1 |z_t(x,t)|^2 dx dt$$
$$= \sum_{m=1}^{\infty} (m\pi)^4 (a_m^2 + b_m^2)$$
$$= \|z_t(0)\|_{L^2(\Omega)}^2 + \|z_{xx}(0)\|_{L^2(\Omega)}^2$$

We conclude that the control

$$u(t) = -\frac{\int_0^1 |y_t(x,t)|^2 dx}{\|y_t\|_{L^2(\Omega)}^2 + \|y_{xx}\|_{L^2(\Omega)}^2} \quad (23)$$

ensures the exponential stabilization of system (22).

3 Stabilization of Semilinear Systems

In this section we study the strong and exponential stabilization of the semilinear system (1) when B is nonlinear.

In other hand, we consider the system

$$\begin{cases} z_{tt}(x,t) = Az(t), \\ z(0) = z_0, z_t(0) = z_1, \end{cases} \quad (24)$$

The following result gives sufficient conditions for strong stabilization of system (1).

Theorem 3 *If B is locally Lipschitz and there exist $T > 0$ and $\delta > 0$ such that the solution (z, z_t) of system (24) verifies the inequality*

$$\int_0^T |\langle Bz_t(t), z_t(t) \rangle| dt \geq \delta \{ \|z_t(0)\|_H^2 + \|z(0)\|_V^2 \}, \tag{25}$$

then the control

$$u(t) = -\langle By_t(t), y_t(t) \rangle \tag{26}$$

strongly stabilizes system (1).

Proof We show first that $h : y_t \mapsto \langle By_t, y_t \rangle By_t$ is locally Lipschitz.
Since B is locally Lipschitz, then for all $R > 0$, there exists $L > 0$ such that

$$\|By_t - Bz_t\|_H \leq L\|y_t - z_t\|_H, \quad \forall y_t, z_t \in H : 0 < \|y_t\|_H \leq \|z_t\|_H \leq R. \tag{27}$$

Let us remark that

$$\begin{aligned}
\langle Bz_t, z_t \rangle Bz_t - \langle By_t, y_t \rangle By_t &= \langle Bz_t, z_t \rangle (Bz_t - By_t) + (\langle Bz_t, z_t \rangle - \langle By_t, y_t \rangle) By_t \\
&= (\langle Bz_t, z_t - y_t \rangle - \langle Bz_t - By_t, y_t \rangle) By_t \\
&\quad + \langle Bz_t, z_t \rangle (Bz_t - By_t)
\end{aligned}$$

Using (27), we have

$$\begin{aligned}
\|h(z_t) - h(y_t)\|_H &\leq L^2 \|z_t\|_H^2 \|z_t - y_t\|_H + L^2 \|z_t - y_t\|_H (\|z_t\|_H + \|y_t\|_H) \|y_t\|_H \\
&\leq \mathcal{L} \|z_t - y_t\|_H
\end{aligned}$$

where $\mathcal{L} = 3LR^2$.
It follows that system (1) has a unique weak solution (y, y_t) (see [11]).
Using (7) and (27), we obtain

$$|\langle Bz_t, z_t \rangle| \leq L\|z_t - y_t\|_H \|z_t\|_H + L\|y_t - z_t\|_H \|y_t\|_H + |\langle By_t, y_t \rangle| \tag{28}$$

Now, we differentiate $E_y(t)$

$$\frac{d}{dt} E_y(t) \leq -\langle By_t, y_t \rangle^2$$

Then

$$E_y(t) \leq -\int_0^t \langle By_s(s), y_s(s) \rangle^2 ds + E_y(0)$$

It follows that

$$E_y(t) \leq E_y(0) \tag{29}$$

Moreover, we have

$$\|y_t\|_H^2 \leq \|(y, y_t)\|_X^2 \leq 2E_y(t) \quad and \quad \|z_t - y_t\|_H^2 \leq \|(\psi - \phi)\|_X^2 \tag{30}$$

where $\psi = (z, z_t)$ and $\phi = (y, y_t)$.

From (28) and combining (29) with (30), we get

$$|\langle Bz_t, z_t \rangle| \leq 2L \|\psi - \phi\|_X \sqrt{2E_y(0)} + |\langle By_t, y_t \rangle| \tag{31}$$

From (3), we have

$$\|\psi - \phi\|_X \leq \sqrt{2E_y(0)} \int_0^t |\langle By_t, y_t \rangle|^2 ds$$

For a fixed $T > 0$, Schwarz inequality, yields

$$\|\psi - \phi\|_X \leq \sqrt{2T E_y(0) g(0)} \tag{32}$$

where $g(t) = \int_t^{t+T} |\langle By_s(s), y_s(s) \rangle|^2 ds$.

Integrating (31) over the interval $[0, T]$ and by (32), we obtain

$$\int_0^T |\langle Bz_t, z_t \rangle| dt \leq (2E_y(0) + 1) L \sqrt{T g(0)}$$

Condition (25) allows

$$\|z_1\|_H^2 + \|z_0\|_V^2 \leq \frac{(2E_y(0) + 1) L \sqrt{T}}{\delta} \sqrt{g(0)}$$

Replacing (z_0, z_1) by (y_0, y_1), we get

$$\|y_1\|_H^2 + \|y_0\|_V^2 \leq \frac{(2E_y(0) + 1) L \sqrt{T}}{\delta} \sqrt{g(0)}$$

Now, replacing (y_0, y_1) by (y, y_t), we obtain

$$\|y_t\|_H^2 + \|y\|_V^2 \leq \beta \sqrt{g(t)}$$

where $\beta = \dfrac{(2E_y(0) + 1) L \sqrt{T}}{\delta}$.

It follows that $\|(y, y_t)\|_X \longrightarrow 0$ as $t \longrightarrow +\infty$, we deduce that the control (26) strongly stabilizes system (1).

Example 3 On $\Omega =]0, 1[$, we consider the following system

$$\begin{cases} y_{tt}(x, t) = y_{xx}(x, t) + u(t) B y_t(x, t), & \Omega \times]0, +\infty[\\ y(x, 0) = y_0(x), \ y_t(x, 0) = y_1(x), & \Omega \\ y(0, t) = y(1, t) = 0, &]0, +\infty[\end{cases} \tag{33}$$

where the operator $B y_t = (\max\{\|y_t\|_{L^2(\Omega)}, 1\}) y_t$ and $u(t) \in L^\infty(0, +\infty)$ is a real valued control.

System (33) has a unique solution on state space $X = H_0^1(\Omega) \times L^2(\Omega)$ [8].

The solution of the uncontrolled system

$$\begin{cases} z_{tt}(x, t) = z_{xx}(x, t), & \Omega \times]0, +\infty[\\ z(x, 0) = z_0(x), z_t(x, 0) = z_1(x), & \Omega \\ z(0, t) = z(1, t) = 0, &]0, +\infty[\end{cases}$$

is given by

$$z(t, x) = \sum_{k=1}^{\infty} (a_k \cos(k\pi t) + b_k \sin(k\pi t)) \sin(k\pi x).$$

The assumption (25) is verified, indeed

$$\int_0^2 \langle B z_t(t), z_t(t) \rangle dt = (\max\{\|z_t\|_{L^2(\Omega)}, 1\}) \int_0^2 \int_0^1 |z_t(x, t)|^2 \, dx dt$$

$$\geq \sum_{k=1}^{\infty} (k\pi)^2 (a_k^2 + b_k^2)$$

$$\geq \|z_t(0)\|_{L^2(\Omega)}^2 + \|z_x(0)\|_{L^2(\Omega)}^2$$

We conclude that

$$u(t) = -(\max\{\|y_t\|_{L^2(\Omega)}, 1\}) \int_0^1 |y_t(x, t)|^2 dx \tag{34}$$

strongly stabilizes system (33).

Theorem 4 *Assume that B is locally Lipschitz and there exist $T > 0$ and $C > 0$ such that the solution (z, z_t) of system (24) satisfy the following condition*

$$\int_0^T |\langle B z_t(t), z_t(t) \rangle| dt \geq C\{\|z_t(0)\|_H^2 + \|z(0)\|_V^2\}, \tag{35}$$

then the control

$$u(t) = \begin{cases} -\dfrac{\langle B y_t(t), y_t(t) \rangle}{\|(y, y_t)\|_X^2}, & (y, y_t) \neq (0, 0) \\ 0, & (y, y_t) = (0, 0) \end{cases} \tag{36}$$

exponentially stabilizes system (1).

Proof Let $\phi = (y, y_t)$ et $\psi = (z, z_t)$. We show first that the operator $f : \phi \mapsto \dfrac{\langle By_t, y_t \rangle}{\|(y, y_t)\|_X^2} By_t$ is locally Lipschitz.

Since B is locally Lipschitz, then for all $R > 0$, there exists $K > 0$ such that

$$\|By_t - Bz_t\|_H \leq K\|y_t - z_t\|_H, \quad \forall y_t, z_t \in H : 0 < \|y_t\|_H \leq \|z_t\|_H \leq R. \quad (37)$$

Then

$$f(\psi) - f(\phi) = \frac{\|\phi\|_X^2 (\langle Bz_t, z_t \rangle Bz_t - \langle By_t, y_t \rangle By_t)}{\|\psi\|_X^2 \|\phi\|_X^2} - \frac{(\|\psi\|_X^2 - \|\phi\|_X^2)\langle By_t, y_t \rangle By_t}{\|\psi\|_X^2 \|\phi\|_X^2}.$$

It follows that

$$\|f(\psi) - f(\phi)\|_X \leq \frac{\|\langle Bz_t, z_t \rangle Bz_t - \langle By_t, y_t \rangle By_t\|_H}{\|\psi\|_X^2} + K^2 |\|\psi\|_X^2 - \|\phi\|_X^2| \cdot \frac{\|\phi\|_X}{\|\psi\|_X^2}$$

$$\leq \frac{\|\langle Bz_t, z_t \rangle (Bz_t - By_t) + (\langle Bz_t, z_t - y_t \rangle}{\|\psi\|_X^2}$$

$$- \frac{\langle By_t - Bz_t, y_t) \rangle By_t\|_H}{\|\psi\|_X^2} + 2K^2 \|\psi - \phi\|_X$$

Furthermore, we have

$$\|z_t - y_t\|_H^2 \leq \|(\psi - \phi)\|_X^2 \quad (38)$$

Using (38), we get $\|f(\psi) - f(\phi)\|_X \leq L\|\psi - \phi\|_X$ where $L = 5K^2$.

We conclude that system (1) has a unique weak solution (y, y_t) (see [11]). From (37) and (7), we get

$$|\langle Bz_t, z_t \rangle| \leq K\|z_t - y_t\|_H \|z_t\|_H + K\|y_t - z_t\|_H \|y_t\|_H + |\langle By_t, y_t \rangle| \quad (39)$$

Moreover, we have

$$\|y_t\|_H^2 \leq \|(y, y_t)\|_X^2 \leq 2E_y(t)$$

Since A is dissipative, then

$$\frac{d}{dt} E_y(t) \leq -\frac{\langle By_t, y_t \rangle^2}{\|y_t\|_H^2 + \|y\|_V^2}$$

Thus

$$E_y(t) = -\int_0^t \frac{\langle By_s(s), y_s(s) \rangle^2}{\|y_s(s)\|_H^2 + \|y(s)\|_V^2} ds + E_y(0)$$

It follows that

$$E_y(t) \leq E_y(0) \quad (40)$$

Using (38), (39) and (40), we obtain

$$|\langle Bz_t, z_t\rangle| \le 2K\|\psi - \phi\|_X \sqrt{2E_y(0)} + |\langle By_t, y_t\rangle| \tag{41}$$

From (3), we deduce

$$\|\psi - \phi\|_X \le K \int_0^t |f(s)|\sqrt{2E_y(s)}ds$$

where $f(s) = \dfrac{\langle By_s(s), y_s(s)\rangle}{2E_y(s)}$.

Schwarz inequality, gives

$$\|\psi - \phi\|_X \le K\sqrt{2T}\left(\int_0^T |f(s)|^2 E_y(s)ds\right)^{\frac{1}{2}} \tag{42}$$

Using (40), we get

$$|\langle By_t, y_t\rangle| \le |f(t)|\sqrt{2E_y(t)}\sqrt{2E_y(0)} \tag{43}$$

Integrating (41) on the interval $[0, T]$ and from (42) and (43), we obtain

$$\int_0^T |\langle Bz_s(s), z_s(s)\rangle|ds \le 2(2K^2 + 1)\sqrt{TE_y(0)}\left(\int_0^T |f(s)|^2 E_y(s)ds\right)^{\frac{1}{2}}$$

It follows from (35), that

$$\|z_t(0)\|_H^2 + \|z(0)\|_V^2 \le \frac{2(2K + 1)}{C}\sqrt{TE_y(0)}\left(\int_0^T |f(s)|^2 E_y(s)ds\right)^{\frac{1}{2}}$$

Taking $(z_0, z_1) = (y_0, y_1)$, we have

$$\|y_1\|_H^2 + \|y_0\|_V^2 \le \frac{2(2K + 1)}{C}\sqrt{TE_y(0)}\left(\int_0^T |f(s)|^2 E_y(s)ds\right)^{\frac{1}{2}}$$

Replacing (y_0, y_1) by (y, y_t), we get

$$\|y_t\|_H^2 + \|y\|_V^2 \le \frac{2(2K + 1)}{C}\sqrt{TE_y(t)}\left(\int_t^{t+T} |f(s)|^2 E_y(s)ds\right)^{\frac{1}{2}}$$

It follows that

$$E_y(t) \le \gamma \int_t^{t+T} |f(s)|^2 E_y(s)ds \tag{44}$$

where $\gamma = \dfrac{2(2K + 1)}{C}\sqrt{T}$.

Integrating from kT to $(k+1)T$ the inequality

$$\frac{d}{dt}E_y(t) \leq -\frac{\langle By_t, y_t \rangle^2}{2E_y(t)}$$

and since $E_y(t)$ decreases, we obtain

$$E_y(kT) - E_y((k+1)T) \geq \int_{kT}^{(k+1)T} |f(s)|^2 E_y(s)ds$$

From (44), we get

$$E_y((k+1)T) \leq rE_y(kT)$$

where $r = \dfrac{1}{\gamma}$.

This implies that

$$E_y((k+1)T) \leq r^k E_y(0) \tag{45}$$

Applying (45) to the integer part $k = E(\dfrac{t}{T})$ of $\dfrac{t}{T}$, we deduce that

$$E_y(t) \leq Me^{-\lambda t}E_y(0)$$

where $M = \frac{2\sqrt{T}}{C}(2K+1)$ and $\lambda = \dfrac{\ln(\frac{2\sqrt{T}}{C}(2K+1))}{T}$.

Then the system (1) is exponentially stabilizable by the control (36).

Example 4 On $\Omega =]0, 1[$, we consider the following system

$$\begin{cases} y_{tt}(x,t) + y_{xxxx}(x,t) = u(t)By_t(x,t), & \Omega \times]0, +\infty[\\ y(x,0) = y_0(x), \ y_t(x,0) = y_1(x), & \Omega \\ y(0,t) = y(1,t) = y_{xx}(0,t) = y_{xx}(1,t) = 0, &]0, +\infty[\end{cases} \tag{46}$$

where the operator $By_t = (\max\{\|y_t\|_{L^2(\Omega)}, 1\})y_t$ and $u(t) \in L^\infty(0, +\infty)$ is a real valued control.

System (46) has a unique solution on the state space $X = (H^2(\Omega) \cap H_0^1(\Omega)) \times L^2(\Omega)$ [8].

The condition (35) is verified, indeed

$$\int_0^2 \langle Bz_t(t), z_t(t) \rangle dt = (\max\{\|z_t\|_{L^2(\Omega)}, 1\}) \int_0^2 \int_0^1 |z_t(x,t)|^2 dx dt$$
$$\geq \sum_{m=1}^\infty (m\pi)^4 (a_m^2 + b_m^2)$$
$$\geq \|z_t(0)\|_{L^2(\Omega)}^2 + \|z_{xx}(0)\|_{L^2(\Omega)}^2$$

where $z(x, t) = \sum_{m=1}^{\infty} (a_m \cos((m\pi)^2 t) + b_m \sin((m\pi)^2 t)) \sin(m\pi x)$ is the solution of the system

$$\begin{cases} z_{tt}(x, t) + z_{xxxx}(x, t) = 0, & \Omega \times]0, +\infty[\\ z(x, 0) = z_0(x), z_t(x, 0) = z_1(x), & \Omega \\ z(0, t) = z(1, t) = z_{xx}(0, t) = z_{xx}(1, t) = 0, &]0, +\infty[\end{cases}$$

We deduce from theorem 4 that the control

$$u(t) = -\frac{(\max\{\|y_t\|_{L^2(\Omega)}, 1\}) \int_0^1 |y_t(x, t)|^2 dx}{\|y_t\|_{L^2(\Omega)}^2 + \|y_{xx}\|_{L^2(\Omega)}^2} \tag{47}$$

exponentially stabilizes the system (46).

4   Simulations Results

In this section we illustrate the previous results by simulations. The developed results show that the stabilizing control is given by (14) for exponential stabilization and by (6) for strong stabilization. The computation of such control may be achieved using the following algorithm:

Step 1: Initial data : threshold $\varepsilon > 0$ and initial condition y_0;
Step 2: Apply the control $u(t_i) = \langle By_t(t_i), y_t(t_i) \rangle$;
Step 3: Solving the system (1) using Lax-Wendroff method given in [9], gives $y(t_{i+1})$;
Step 4: If $\|y(t_{i+1})\| < \varepsilon$ stop, otherwise;
Step 5: $i = i + 1$ and go to step 2.

Example 5 On $\Omega =]0, 1[$, consider the following system

$$\begin{cases} y_{tt}(x, t) - y_{xx}(x, t) = u(t)y_t(x, t), & \Omega \times]0, \infty[\\ y(x, 0) = \sin(2\pi x), y_t(x, 0) = 0, & \Omega \\ y(0, t) = y(1, t) = 0, &]0, \infty[\end{cases} \tag{48}$$

where the control

$$u(t) = -\int_0^1 |y_t(x, t)|^2 dx.$$

System (48) has a unique solution on the state space $H_0^1(\Omega) \times L^2(\Omega)$ [8].

Fig. 1 Initial position

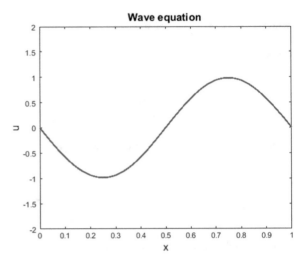

Fig. 2 The position at t = 5

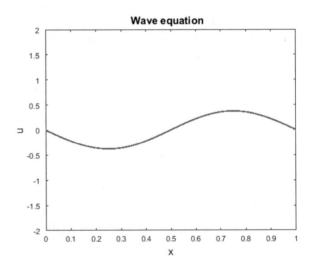

Figures 1 and 2 represent the evolution of the state.

Fig. 3 The position at t = 8

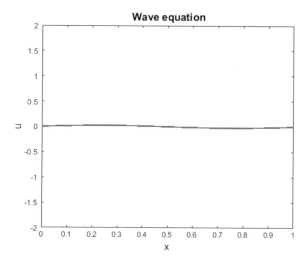

Fig. 4 The energy of the system

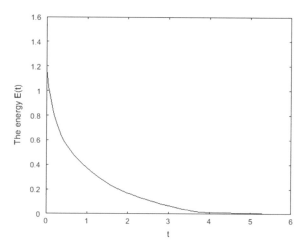

Figure 3 shows that the state is strongly stabilized on $\Omega =]0, 1[$ with stabilization error equals to 1.03×10^{-4}.

Figure 4 shows that the energy of the system decreases to zero.

Fig. 5 Evolution of the control

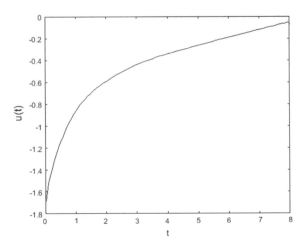

Figure 5 represent the evolution of the control.

5 Conclusion

We explore controls that stabilize strongly and exponentially a second order semi-linear systems where the control operator is possibly nonlinear. The obtained results were illustrated by many examples and simulations. This work gives an opening to others questions : establishing similar results for regional case.

References

1. Bardos, C., Lebeau, G., Rauch, J.: Sharp sufficient conditions for the observation, control and stabilization of waves from the boundary. SIAM J. Control Optim. **30**(5), 1024–1065 (1992)
2. Chen, G., Fulling, A., Narcowich, F.J., Sun, S.: Exponential decay of energy of evolution equations with locally distributed damping. SIAM J. Appl. Math. **51**, 266–301 (1991)
3. Couchoroun, J.M.: Stabilization of controlled vibrating systems. ESAIM: COCV **17**, 1144–1157 (2011)
4. Haraux, A.: Une remarque sur la stabilisation de certains systmes du deuxime ordre en temps. Port. Math. **46**, 245–258 (1989)
5. Haraux, A.: Remarks on weak stabilization of semilinear wave equations. ESAIM Control Optim. Calc. Var. **6**, 553–560 (2001)
6. Ghisi, M., Gobbino, M., Haraux, A.: The remarkable effectiveness of time-dependent damping terms for second order evolution equations. SIAM J. Control Optim. **54**(3), 1266–1294 (2016)
7. Liu, Kangsheng, Rao, Bopeng, Zhang, Xu: Stabilization of the wave equations with potential and indefinite damping. J. Math. Anal. Appl. **269**, 747–769 (2002)
8. Komornik, V., Loreti, P.: Fourier Series in Control Theory. Springer Monographs in Mathematics (2005)

9. Lax, P., Wendroff, B.: Systems of conservation laws. Commun. Pure Appl. Math. **13**(2), 217–237 (1960)
10. Martinez, P., Vancostenoble, J.: Exponential stability for the wave equation with weak non-monotone damping. Port. Math. **57**(3) (2000)
11. Pazy, A.: Semi-Groups of Linear Operators and Applications to Partial Differential Equations. Springer, New York (1983)
12. Tebou, L.: Equivalence between observability and stabilization for a class of second order semilinear evolution equation. Discret. Contin. Dyn. Syst. 744–752 (2009)

On the Fractional Output Stabilization for a Class of Infinite Dimensional Linear Systems

Hanaa Zitane, Rachid Larhrissi and Ali Boutoulout

Abstract The aim of this paper is to investigate the stability and the stabilization of the fractional spatial derivative of order $\alpha \in]0, 1[$ for a class of infinite dimensional linear systems. Firstly, we explore some properties of the fractional derivative stability. Then, we establish the characterization of the fractional spatial derivative stabilization with an approach based on the steady state Riccati equation. Moreover, we discuss a minimization fractional problem. Finally, we illustrate the developed results by numerical simulations.

1 Introduction

The stability concept is one of the most interesting notion in systems theory. Indeed, the stability problem of linear systems and related questions made the objective of many works [2, 3, 13]. Several results have been explored. In [9], the authors have been studied the relationship between the asymptotic behavior of a linear system, the spectral properties of its dynamics and the existence of a Lyapunov function. Moreover, various approaches have been developed to characterize different kinds of stabilizability. In [11], the exponential stabilization is characterized using a specific decomposition of the state space. Hence, the strong stabilization is studied by means of Riccati Equation [1].

Recently, in [12], the notion of the state gradient stabilization for a distributed linear systems has been introduced. Sufficient conditions for the gradient state stability for such systems, using the spectrum properties are given. Also, the gradient

H. Zitane (✉) · R. Larhrissi · A. Boutoulout
MACS Laboratory, Department of Mathematics, Faculty of Sciences,
Moulay Ismail University, Meknes, Morocco
e-mail: hanaa.zit@gmail.com

R. Larhrissi
e-mail: r.larhrissi@gmail.com

A. Boutoulout
e-mail: boutouloutali@yahoo.fr

© Springer Nature Switzerland AG 2020
E. H. Zerrik et al. (eds.), *Recent Advances in Modeling, Analysis and Systems Control: Theoretical Aspects and Applications*, Studies in Systems, Decision and Control 243,
https://doi.org/10.1007/978-3-030-26149-8_18

weak stabilization and the gradient strong stabilization were assured under a dissi-pation hypothesis. Moreover, the gradient exponential stabilization is obtained via the resolution of a Riccati equation.

The fractional derivative has several interesting properties that are useful for mod-eling a wide array of phenomena in many branches of science. For instance, the derivative is able to describe substance heterogeneities and configurations with dif-ferent scales. It has become very popular in recent years due to its demonstrated applications in various fields of sciences, engineering, mathematics, viscoelasticity, biology, image processing, economics, physics, control theory and in other fields. Some of the most applications are given in the book of Kilbas, Srivastava and Trujillo [4], the book of Oldham and Spanier [5], the book of Ross [10], and the book of Podlubny [8].

Motivated by the above discussion, in this work, we propose to extend the above results established on the integer cases: $\alpha = 0$ and $\alpha = 1$ that correspond, respec-tively, to the state stabilization and the state gradient stabilization, to the fractional order $\alpha \in]0, 1[$ case, that corresponds to the state fractional spatial derivative sta-bilization. Thus, our main contribution is to study the stability and the stabilization of the state Riemann Liouville fractional spatial derivative of order $\alpha \in]0, 1[$, for a class of infinite dimensional linear systems.

The rest of the present paper is organized as follows. In the second section, we characterize the fractional derivative stability. The third section is devoted to the characterization of the state fractional spatial derivative stabilization and the min-imization of a quadratic performance given cost. Finally, the obtained results are illustrated by numerical simulations.

2   Fractional Derivative Stability

The goal of this section is to establish the characterization of the state fractional spatial derivative stability, with the fractional order $\alpha \in]0, 1[$, for a class of infinite dimensional linear systems.

2.1   Considered System and Definitions

Let Ω be an open bounded subset of $\mathbb{R}$ and let us consider the state-space system

$$\begin{cases} \dot{z}(t) = Az(t) \ ]0, +\infty[\\ z(0) = z_0 \qquad z_0 \in H \end{cases} \tag{1}$$

where $A : D(A) \subset H \longrightarrow H$ is a linear operator generating a strongly continuous semi-group $S(t)_{t\geq0}$, on the Hilbert state space $H \subset H^1(\Omega)$ endowed with its usual inner product, denoted by $\langle ., . \rangle$ and the corresponding norm $\|.\|$.

The mild solution of the system (1) is given by $z(t) = S(t)z_0$, where $z \in \mathcal{C}(0, T, H)$, $T > 0$ [6].

Consider that the system (1) is augmented with the following fractional output

$$y(x, t) = D_x^\alpha z(x, t) \tag{2}$$

where D_x^α denotes the Riemann Liouville fractional spatial derivative of order $\alpha \in]0, 1[$, defined by [4]

$$D_x^\alpha : H \longrightarrow L^2(\Omega)$$

$$z \longrightarrow D_x^\alpha z = \frac{1}{\Gamma(1 - \alpha)} \frac{d}{dx} \int_0^x (x - \tau)^{-\alpha} z(\tau, t) \, d\tau$$

where $\Gamma(1 - \alpha)$ is the Eulers Gamma function. The Hilbert space $L^2(\Omega)$ is endowed with its usual inner product, denoted by $\langle ., . \rangle_{L^2(\Omega)}$ and the corresponding norm $\|.\|_{L^2(\Omega)}$.

In what follows, we denote by $(D_x^\alpha)^*$ the adjoint operator of D_x^α [7] and we define the operator $C^\alpha = (D_x^\alpha)^* D_x^\alpha$ which is an operator mapping H into itself.

Let us recall the following definitions

Definition 1 Let $z_0 \in H$, the output (2) of the system (1) is said to be

1. Weakly stable, if the corresponding solution $z(x, t)$ of (1) satisfies

$$\langle D_x^\alpha z(x, t), y \rangle_{L^2(\Omega)} \longrightarrow 0 \ as \ t \longrightarrow +\infty, \forall y \in L^2(\Omega)$$

2. Strongly stable, if the corresponding solution $z(x, t)$ of (1) satisfies

$$\|D_x^\alpha z(x, t)\|_{L^2(\Omega)} \longrightarrow 0 \ as \ t \longrightarrow +\infty$$

3. Exponentially stable, if there exist M and $\sigma > 0$ such that

$$\|D_x^\alpha z(x, t)\|_{L^2(\Omega)} \leq Me^{-\sigma t}\|z_0\|, \forall t \geq 0, \forall z_0 \in H$$

Remark 1　1. It is clear that the exponential stability of (2) $\Rightarrow$ the strong stability of (2) $\Rightarrow$ the weak stability of (2).
2. On $H = H_0^1(\Omega)$, due to the continuity property of the fractional derivation operator D_x^α from the space $H_0^1(\Omega)$ into $L^2(\Omega)$, if the system (1) is stable then its fractional output (2) is also stable.
3. If the system (1) is gradient stable then its output (2) is also stable on $H = H_0^1(\Omega)$.

The following lemma plays a significant role to obtain our results.

Lemma 1 *Let* $(S(t))_{t \geq 0}$ *be a strongly continuous semi-group generated by A on H. Suppose that there exists a function* $R(t) \in L^2(0, +\infty; \mathbb{R}^+)$ *such that*

$$\|D_x^\alpha S(t+s)z\|_{L^2(\Omega)} \leq R(t)\|D_x^\alpha S(s)z\|_{L^2(\Omega)}, \; \forall t, s \geq 0, \;\; \forall z \in H \qquad (3)$$

holds, then the operators $(D_x^\alpha S(t))_{t \geq 0}$ *are uniformly bounded.*

Proof Let us show that $\forall z \in H \; \sup_{t \geq 0}\|D_x^\alpha S(t)z\|_{L^2(\Omega)} < \infty$.

Otherwise there exists a sequence $(t_1 + \tau_k)$, $t_1 > 0$ and $\tau_k \longrightarrow +\infty$ such that

$$\|D_x^\alpha S(t_1 + \tau_k)z\|_{L^2(\Omega)} \longrightarrow +\infty \text{ as } k \longrightarrow +\infty$$

In the other-hand, one has

$$\int_0^{+\infty} \|D_x^\alpha S(s + \tau_k)z\|_{L^2(\Omega)}^2 \, ds = \int_{\tau_k}^{+\infty} \|D_x^\alpha S(s)z\|_{L^2(\Omega)}^2 \, ds$$

and the right-hand side goes to zero when $k \longrightarrow +\infty$. By Fatou's lemma

$$\liminf_{k \longrightarrow +\infty} \|D_x^\alpha S(s + \tau_k)z\|_{L^2(\Omega)} = 0 \; \forall s > 0$$

Then, for some $s_0 < t_1$ we may find a subsequence τ_{k_n} such that

$$\lim_{n \longrightarrow +\infty} \|D_x^\alpha S(s_0 + \tau_{k_n})z\|_{L^2(\Omega)} = 0$$

But, from the assumption (3) it follows that

$$\|D_x^\alpha S(t_1 + \tau_{k_n})z\|_{L^2(\Omega)} \leq R(t_1 - s_0)\|D_x^\alpha S(s_0 + \tau_{k_n})z\|_{L^2(\Omega)} \xrightarrow[n \longrightarrow +\infty]{} 0$$

which is a contradiction.

By applying the uniform boundedness principal, we deduce that the operators $(D_x^\alpha S(t))_{t \geq 0}$ are uniformly bounded.

2.2 Characterization of the Fractional Derivative Stability

We begin with the following result which links the stability of the fractional output (2) to spectrum properties of the operator A.

Let

$$\sigma^1(A) = \{\lambda \in \sigma(A) : Re(\lambda) \geq 0, \; N(A - \lambda I) \text{ is not included in } N(C^\alpha)\}$$

and

$$\sigma^2(A) = \{\lambda \in \sigma(A) : Re(\lambda) < 0, \ N(A - \lambda I) \text{ is not included in } N(C^\alpha)\}$$

where $\sigma(A)$ and $N(A)$ indicate, respectively, the points spectrum and the kernel of the operator A.

Proposition 1 *Let A generate a strongly continuous semi-group $(S(t))_{t\geq 0}$ on H.*

1. If the output (2) of the system (1) is weakly stable, then

$$\sigma^1(A) = \varnothing$$

2. Assume that $(\phi_n)_{n\geq 0}$ is an orthonormal basis of eigenfunctions of A in H. If

$$\sigma^1(A) = \varnothing \text{ and } \forall \lambda \in \sigma^2(A) \quad Re(\lambda) \leq -\beta \text{ for some } \beta > 0$$

then, the output (2) of the system (1) is exponentially stable.

Proof 1. By absurd, we suppose that there exist $\lambda \in \sigma^1(A)$ and $\phi \in H$ such that $A\phi = \lambda\phi$.

For $z_0 = \phi$, the solution of the system (1) is $S(t)z_0 = e^{\lambda t}\phi$, therefore

$$\mid \langle D_x^\alpha S(t)\phi, D_x^\alpha \phi\rangle_{L^2(\Omega)} \mid = e^{Re(\lambda)t}\|D_x^\alpha\phi\|^2_{L^2(\Omega)}$$

This implies that

$$\mid \langle D_x^\alpha S(t)\phi, D_x^\alpha \phi\rangle_{L^2(\Omega)} \mid \geq \|D_x^\alpha\phi\|^2_{L^2(\Omega)} \neq 0$$

Thus, the output (2) of the system (1) isn't weakly stable, which is a contradiction.

2. Without loss of generality, we can suppose that the eigenvalues of A are simple. For $z_0 \in H$, one has

$$D_x^\alpha S(t)z_0 = \sum_{i\geq 0} e^{\lambda_i t}\langle z_0, \phi_i\rangle D_x^\alpha \phi_i$$

Moreover, if for some $\beta > 0$, $Re(\lambda) \leq -\beta$ for all $\lambda \in \sigma^2(A)$, then we have

$$\|D_x^\alpha S(t)z_0\|^2_{L^2(\Omega)} \leq e^{-2\beta t}\sum_{i\geq 0}\langle z_0, \phi_i\rangle^2\|D_x^\alpha \phi_i\|^2_{L^2(\Omega)}$$

Using Parsevals identity, we deduce that

$$\|D_x^\alpha S(t)z_0\|_{L^2(\Omega)} \leq Me^{-\beta t}\|z_0\|, \forall t \geq 0 \ \text{ for some } M > 0$$

Example 1 Let us consider, on $\Omega =]0, 1[$, the following heat equation

$$\begin{cases} \dfrac{\partial z}{\partial t}(x, t) = \dfrac{\partial^2 z}{\partial x^2}(x, t) & \Omega \times]0, +\infty[\\ \dfrac{\partial z}{\partial x}(0, t) = \dfrac{\partial z}{\partial x}(1, t) = 0 \ \forall t > 0 \\ z(x, 0) = z_0 & \Omega \end{cases} \tag{4}$$

where the dynamic is defined by $A = \dfrac{\partial^2}{\partial x^2}$, with spectrum given by the eigenvalues $\lambda_i = -(i\pi)^2$, $i \in \mathbb{N}$ and the corresponding eigenfunctions $\phi_i(x) = \sqrt{2}cos(i\pi x)$, $\forall i \geq 1$ and $\phi_0(x) = 1$.

We have $\sigma^1(A) = \varnothing$ and $\forall \lambda \in \sigma^2(A)$, $Re(\lambda) \leq -(\pi)^2$.

Then, the output (2) of the system (4) is exponentially stable.

In the following result, we establish an important criterion for the exponential stability of the fractional output (2).

Proposition 2 *Assume that the semi-group* $(S(t))_{t\geq 0}$ *satisfies the condition* (3). *Moreover, if*

$$\forall z \in H \quad \|D_x^\alpha S(t + s)z\|_{L^2(\Omega)} \leq \|D_x^\alpha S(t)z\|_{L^2(\Omega)}.\|D_x^\alpha S(s)z\|_{L^2(\Omega)}, \quad \forall t, s \geq 0 \tag{5}$$

holds, then the output (2) *of the system* (1) *is exponentially stable, if and only if*

$$\forall z \in H \quad \int_0^{+\infty} \|D_x^\alpha S(t)z\|_{L^2(\Omega)}^2 \, dt < \infty$$

Proof We pose $\sigma_0 = \underset{t \geq 0}{\inf} \dfrac{ln\|D_x^\alpha S(t)\|_{L^2(\Omega)}}{t}$. We have

$$t\|D_x^\alpha S(t)z\|_{L^2(\Omega)}^2 = \int_0^t \|D_x^\alpha S(t)z\|_{L^2(\Omega)}^2 \, ds$$

$$= \int_0^t \|D_x^\alpha S(t - s + s)z\|_{L^2(\Omega)}^2 \, ds$$

From condition (3) and Lemma 1, it follows that

$$t\|D_x^\alpha S(t)z\|_{L^2(\Omega)}^2 \leq \int_0^t R^2(s)\|D_x^\alpha S(t - s)z\|_{L^2(\Omega)}^2 \, ds$$

$$\leq \omega^2 \|z\|_{L^2(\Omega)}^2$$

for some $\omega > 0$ independent of t.

This implies that $\|D_x^\alpha S(t)\|_{L^2(\Omega)} < 1$, for t sufficiently large.

Yields $ln\|D_x^\alpha S(t)\|_{L^2(\Omega)} < 0, \quad \forall t \geq t_0$ for some $t_0 > 0$. Then

$$\sigma_0 = \inf_{t \geq 0} \frac{ln\|D_x^\alpha S(t)\|_{L^2(\Omega)}}{t} < 0$$

Now we prove that

$$\sigma_0 = \lim_{t \longrightarrow +\infty} \frac{ln\|D_x^\alpha S(t)\|_{L^2(\Omega)}}{t} \tag{6}$$

Let $t_1 > 0$, we pose $N' = \sup_{t \in [0,t_1]} \|D_x^\alpha S(t)\|_{L^2(\Omega)}$. For each $t > t_1$ there exists $m \in \mathbb{N}$ such that $mt_1 \leq t \leq (m+1)t_1$. Then

$$\frac{ln\|D_x^\alpha S(t)\|_{L^2(\Omega)}}{t} \leq \frac{ln\|D_x^\alpha S(mt_1)\|_{L^2(\Omega)}}{t} + \frac{ln\|D_x^\alpha S(t - mt_1)\|_{L^2(\Omega)}}{t}$$

Using (5), that implies

$$\frac{ln\|D_x^\alpha S(t)\|_{L^2(\Omega)}}{t} \leq \frac{mt_1}{t} \frac{ln\|D_x^\alpha S(t_1)\|_{L^2(\Omega)}}{t_1} + \frac{ln\|D_x^\alpha N'\|_{L^2(\Omega)}}{t}$$

Since $\dfrac{mt_1}{t} \leq 1$ and t_1 is arbitrary, it follows that

$$\limsup_{t \longrightarrow +\infty} \frac{ln\|D_x^\alpha S(t)\|_{L^2(\Omega)}}{t} \leq \inf_{t > 0} \frac{ln\|D_x^\alpha S(t)\|_{L^2(\Omega)}}{t} \leq \liminf_{t \longrightarrow +\infty} \frac{ln\|D_x^\alpha S(t)\|_{L^2(\Omega)}}{t}$$

Then, (6) holds. So we deduce that for all $\sigma \in]0, -\sigma_0[$, there exists $M > 0$ such that

$$\forall z \in H \qquad \|D_x^\alpha S(t)z\|_{L^2(\Omega)} \leq Me^{-\sigma t}\|z\|_{L^2(\Omega)} \quad \forall t \geq 0$$

The converse is immediate.

Corollary 1 *Suppose that the conditions (3) and (5) are satisfied. Moreover, if there exists a linear self-adjoint and positive operator $P \in \mathcal{L}(H)$ satisfying*

$$\langle Az, Pz \rangle + \langle Pz, Az \rangle + \langle Rz, z \rangle = 0, \quad \forall z \in D(A) \tag{7}$$

where $R \in \mathcal{L}(H)$ is a linear self-adjoint operator such that

$$\forall z \in H \qquad \langle Rz, z \rangle \geq \beta\|D_x^\alpha z\|_{L^2(\Omega)}^2, \text{ for some } \beta \geq 0 \tag{8}$$

then, the output (2) of the system (1) is exponentially stable.

Proof Let us consider the functional $F(z) = \langle Pz, z \rangle$ for all $z \in H$.

Let $z_0 \in D(A)$, we have

$$
\frac{d}{dt}F(S(t)z_0) = \langle AS(t)z_0, PS(t)z_0 \rangle + \langle PS(t)z_0, AS(t)z_0 \rangle
$$
$$
= -\langle RS(t)z_0, S(t)z_0 \rangle
$$

Therefore

$$
\int_0^{+\infty} \langle RS(s)z_0, S(s)z_0 \rangle \, ds \le F(z_0)
$$

By the assumption (8), it follows that

$$
\int_0^{+\infty} \|D_x^\alpha S(t)z_0\|_{L^2(\Omega)}^2 \, dt \le \frac{F(z_0)}{\beta} \qquad \forall z_0 \in D(A) \tag{9}
$$

From the density of $D(A)$ in H and the continuity of the function F, the inequality (9) holds for all $z_0 \in H$. Then

$$
\int_0^{+\infty} \|D_x^\alpha S(t)z_0\|_{L^2(\Omega)}^2 \, dt < \infty \qquad \forall z_0 \in H
$$

Hence, from the Proposition 2, we get the exponential stability of the output (2) of the system (1).

For the strong stability of the state fractional spatial derivative (2), we have the following proposition.

Proposition 3 *Assume that there exists a linear self-adjoint and positive operator $P \in \mathscr{L}(H)$ such that the Eq. (7) holds. If, in addition, the assumption*

$$
\langle C^\alpha A z, z \rangle_{L^2(\Omega)} \le 0, \quad \forall z \in D(A) \tag{10}
$$

is satisfied, then the output (2) of the system (1) is strongly stable.

Proof Suppose that the Eq. (7) has a positive solution $P \in \mathscr{L}(H)$ and let us consider the functional $F(z) = \langle Pz, z \rangle, \forall z \in H$.

Let $z_0 \in D(A)$, one has

$$
\frac{d}{dt}F(S(t)z_0) = \langle AS(t)z_0, PS(t)z_0 \rangle + \langle PS(t)z_0, AS(t)z_0 \rangle
$$
$$
= -\langle RS(t)z_0, S(t)z_0 \rangle
$$

Therefore

$$
\int_0^{+\infty} \langle RS(s)z_0, S(s)z_0 \rangle \, ds \le F(z_0)
$$

By the condition (8), we get

$$\int_0^{+\infty} \|D_x^\alpha S(s)z_0\|_{L^2(\Omega)}^2 \, ds < \infty \quad \forall z_0 \in D(A)$$

On the other-hand, from the assumption (10), we have

$$\frac{d}{dt}\|D_x^\alpha S(t)z_0\|_{L^2(\Omega)}^2 = \langle \frac{d}{dt}D_x^\alpha S(t)z_0, D_x^\alpha S(t)z_0\rangle_{L^2(\Omega)} + \langle D_x^\alpha S(t)z_0, \frac{d}{dt}D_x^\alpha S(t)z_0\rangle_{L^2(\Omega)}$$
$$= 2\langle C_\alpha AS(t)z_0, S(t)z_0\rangle_{L^2(\Omega)} \le 0$$

This implies that

$$t\|D_x^\alpha S(t)z_0\|_{L^2(\Omega)}^2 = \int_0^t \|D_x^\alpha S(t)z_0\|_{L^2(\Omega)}^2 \, ds \le \int_0^t \|D_x^\alpha S(s)z_0\|_{L^2(\Omega)}^2 \, ds$$

So, we deduce that

$$\|D_x^\alpha S(t)z_0\|_{L^2(\Omega)}^2 \le \frac{f(., z_0)}{t} \quad \forall z_0 \in D(A), \ t > 0$$

where $f(., z_0) = \int_0^t \|D_x^\alpha S(s)z_0\|_{L^2(\Omega)}^2 \, ds$. From the density of $D(A)$ in H and the continuity of $f(., z_0)$, we conclude that

$$\lim_{t\to+\infty} \|D_x^\alpha S(t)z_0\|_{L^2(\Omega)} = 0, \quad \forall z_0 \in H$$

3 Fractional Derivative Stabilization

Let Ω be an open bounded subset of $\mathbb{R}$, we consider the following linear system

$$\begin{cases} \dot{z}(t) = Az(t) + Bu(t) &]0, +\infty[\\ z(0) = z_0 & z_0 \in H \end{cases} \tag{11}$$

with the same assumptions on A and $B \in \mathcal{L}(U, H)$, where U is the space of controls.

The mild solution $z(t)$ of the system (11) is given by the following variation of constants formula

$$z(t) = S(t)z_0 + \int_0^t S(t - s)Bu(s) \, ds \tag{12}$$

where $z \in \mathcal{C}(0, T, H)$, $T > 0$ [6].

Definition 2 The output (2) of the system (11) is said to be weakly (respectively strongly, exponentially) stabilizable if there exists a bounded operator $K \in \mathcal{L}(H, U)$

such that the output (2) of the system

$$\begin{cases} \dot{z}(t) = (A + BK)z(t) \ ]0, +\infty[\\ z(0) = z_0 \qquad\qquad z_0 \in H \end{cases} \tag{13}$$

is weakly (respectively strongly, exponentially) stable.

Remark 2 It is clear that the exponential stabilization of (2) $\Rightarrow$ the strong stabilization of (2) $\Rightarrow$ the weak stabilization of (2).

In what follows, we denote by $S_K(t)$, $t \geq 0$, the strongly continuous semi-group generated by $A + BK$, where $K \in \mathscr{L}(H, U)$ is the feedback operator.

3.1 Riccati Method

In this paragraph, we propose an approach characterizing the stabilization of the state fractional spatial derivative (2) of the system (11).

We consider the system (13) with the same assumptions on A and B. Let $R \in \mathscr{L}(H)$ be a linear and self-adjoint operator satisfying (8) and assume that the following Riccati equation

$$\langle Az, Pz \rangle + \langle Pz, Az \rangle - \langle B^*Pz, B^*Pz \rangle + \langle Rz, z \rangle = 0, \qquad \forall z \in D(A) \tag{14}$$

has a linear self-adjoint positive solution $P \in \mathscr{L}(H)$ and let $K = -B^*P$.

Proposition 4 *1. Assume that the semi-group $S_K(t)$ satisfies the conditions (3) and (5). Then, the output (2) of the system (11) is exponentially stabilizable by the feedback control*

$$u(t) = Kz(x, t)$$

2. If the following condition

$$\langle C^\alpha(A + BK)z, z \rangle_{L^2(\Omega)} \leq 0 \qquad \forall z \in D(A) \tag{15}$$

holds, then the output (2) of the system (11) is strongly stabilizable.

Proof The proof is deduced directly from the Corollary 1 and the Proposition 3.

Proposition 5 *Assume that the output (2) of the system (11) is exponentially stabilizable. Moreover, if the condition*

$$\langle C^\alpha z, z \rangle_{L^2(\Omega)} \geq d \langle C^\alpha(A + BK)z, z \rangle_{L^2(\Omega)} \quad \forall z \in D(A), \ for\ some\ d > 0 \tag{16}$$

holds, then the state fractional derivative of the system (13) remains bounded.

Proof Let $z_0 \in D(A)$, we have

$$\frac{d}{dt}\|D_x^\alpha z(x, t)\|_{L^2(\Omega)}^2 = 2\langle C^\alpha(A + BK)z, z\rangle_{L^2(\Omega)}, \quad z_0 \in D(A)$$

Using (16) and integrating this inequality, it follows that

$$\|D_x^\alpha z(x, t)\|_{L^2(\Omega)}^2 - \|D_x^\alpha z_0\|_{L^2(\Omega)}^2 \le \frac{2}{d} \int_0^t \|D_x^\alpha z(x, s)\|_{L^2(\Omega)}^2 \, ds \qquad (17)$$

And since the output (2) of the system (11) is exponentially stabilizable, one has

$$\int_0^{+\infty} \|D_x^\alpha z(x, t)\|_{L^2(\Omega)}^2 \, dt < +\infty$$

Therefore, the inequality (17) implies that there exists $R > 0$ such that

$$\|D_x^\alpha z(x, t)\|_{L^2(\Omega)} \le R, \quad t \ge 0, \forall \, z_0 \in D(A)$$

Thus, by the density of $D(A)$ in H, we get the stated result.

3.2 Stabilization Control Problem

The aim of this paragraph is to characterize the control that stabilizes the state fractional spatial derivative of order $\alpha \in]0, 1[$ of the system (11) minimizing the following functional cost

$$\begin{cases} min \, J(z_0, u) = \int_0^{+\infty} \|D_x^\alpha z(x, t)\|_{L^2(\Omega)}^2 \, dt + \int_0^{+\infty} \|u(t)\|^2 \, dt, \\ u \in U_{ad} = \{u \in L^2(0, +\infty; U) | J(z_0, u) < +\infty\} \end{cases} \qquad (18)$$

Proposition 6 *Assume that the semi-group $S_K(t)$, $t \ge 0$, satisfies the conditions (3) and (5). If, in addition, there exists a self-adjoint and positive operator $P \in \mathcal{L}(H)$ such that*

$$\langle Az, Pz\rangle + \langle Pz, Az\rangle - \langle B^*Pz, B^*Pz\rangle + \|D_x^\alpha z(x, t)\|_{L^2(\Omega)}^2 = 0, \quad \forall \, z \in D(A) \quad (19)$$

Then, the feedback control
$$u^*(t) = -B^*Pz(x, t) \qquad (20)$$

is the unique solution of problem (18) which stabilizes exponentially the output (2) of the system (11).

Proof 1. Let $z_0 \in D(A)$ and $u \in L^2([0, +\infty[; U)$, we consider

$$J(z_0, u, k) = \int_0^k \|D_x^\alpha z(x, t)\|_{L^2(\Omega)}^2 \, ds + \int_0^k \|u(t)\|^2 \, ds \leq J(z_0, u), \text{ for some } k \geq 0$$

Since the operator P satisfies the Eq. (19), we obtain

$$J(z_0, u, k) = \langle z_0, P z_0 \rangle - \langle z(x, k), P z(x, k) \rangle +$$
$$\int_0^k \langle [u(s) + B^* P z(x, s)], [u(s) + B^* P z(x, s)] \rangle \, ds \quad \forall z_0 \in D(A)$$
$$(21)$$

From the density of $D(A)$ in the space H, (21) holds for all $z_0 \in H$.
Using the fact that $P \in \mathcal{L}(H)$ is a positive operator, it follows that

$$J(z_0, u, k) \leq \langle z_0, P z_0 \rangle + \int_0^k \langle [u(s) + B^* P z(x, s)], [u(s) + B^* P z(x, s)] \rangle \, ds$$
$$(22)$$

If we chose $u(s) = u^*(s) = -B^* P z(x, s)$, it follows that

$$J(z_0, u^*, k) \leq \langle z_0, P z_0 \rangle. \tag{23}$$

Since the right-hand side of (23) does not depend on k, we tend k to $+\infty$, we obtain that

$$J(z_0, u^*) = \lim_{k \to \infty} J(z_0, u^*, k) \leq \langle z_0, P z_0 \rangle \quad \forall \, z_0 \in H \tag{24}$$

Thus $J(z_0, u^*)$ is finite for all $z_0 \in H$ and $U_{ad} \neq 0$.
On the other hand, for all $u \in U_{ad}$ [3], we have

$$J(z_0, u, k) = \int_0^k \|D_x^\alpha z(x, t)\|_{L^2(\Omega)}^2 \, ds + \int_0^k \|u(t)\|^2 \, ds \leq J(z_0, u)$$

So

$$\inf_{u \in L^2([0, +\infty]; U)} J(z_0, u) \geq \inf_{u \in L^2([0, +\infty]; U)} J(z_0, u, k)$$
$$= \inf_{u \in L^2([0, k]; U)} J(z_0, u)$$
$$= \langle z_0, P(k) z_0 \rangle \quad \forall k \geq 0$$

where $P(k)$, $k \in N$ is a uniformly bounded sequence and satisfies
$\lim_{k \to \infty} P(k) z_0 = P z_0$, $\forall k \geq 0$, with $P \in \mathcal{L}(H)$ is the solution of (19) [3].
Let us pass to the limit on k, we get

$$\inf_{u \in L^2([0,+\infty];U)} J(z_0, u) \geq \langle z_0, Pz_0 \rangle \tag{25}$$

Combining (25) with (24), we obtain

$$\langle z_0, Pz_0 \rangle \leq \inf_{u \in L^2([0,+\infty];U)} J(z_0, u) \leq J(z_0, u^*) \leq \langle z_0, Pz_0 \rangle$$

Then, we deduce that

$$\min_{u \in L^2([0,+\infty];U)} J(z_0, u) = J(z_0, u^*) = \langle z_0, Pz_0 \rangle$$

This means that the feedback control (20) is the optimal one that is solution of the problem (18).
The uniqueness of the optimal control u^* follows from the uniqueness of $z(x, t)$ solution of system (11).

2. From Proposition 4, we deduce the exponential stabilization of the fractional output (2) of the system (11) by control (20).

4   Simulation Results

Here, we give an algorithm that calculates the control u^* solution of the problem (18) stabilizing the fractional output of order $\alpha \in]0, 1[$ of the system (11). From the previous result, this control can be obtained by solving the Riccati Equation (19).

Let $H_n = span\{\phi_i, i = 1, 2, \ldots, n\}$ be the subspace of H where $\{\phi_i, i \in \mathbb{N}^*\}$ is an orthonormal basis of $L^2(\Omega)$. H_n is endowed with the restriction of the inner product of $L^2(\Omega)$.

We define the projection operator

$$\Pi_n(z) : H \longrightarrow H_n$$
$$z \longrightarrow \sum_{i=1}^{n} \langle z, \phi_i \rangle \phi_i \qquad \forall z \in H$$

The projection of the Eq. (19) on the space H_n is given by the following matrix algebraic Riccati equation

$$P_n A_n + A_n P_n - P_n B_n^* B_n^* P_n + (D_x^\alpha)_n^* (D_x^\alpha)_n = 0 \tag{26}$$

where A_n, P_n, B_n and $(D_x^\alpha)_n$ are respectively the projections of A, P, B and D_x^α in H_n.
Also, the projection of the system (11) is given by

$$\begin{cases} \dot{z}_n(t) = A_n z_n(t) - B_n B_n^* P_n z_n(t) \\ z(0) = z_{0_n} \end{cases} \tag{27}$$

we perform the following algorithm steps

Algorithm:

Step 1: Let $\varepsilon > 0$ the Threshold accuracy, α the order of the derivative, $t_i = ih$ the time sequence with $i \in \mathbb{N}$ and $h > 0$ small enough, n the dimension of the projection space and $z(0) = z_{0_n}$ the initial condition.

Step 2: Solve the Riccati Equation (26).

Step 3: Apply the feedback control $U_n(t_i) = -B_n^* P_n z_n(x, t_i)$.

Step 4: Solving the system (27).

Step 5: Calculate $D_x^\alpha z_n(x, t_i) = \sum_{j \geq 1} \langle z_n(x, t_i), \phi_j(x) \rangle D_x^\alpha \phi_j(x)$.

Step 6: If $\|D_x^\alpha z_n(x, t_i)\|_{L^2(\Omega)} < \varepsilon$ stop, else $i = i + 1$ and go to 3.

To illustrate these steps, we consider the example

Example 2 Let us consider, on $\Omega =]0, 1[$, the following system

$$
\begin{cases}
\dfrac{\partial z}{\partial t}(x, t) = (0.05)\dfrac{d^2 z}{dx^2}(x, t) + z(x, t) + u(t) & \Omega \times]0, +\infty[\\
z(0, t) = z(1, t) = 0 & \forall t > 0 \\
z(x, 0) = \dfrac{x}{2}(1 - x) & \Omega
\end{cases}
\tag{28}
$$

where the dynamic $Az = (0.05\dfrac{d^2}{dx^2} + 1)z$ with spectrum given by the simple eigenvalues $\lambda_i = 1 - 0.05(i\pi)^2$, $i \geq 1$ and the corresponding eigenfunctions $\phi_i(x) = \sqrt{\dfrac{2}{1+(i\pi)^2}} \sin(i\pi x)$, $i \geq 1$.

A generates a strongly continuous semi-group $S(t)_{t \geq 0}$ defined by

$$
S(t)z_0 = \sum_{i \geq 1} e^{\lambda_i t} \langle z_0, \phi_i \rangle \phi_i
$$

Let the system (28) be augmented with the following fractional outputs

$$
y(x, t) = D_x^{\frac{1}{3}} z(x, t)
\tag{29}
$$

$$
y(x, t) = D_x^{\frac{1}{2}} z(x, t)
\tag{30}
$$

$$
y(x, t) = D_x^{\frac{5}{6}} z(x, t)
\tag{31}
$$

We consider $H_n = Span\{\beta_i \sin(i\pi x), \beta_i = \sqrt{\dfrac{2}{1+(i\pi)^2}}, 1 \leq i \leq n, x \in \Omega\}$ and we apply the previous algorithm taking the truncation at $n = 5$. Therefore, we obtain the following simulation results:

For the output (29) (Figs. 1, 2, 3)

Figure 2 indicates that the $D_x^{\frac{1}{3}} z(x, t)$ is stabilized on Ω at time $t = 12$ with stabilization error equals to 3.1423×10^{-06} and a stabilization cost equals to $J = 0.0239$.

Fig. 1 The output $D_x^{\frac{1}{3}} z(x, t)$ evolution on Ω

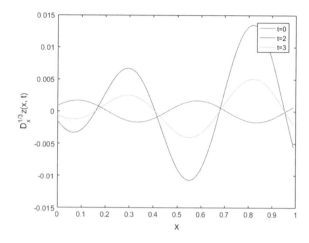

Fig. 2 The fractional output $D_x^{\frac{1}{3}} z(x, t)$ stabilization at $t = 12$

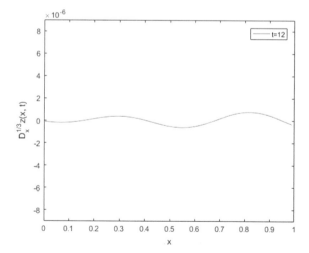

Fig. 3 The control evolution using $\alpha = \frac{1}{3}$

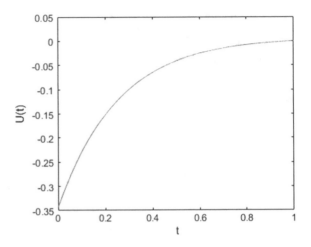

Fig. 4 The output $D_x^{\frac{1}{2}} z(x, t)$ evolution on Ω

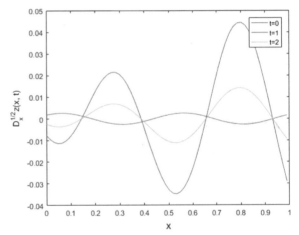

For the output (30) (Figs. 4, 5, 6)

Figure 5 shows how the $D_x^{\frac{1}{2}} z(x, t)$ evolves close to zero when the time t increases. The $D_x^{\frac{1}{2}}$ is stabilized on Ω at time $t = 11$ with error equals to 2.0040×10^{-06} and stabilization cost equals to $J = 0.0434$.

Fig. 5 The fractional output $D_x^{\frac{1}{2}} z(x, t)$ stabilization at $t = 11$

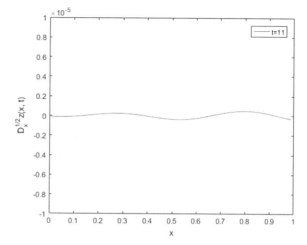

Fig. 6 The control evolution using $\alpha = \frac{1}{2}$

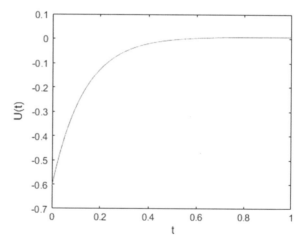

For the output (31) (Figs. 7, 8, 9)

Figure 8 shows how the $D_x^{\frac{5}{6}} z(x, t)$ of system (27) is stabilized on Ω at $t = 1$. With a stabilization error equals to 3.2695×10^{-07} and a stabilization cost equals to $J = 0.1162$.

Fig. 7 The output $D_x^{\frac{5}{6}} z(x, t)$
evolution on Ω

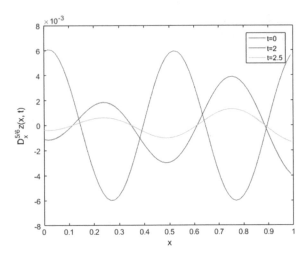

Fig. 8 The fractional output
$D_x^{\frac{5}{6}} z(x, t)$ stabilization at
$t = 7$

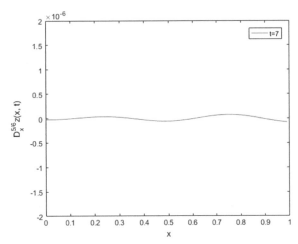

5 Conclusion

The state fractional spatial derivative stability of order $\alpha \in]0, 1[$, for infinite dimensional linear systems, is studied using the spectrum proprieties of the system dynamics. Also, the fractional output stabilization is characterized via Riccati approach. Moreover, a control optimal problem is solved. Hence, the obtained results are successfully illustrated by numerical simulations.

This work gives an opening to others questions, this is the case of extending these results to infinite dimensional bilinear systems.

Fig. 9 The control evolution using $\alpha = \frac{5}{6}$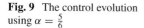

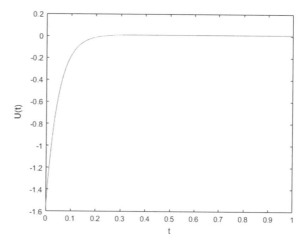

References

1. Balakrishnan, A.V.: Strong stabilizability and the steady state Riccati equation. Appl. Math. Optim. **7**(1), 335–345 (1981)
2. Curtain, R.F., Pritchard, A.J.: Infinite Dimensional Linear Systems Theory. Springer, Berlin (1978)
3. Curtain, R.F., Zwart, H.J.: An Introduction to Infinite Dimensional Linear Systems Theory. Springer, New York (1995)
4. Kilbas, A.A., Srivastava, H.M., Trujillo, J.J.: Theory and Applications of Fractional Differential Equations. Elsevier Science Limited (2006)
5. Oldham, K.B., Spanier, J.: The Fractional Calculus: Theory and Application of Differentiation and Integration to Arbitrary Order. Academic, New York and London (1974)
6. Pazy, A.: Semigroups of Linear Operators and Applications to Partial Differential Equations. Springer, New York (1983)
7. Podlubny, I., Chen, Y.: Adjoint fractional differential expressions and operators. In: ASME 2007 International Design Engineering Technical Conferences and Computers and Information in Engineering Conference, American Society Of Mechanical Engineers, pp. 1385–1390 (2007)
8. Podlubny, I.: Fractional Differential Equations. Academic, San Diego (1999)
9. Pritchard, A.J., Zabczyk, J.: Stability and stabilizability of infinite dimensional systems. SIAM Rev. **23**(1), 25–51 (1981)
10. Ross. B.: Fractional Calculus and Its Applications, Springer Verlag, in Lecture Notes in Mathematics, Proc. int. conf. New Haven, **457** (1974)
11. Triggiani, R.: On the stabilizability problem in Banach space. J. Math. Anal. Appl. **52**(3), 383–403 (1975)
12. Zerrik, E., Benslimane, Y.: An output gradient stabilization of distributed linear systems approaches and simulations. Intell. Control Autom. **3**(2), 159–167 (2012)
13. Zerrik, E., Ouzahra, M.: Regional stabilization for infinite dimensional systems. Int. J. Control, **76**(1), 73–81 (2003)

Regional Boundary Observability with Constraints on the State of Semilinear Parabolic Systems

Hayat Zouiten, Ali Boutoulout and Fatima-Zahrae El Alaoui

Abstract The purpose of this paper is to study the concept of regional boundary observability with constraints on the state of distributed parameter systems governed by semilinear parabolic ones, which consists in the reconstruction of the initial state between two prescribed functions on a part Γ of the boundary $\partial\Omega$, without the knowledge of the state using the Lagrangian multiplier approach. Finally, this approach leads to an algorithm which is illustrated through numerical examples and simulations.

1 Introduction

Control theory is one of the most interdisciplinary areas of research that brings together a set of notions such as controllability, observability, stability, etc., which lead to various reasoning techniques and mathematical tools for better understanding and knowledge of the system. For this, it is necessary to establish a precise description of the system, which supposes a rather fine knowledge of its different components, their behaviors, and their interactions. This description can be represented in the form of differential equations or partial differential equations, showing time and space variables in the case of Distributed Parameter Systems. DPSs represent dynamical systems in which the states depend on not only time but also space or spatial variables, which makes the system infinite dimensional. In the literature, DPSs are also called spatio–temporal dynamic systems, and usually, they are augmented with input–output equations that can be dependent in their turn of time and space variables.

H. Zouiten (✉) · A. Boutoulout · F.-Z. El Alaoui
TSI Team, MACS Laboratory, Faculty of Sciences, Moulay Ismail University, Meknes, Morocco
e-mail: hayat.zouiten@yahoo.fr

A. Boutoulout
e-mail: boutouloutali@yahoo.fr

F.-Z. El Alaoui
e-mail: fzelalaoui2011@yahoo.fr

© Springer Nature Switzerland AG 2020
E. H. Zerrik et al. (eds.), *Recent Advances in Modeling, Analysis and Systems Control: Theoretical Aspects and Applications*, Studies in Systems, Decision and Control 243, https://doi.org/10.1007/978-3-030-26149-8_19

261

In a DPS, the notion of observability is linked to the possibility to reconstruct the initial state of the system in a finite duration using sensor measurements. The notion of observability was introduced at the beginning of the sixties by R. E. Kalman [7] for the case of the finite dimension, and has been generalized to the infinite–dimensional context by D. L. Russell, S. Dolecki, G. Weiss, M. C. Delfour and S. K. Mitter [3, 4, 12].

Recently, the concept of regional analysis has been introduced by El jai and his co-workers, which the target of interest is not fully the geometrical evolution domain Ω, but just in an internal subregion ω of Ω (see [1, 6]) or on a part of the boundary subregion Γ of $\partial\Omega$ (see [14, 15]). It should be pointed out that in the literature there are very little results concerning regional observability for the class of semilinear systems and this field of research is still in development.

In this paper, we introduce a new concept which is the regional boundary observability with constraints on the state (also called boundary enlarged observability) for semilinear parabolic systems. The goal of this problem is to rebuild the initial state between two prescribed functions on a part Γ of the boundary $\partial\Omega$. The reasons for studying this kind of notion are so many. For instance, we find that the mathematical model of real systems is obtained from measures or approximations. Then, the solution for such systems is approximately known and required to be only between two bounds. This is the case, for example, of the biological treatment of wastewater using a fixed bed bioreactor. The process has to regulate the substrate concentration of the bottom of the reactor between two prescribed levels, for more details (see [16]).

This paper is organized as follows. The studied problem and some preliminaries are introduced in Sect. 2, In Sect. 3 we characterize the regional boundary enlarged observability of the system. Section 4, is focused on the regional boundary reconstruction of the initial state using the Lagrangian multiplier approach. In Sect. 5 we give an algorithm which is successfully implemented numerically and illustrated by an example which led to successful results in Sect. 6. We end with Sect. 7 of conclusions.

2   Problem Statement and Preliminaries

Let Ω be an open subset of $\mathbb{R}^n$ ($n = 1, 2, 3$), with sufficiently smooth boundary $\partial\Omega$. For $T > 0$, let $Q_T = \Omega \times]0, T[$ and $\Sigma_T = \partial\Omega \times]0, T[$. We consider the following semilinear parabolic system:

$$
\begin{cases}
\dfrac{\partial u}{\partial t}(x, t) = Au(x, t) + Nu(x, t) \ in & Q_T \\[2mm]
\dfrac{\partial u}{\partial \nu_A}(\xi, t) = 0 & on & \Sigma_T \\[2mm]
u(x, 0) = u_0(x) & in & \Omega,
\end{cases}
\tag{1}
$$

where A is a second-order linear differential operator with compact resolvent which generates a strongly continuous semigroup $(S(t))_{t \geq 0}$ on a Hilbert space $L^2(\Omega)$ and $\dfrac{\partial u}{\partial \nu_A}$ denotes the co-normal derivation with respect to A. We suppose that $u_0 \in H^1(\Omega)$ is unknown and $\mathcal{N}$ is a nonlinear continuous operator.

We associate to the system (1) the linear one defined by

$$
\begin{cases}
\dfrac{\partial u}{\partial t}(x, t) = Au(x, t) \ in & Q_T \\
\dfrac{\partial u}{\partial \nu_A}(\xi, t) = 0 & on \quad \Sigma_T \\
u(x, 0) = u_0(x) & in \quad \Omega,
\end{cases}
\tag{2}
$$

without loss of generality we denote $u(x, t) := u(t)$. System (2) admits a unique solution $u \in L^2(0, T; H^1(\Omega)) \cap C^0(0, T; L^2(\Omega))$ (see [8]).

The measurements (possibly unbounded) are given depending on the number and the structure of sensors with dense domain in $H^1(\Omega)$ and rang in O as follows:

$$
y(t) = Cu(t), \quad t \in]0, T[,
\tag{3}
$$

where $O = L^2(0, T; \mathbb{R}^q)$ denotes the observation space and $q \in \mathbb{N}$ is the finite number of the considered sensors.

It is well known that the global Lipschitz continuity of the nonlinear $\mathcal{N}$ implies that the problem (1) admits a unique mild solution (see [10, 13]) given by the variation of parameters formula

$$
u(t) = S(t)u_0 + \int_0^t S(t - \tau)\mathcal{N}u(\tau)d\tau,
\tag{4}
$$

with $u \in L^2(0, T; H^1(\Omega))$. We defined the following operator:

$$
\mathcal{X} : L^2(0, T; H^1(\Omega)) \longrightarrow L^2(0, T; H^1(\Omega))
$$
$$
u(\cdot) \longmapsto \int_0^t S(t - \tau)\mathcal{N}u(\tau)d\tau,
$$

We define a nonlinear semigroup $(F(t))_{t \geq 0}$ associated to the solution of (1) by

$$
F(t)u_0 = u(t), \quad t \in]0, T[.
\tag{5}
$$

Hence the output function is formally given by

$$
y(t) = CF(t)u_0, \quad t \in]0, T[.
$$

Moreover, if a semigroup on $L^2(\Omega)$ provides the mild solution of (1) in the sense of (5) (form more details see [2]), we call it the nonlinear semigroup on $L^2(\Omega)$ associated with the semilinear system (1) and we have

$$F(t)u_0 = S(t)u_0 + \mathcal{X}u(\cdot), \quad t \in]0, T[, \ u_0 \in H^1(\Omega) \tag{6}$$

The output function is only well defined if C is bounded. However, in case of unbounded observation operator, we have $Cu(t)$ is not defined, then we introduce the notion of admissible operator and we have that C is an admissible operator for the nonlinear semigroup $(F(t))_{t \geq 0}$ in the sense of Definition 1, if we can extend the mapping

$$u \longmapsto CF(\cdot)u$$

to a bounded operator from $H^1(\Omega)$ to O.

Definition 1 [2]

Let $C \in \mathcal{L}(D(C), O)$, we say that C is an admissible observation operator for $(F(t))_{t \geq 0}$, if there exists a constant $\lambda > 0$ such that

$$\int_0^T \|CF(\cdot)u - CF(\cdot)v\|_{\mathbb{R}^q}^2 \leq \lambda \|u - v\|^2,$$

for any $u, v \in D(C)$.

3 Regional Boundary Enlarged Observability

Let us consider the observability operator defined by

$$K : H^1(\Omega) \longrightarrow O$$
$$u \longmapsto CS(\cdot)u,$$

its adjoint K^* is given by

$$K^* : O \longrightarrow H^1(\Omega)$$
$$u \longmapsto \int_0^T S^*(t)C^*u(t)dt.$$

We consider the following operators:

- $\gamma_0 : H^1(\Omega) \longrightarrow H^{1/2}(\partial\Omega)$ the trace operator of order zero, which is linear continuous and surjective, γ_0^* denotes its adjoint.
- The restriction operator

$$\chi_\Gamma : H^{1/2}(\partial\Omega) \longrightarrow H^{1/2}(\Gamma)$$
$$u \longmapsto \chi_\Gamma u = u_{|\Gamma},$$

where $u_{|_\Gamma}$ is the restriction of the state u to Γ, the adjoint operator χ_Γ^* can be given by

$$\chi_\Gamma^* u(x) := \begin{cases} u(x) \,, x \in \Gamma \\ 0 \quad\quad, x \in \partial\Omega \backslash \Gamma. \end{cases}$$

- For $\omega \subset \Omega$, we define χ_ω by

$$\chi_\omega : H^1(\Omega) \longrightarrow H^1(\omega)$$
$$u \longmapsto \chi_\omega u = u_{|_\omega}.$$

Let $\alpha(\cdot)$ and $\beta(\cdot)$ be two functions defined in $H^{1/2}(\Gamma)$ such that $\alpha(\cdot) \leq \beta(\cdot)$ a.e. on Γ. Throughout the paper we set

$$\mathcal{E} = \{u \in H^{1/2}(\Gamma) \,|\, \alpha(\cdot) \leq u(\cdot) \leq \beta(\cdot) \text{ a.e. on } \Gamma\}.$$

Definition 2

We say that the linear system (2) together with the output (3) is $\mathcal{E}$−observable on Γ if

$$Im(\chi_\Gamma \gamma_0 K^*) \cap \mathcal{E} \neq \emptyset$$

Definition 3

The system (1) together with the output (3) is said to be $\mathcal{E}$−observable on Γ if it is possible to reconstruct the state $u(t)_{|_\Gamma}$ between $\alpha(\cdot)$ and $\beta(\cdot)$ at any time t and the reconstructed state depends continuously on y.

The study of regional boundary enlarged observability of a distributed parameter system governed by a semilinear parabolic equation amounts to solve the following problem.

Problem 1

Since Γ is an open set of $\partial\Omega$, then there exists $u_0^1 = \chi_\Gamma \gamma_0 u_0$ with $u_0^1 \in H^{1/2}(\Gamma)$, such that

$$u_0 = \gamma_0^* \chi_\Gamma^* u_0^1 + \bar{u}_0,$$

where $\bar{u}_0$ is the residual part and $\bar{u}_0 = 0$ on Γ.
Given the system (1) together with the output (3) on Γ at time $t \in [0, T]$, is it possible to reconstruct u_0^1 between $\alpha(\cdot)$ and $\beta(\cdot)$ on Γ?

4 Lagrangian Multiplier Approach

The aim of this section is to use the Lagrangian multiplier approach to solve the Problem 1.
Indeed, this problem is equivalent to minimize the reconstruction error between the output function and the observation model, which can be formulated as follows:

$$\begin{cases} \inf \| Ku + Gu(\cdot) - y\|_O^2 \\ u \in U, \end{cases} \qquad (7)$$

where $G = C\mathcal{X}$ and $U = \{u \in H^1(\Omega) \mid \chi_\Gamma \gamma_0 u \in \mathcal{E}\}$.

We consider the system (1) observed by an internal zone sensor and from the definition of the $\mathcal{E}$−observable on Γ, there exists at least $\theta \in O$ that allows to observe a state $y \in \mathcal{E}$ on Γ of the form $K^*\theta$ such that $\chi_\Gamma \gamma_0 K^*\theta \in \mathcal{E}$ with $\theta \in O$.

Hence the problem (7) can be rewritten as follows:

$$\begin{cases} \inf \| KK^*\theta + Gu(\cdot) - y\|_O^2 \\ \theta \in \mathcal{J} = \{\theta \in O \mid \chi_\Gamma \gamma_0 K^*\theta \in \mathcal{E}\}. \end{cases} \qquad (8)$$

Let us consider $r > 0$ small enough, we set $F_r = \bigcup_{z \in \Gamma} B(z, r)$ and $\omega_r = F_r \cap \Omega$ with $B(z, r)$ is the ball of radius r centered in z, and let $\alpha'(\cdot)$ and $\beta'(\cdot)$ be two functions defined in $H^1(\omega_r)$ such that $\alpha'(\cdot) \le \beta'(\cdot)$ in ω_r, we have

$$\mathcal{E}' = \{u \in H^1(\omega_r) \mid \alpha'(\cdot) \le u(\cdot) \le \beta'(\cdot) \text{ a.e. in } \omega_r\}.$$

Then we have the following results.

Proposition 1

If $\alpha(\cdot)$ (respectively $\beta(\cdot)$) is the restriction of the trace of $\alpha'(\cdot)$ (respectively $\beta'(\cdot)$) and if the system (2) together with output (3) is $\mathcal{E}'$−observable in ω_r, then it is $\mathcal{E}$−observable on Γ.

Proof For the details of the proof, we refer the reader to (see [17]). □

Theorem 1

If the linear system (2) together with the output (3) is exactly observable in ω_r, then the solution of (8) verify

$$KK^*KK^*\theta^* = -KK^*Gu(\cdot) + KK^*y - \frac{1}{2}H_\Gamma^*\lambda^*, \qquad (9)$$

and the solution of the problem (7) u^* is unique, where λ^* is the solution of

$$\begin{cases} \lambda^* = (H_\Gamma H_\Gamma^*)^{-1}\left[-2H_\Gamma(KK^*KK^*)H_\Gamma^\dagger u^* - 2H_\Gamma KK^*Gu(\cdot) + 2H_\Gamma KK^*y\right] \\ u^* = P_{\mathcal{E}}(\rho\lambda^* + u^*), \end{cases} \qquad (10)$$

while $P_{\mathcal{E}} : H^{1/2}(\Gamma) \longrightarrow \mathcal{E}$ denotes the projection operator, $\rho > 0$, $H_\Gamma = \chi_\Gamma \gamma_0 K^*$ and $H_\Gamma^\dagger = (H_\Gamma^* H_\Gamma)^{-1} H_\Gamma^*$ is the pseudo-inverse of H_Γ.

Proof If the linear system (2) together with the output (3) is exactly observable in ω_r, thus it is $\mathcal{E}'$−observable in ω_r. By Proposition 1 we see that the system (2)–(3) is $\mathcal{E}$−observable on Γ, consequently $\mathcal{J} \neq \emptyset$ and (8) has a solution. The problem (8) is equivalent to the following saddle point problem:

$$\begin{cases} \inf \|KK^*\theta + Gu(\cdot) - y\|_O^2 \\ (\theta, u) \in \mathcal{W}, \end{cases} \tag{11}$$

where $\mathcal{W} = \{ (\theta, u) \in O \times \mathcal{E} \mid \chi_\Gamma \gamma_0 K^* \theta - u = 0 \}$.

To study this constraint problem, we shall use a Lagrangian functional and steer the problem (11) to a saddle point problem.

We associate to the problem (11) the Lagrangian functional defined by

$$L(\theta, u, \lambda) = \|KK^*\theta + Gu(\cdot) - y\|_O^2 + \langle \lambda, \chi_\Gamma \gamma_0 K^*\theta - u \rangle_{H^{1/2}(\Gamma)},$$

for all $(\theta, u, \lambda) \in O \times \mathcal{E} \times H^{1/2}(\Gamma)$.

The proof is divided into in three steps.

I– Let us prove that L admits a saddle point.

The set $O \times \mathcal{E}$ is non empty, closed and convex, and the functional L satisfies the conditions

$- (\theta, u) \longmapsto L(\theta, u, \lambda)$ is convex and lower semi-continuous for all $\lambda \in H^{1/2}(\Gamma)$.

$- \lambda \longmapsto L(\theta, u, \lambda)$ is concave and upper semi-continuous for all $(\theta, u) \in O \times \mathcal{E}$.

Furthermore, there exists $\lambda_0 \in H^{1/2}(\Gamma)$ such that

$$\lim_{\|(\theta, u)\| \to +\infty} L(\theta, u, \lambda_0) = +\infty,$$

and there exists $(\theta_0, u_0) \in O \times \mathcal{E}$ such that

$$\lim_{\|\lambda\| \to +\infty} L(\theta_0, u_0, \lambda) = -\infty.$$

Then, the functional L admits a saddle point. (For more details see e.g., [9, 11]).

II– Let us consider $(\theta^*, u^*, \lambda^*)$ a saddle point of L and prove that $u^* = \chi_\Gamma \gamma_0 K^* \theta^*$ is there restriction on Γ of the solution of (7). We have

$$L(\theta^*, u^*, \lambda) \leq L(\theta^*, u^*, \lambda^*) \leq L(\theta, u, \lambda^*), \tag{12}$$

for all $(\theta, u, \lambda) \in O \times \mathcal{E} \times H^{1/2}(\Gamma)$.

From the first inequality of (12), it follows that

$$\|KK^*\theta^* + Gu(\cdot) - y\|_O^2 + \langle \lambda, \chi_\Gamma \gamma_0 K^*\theta^* - u^* \rangle_{H^{1/2}(\Gamma)} \leq \|KK^*\theta^* + Gu(\cdot) - y\|_O^2$$
$$+ \langle \lambda^*, \chi_\Gamma \gamma_0 K^*\theta^* - u^* \rangle_{H^{1/2}(\Gamma)}$$

$$\forall \lambda \in H^{1/2}(\Gamma),$$

which leads to

$$\langle \lambda, \chi_\Gamma \gamma_0 K^*\theta^* - u^* \rangle_{H^{1/2}(\Gamma)} \leq \langle \lambda^*, \chi_\Gamma \gamma_0 K^*\theta^* - u^* \rangle_{H^{1/2}(\Gamma)} \quad \forall \lambda \in H^{1/2}(\Gamma),$$

it implies that $\chi_\Gamma \gamma_0 K^* \theta^* = u^*$, hence $\chi_\Gamma \gamma_0 K^* \theta^* \in \mathcal{E}$.
From the second inequality of (12), we have

$$\|KK^*\theta^* + Gu(\cdot) - y\|_O^2 + \langle \lambda^*, \chi_\Gamma \gamma_0 K^* \theta^* - u^* \rangle_{H^{1/2}(\Gamma)} \leq \|KK^*\theta + Gu(\cdot) - y\|_O^2 + \langle \lambda^*, \chi_\Gamma \gamma_0 K^* \theta - u \rangle_{H^{1/2}(\Gamma)}$$

$\forall (\theta, u) \in O \times \mathcal{E}$.
Since $\chi_\Gamma \gamma_0 K^* \theta^* = u^*$, which leads to

$$\|KK^*\theta^* + Gu(\cdot) - y\|_O^2 \leq \|KK^*\theta + Gu(\cdot) - y\|_O^2 + \langle \lambda^*, \chi_\Gamma \gamma_0 K^*\theta - u \rangle_{H^{1/2}(\Gamma)}$$

$$\forall (\theta, u) \in O \times \mathcal{E}.$$

Taking $\chi_\Gamma \gamma_0 K^* \theta = u$, we obtain

$$\left\| KK^*\theta^* + Gu(\cdot) - y \right\|_O^2 \leq \left\| KK^*\theta + Gu(\cdot) - y \right\|_O^2,$$

which implies that θ^* is the solution of (8) and we deduce that $u^* = \chi_\Gamma \gamma_0 K^*\theta^*$ is the solution of (7) for all the states which are of the form $K^*\theta$ with $\theta \in O$.

III– If $(\theta^*, u^*, \lambda^*)$ is a saddle point of L, then the following assumptions hold:

$$2\langle KK^*\theta^* + Gu(\cdot) - y, KK^*(\theta - \theta^*) \rangle + \langle \lambda^*, \chi_\Gamma \gamma_0 K^*(\theta - \theta^*) \rangle = 0 \quad \forall \theta \in O, \tag{13}$$

$$\langle \lambda^*, u - u^* \rangle \leq 0 \quad \forall u \in \mathcal{E}, \tag{14}$$

$$\langle \lambda - \lambda^*, \chi_\Gamma \gamma_0 K^* \theta^* - u^* \rangle = 0 \quad \forall \lambda \in H^{1/2}(\Gamma). \tag{15}$$

For more details about the saddle point and its theory, we refer to [5, 11].
From (13), we obtain

$$\langle 2(KK^*)^*(KK^*\theta^* + Gu(\cdot) - y), (\theta - \theta^*) \rangle + \langle (\chi_\Gamma \gamma_0 K^*)^*\lambda^*, (\theta - \theta^*) \rangle = 0 \quad \forall \theta \in O,$$

which leads to

$$2KK^*KK^*\theta^* + 2KK^*Gu(\cdot) - 2KK^*y + H_\Gamma^*\lambda^* = 0. \tag{16}$$

Since the linear system (2)–(3) is exactly observable in ω_r, then it is exactly observable on Γ (see [14]), thus the operator $H_\Gamma^* H_\Gamma$ is invertible, hence we have $\theta^* = H_\Gamma^\dagger u^*$, consequently (16) is equivalent to

$$(KK^*KK^*)H_\Gamma^\dagger u^* = -KK^*Gu(\cdot) + KK^*y - \frac{1}{2}H_\Gamma^*\lambda^*,$$

where $H_\Gamma = \chi_\Gamma \gamma_0 K^*$, $H_\Gamma^\dagger$ the pseudo-inverse of H_Γ, and we obtain

$$\lambda^* = (H_\Gamma H_\Gamma^*)^{-1} \left[-2H_\Gamma(KK^*KK^*)H_\Gamma^\dagger u^* - 2H_\Gamma KK^*Gu(\cdot) + 2H_\Gamma KK^*y \right].$$

From the inequality (14), we have

$$\langle \lambda^*, u - u^* \rangle \leq 0 \quad \forall u \in \mathcal{E},$$

which gives for $\rho > 0$ and $u \in \mathcal{E}$

$$\langle (\rho \lambda^* + u^*) - u^*, u - u^* \rangle \leq 0,$$

then

$$u^* = P_{\mathcal{E}}(\rho \lambda^* + u^*),$$

which prove the second part of (10).
The proof is complete.

$\square$

To prove the uniqueness of the problem (10), we have the following lemma.

Lemma 1

If the linear system (2) together with the output (3) is exactly observable in ω_r and the function

$$R_\Gamma = (H_\Gamma H_\Gamma^*)^{-1} H_\Gamma (K K^* K K^*) H_\Gamma^\dagger,$$

is coercive, then for ρ suitably chosen the system (10) has a unique solution (λ^*, u^*).

Proof We have

$$\lambda^* = -2(H_\Gamma H_\Gamma^*)^{-1} H_\Gamma (K K^* K K^*) H_\Gamma^\dagger u^* - 2(H_\Gamma H_\Gamma^*)^{-1} H_\Gamma K K^* Gu(\cdot) + 2(H_\Gamma H_\Gamma^*)^{-1} H_\Gamma K K^* y.$$

So if $(\theta^*, u^*, \lambda^*)$ is a saddle point of L the system (10) is equivalent to

$$\begin{cases} \lambda^* = -2 R_\Gamma u^* - 2(H_\Gamma H_\Gamma^*)^{-1} H_\Gamma K K^* Gu(\cdot) + 2(H_\Gamma H_\Gamma^*)^{-1} H_\Gamma K K^* y \\ u^* = P_{\mathcal{E}} \left(-2\rho R_\Gamma u^* - 2\rho(H_\Gamma H_\Gamma^*)^{-1} H_\Gamma K K^* Gu(\cdot) + 2\rho(H_\Gamma H_\Gamma^*)^{-1} H_\Gamma K K^* y + u^* \right). \end{cases}$$

It follows that u^* is a fixed point of the function

$$F_\rho : \mathcal{E} \longrightarrow \mathcal{E}$$
$$u \longmapsto P_{\mathcal{E}} \left(-2\rho R_\Gamma u - 2\rho(H_\Gamma H_\Gamma^*)^{-1} H_\Gamma K K^* Gu(\cdot) + 2\rho(H_\Gamma H_\Gamma^*)^{-1} H_\Gamma K K^* y + u \right).$$

The operator R_Γ is coercive, i.e.

$$\exists \mu > 0 \quad \text{such that} \quad \langle R_\Gamma u, u \rangle \geq \mu \|u\|^2, \quad \forall u \in H^{1/2}(\Gamma).$$

It follows that

$$\begin{aligned} \|F_\rho(u_1) - F_\rho(u_2)\|^2 &\leq \| -2\rho R_\Gamma(u_1 - u_2) + (u_1 - u_2)\|^2 \\ &\leq 4\rho^2 \|R_\Gamma\|^2 \|u_1 - u_2\|^2 + \|u_1 - u_2\|^2 - 4\rho \langle R_\Gamma(u_1 - u_2), (u_1 - u_2) \rangle \\ &\leq 4\rho^2 \|R_\Gamma\|^2 \|u_1 - u_2\|^2 + \|u_1 - u_2\|^2 - 4\rho\mu \|u_1 - u_2\|^2 \\ &\leq (4\rho^2 \|R_\Gamma\|^2 + 1 - 4\rho\mu) \|u_1 - u_2\|^2 \quad \forall u_1, u_2 \in \mathcal{E}, \end{aligned}$$

and we deduce that if

$$0 < \rho < \frac{\mu}{\|R_\Gamma\|^2},$$

then F_ρ is contractant, which implies the uniqueness of u^* and λ^*.
The proof is complete.   □

5 Numerical Approach

From the Theorem 1 we proved that the solution of the problem (7) arises to compute
the saddle point of L, which is equivalent to solving the problem

$$\inf_{\theta \in O\, u \in \mathcal{E}} \left(\sup_{\lambda \in H^{1/2}(\Gamma)} L(\theta, u, \lambda) \right). \tag{17}$$

To attain this the implementation can be based on the algorithm of Uzawa (see [5]).

Algorithm.

Step 1 : Initialization

⊙ The subregions ω_r and Γ, time T.

⊙ The domain D and the function of distribution f.

⊙ Two functions $(u_0, \lambda_1) \in \mathcal{E} \times H^{1/2}(\Gamma)$.

⊙ Threshold accuracy ε small enough and $\rho > 0$.

Step 2 : Until $\|u_{n+1} - u_n\|_{H^{1/2}(\omega_r)} \le \varepsilon$, repeat

$\triangleright$ Solve $(KK^*KK^*)\theta_n^* = -KK^*Gu(\cdot) + KK^*y - \frac{1}{2}H_\Gamma^*\lambda^*.$
$\triangleright$ Calculate $u_n = P_\mathcal{E}(\rho\lambda_n + u_{n-1}).$
$\triangleright$ Calculate $\lambda_{n+1} = \lambda_n + (\chi_\Gamma \gamma_0 K^* \theta_n - u_n).$

Step 3 : Let $(\theta^*, u^*, \lambda^*)$ be a saddle point of L, then the sequence θ_n converge to
θ^* solution of the system (8) and u_n converge to the initial state u^* to be
reconstructed on Γ.

6 Simulation Results

Here we present a numerical example that leads to some results related to the choice of the subregion, the constraints functions, and the sensor location.

Let us consider the system defined in $\Omega =]0, 1[\times]0, 1[$ by the following two-dimensional semilinear parabolic equation observed by a pointwise sensor:

$$
\begin{cases}
\dfrac{\partial u}{\partial t}(x, t) = c \displaystyle\sum_{i=1}^{2} \dfrac{\partial^2 u}{\partial x_i^2}(x, t) + \displaystyle\sum_{k,l=0}^{\infty} \left| \langle u(t), \varphi_{kl} \rangle \right| \langle u(t), \varphi_{kl} \rangle \varphi_{kl}(x) & in \quad \Omega \times]0, T[\\
\dfrac{\partial u}{\partial \nu}(\xi, t) = 0 & on \quad \partial\Omega \times]0, T[\\
u(x, 0) = u_0(x) & in \quad \Omega,
\end{cases}
$$

$$\text{(18)}$$

where $x = (x_1, x_2)$, $c = 0.01$ and the operator $A = \displaystyle\sum_{i=1}^{2} \dfrac{\partial^2}{\partial x_i^2}$ has a complete set of eigenfunctions $(\varphi_{kl})_{k,l\in\mathbb{N}}$ in $H^1(\Omega)$.

The system (18) is augmented with the output function given by

$$y(t) = u(b_1, b_2, t), \quad t \in]0, T[,$$

$$\text{(19)}$$

Where

- $T = 2$, the pointwise sensor located in $b = (b_1, b_2)$ with $b_1 = 0.9$ and $b_2 = 0.85$.
- The subregion $\omega_r =]0, 0.45[\times]0, 9[$.
- The boundary subregion $\Gamma = \{0\} \times]0, 1[$.
- The initial state to be observed on Γ

$$u_0(x_1, x_2) = 2\left(x_1^3 - x_1^2 + 0.1\right)\left(x_2^3 - \cos(x_2)^2 + 0.1\right).$$

- The constraints functions

$$
\alpha(x_1, x_2) = 2\left(\dfrac{x_1^3}{3} - \cos(x_1)^2 \mid \dfrac{1}{4}\right)\left(\dfrac{x_2^3}{3} - \dfrac{x_2^2}{2} + \dfrac{1}{4}\right)
$$
$$
\beta(x_1, x_2) = 2\left(\dfrac{x_1^3}{3} - \cos(x_1)^2 + 0.01\right)\left(\dfrac{x_2^3}{3} - \dfrac{x_2^2}{2} + 0.01\right)
$$

Applying the previous algorithm we obtain the following results (Figs. 1, 2, 3, 4 and 5):

The initial state u_{oe} is estimated with a reconstruction error

$$\|u_0 - u_{oe}\|^2 = 3.03 \times 10^{-3}.$$

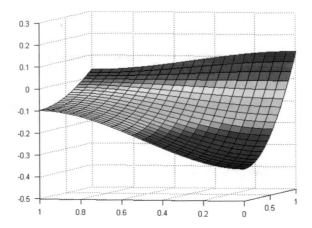

Fig. 1 The constrained function $\alpha(\cdot)$ in ω_r

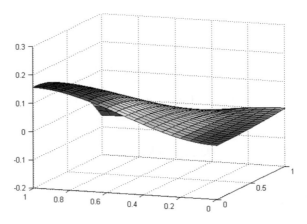

Fig. 2 The constrained function $\beta(\cdot)$ in ω_r

Figure 4 shows that the initial state estimated u_{oe} is between $\alpha(\cdot)$ and $\beta(\cdot)$ on the subregion Γ, then the observed system (18)–(19) is $\mathcal{E}-$observable on Γ.

• If the sensor is located in $(b_1, b_2) = (0.2, 0.65)$, we obtain
Figure 5 shows that the initial state estimated u_{oe} is not between $\alpha(\cdot)$ and $\beta(\cdot)$ on Γ, then the observed system (18)–(19) is not $\mathcal{E}-$observable on Γ.
• The following results show the evolution of the observed error with respect to the sensor location.
Figures 6 and 7 show that the worst locations of the sensor correspond to a great error.

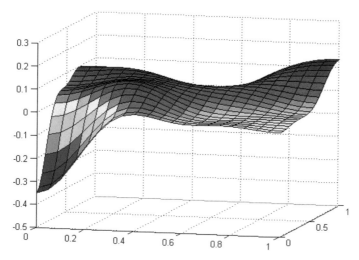

Fig. 3 The estimated initial state u_{0e} in ω_r

Fig. 4 The constraints functions $\alpha(\cdot)$, $\beta(\cdot)$ and the estimated initial state u_{0e} on Γ

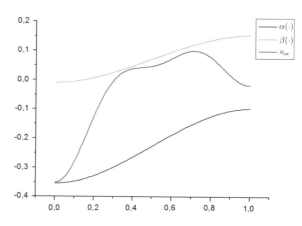

Fig. 5 The constraints functions $\alpha(\cdot)$, $\beta(\cdot)$ and the estimated initial state u_{0e} on Γ

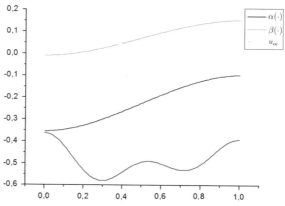

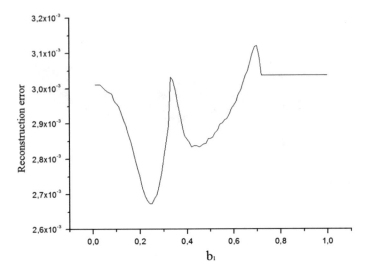

Fig. 6 Evolution of the reconstruction error with respect to the sensor location b_1

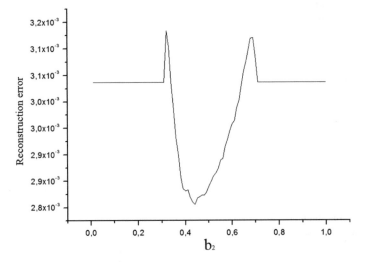

Fig. 7 Evolution of the reconstruction error with respect to the sensor location b_2

7 Conclusion

The regioanl boundary enlarged observability of semilinear parabolic system is considered. The developed method, based on regional observability tools in connection with the Lagrangian multiplier approach leads to an algorithm which is numerically implemented. Many questions remain open, this is the case of studying the problem of regional enlarged gradient observability of semilinear parabolic systems. This question is under consideration and the work shall be the subject of a future paper.

Acknowledgements This work has been carried out with a grant from Hassan II Academy of Sciences and Technology project 630/2016.

References

1. Amouroux, M., EL Jai, A., Zerrik, E.: Regional observability of distributed systems. Int. J. Syst. Sci. **25**, 301–313 (1994)
2. Baroun, M., Jacob, B.: Admissibility and observability of observation operators for semilinear problems. Integr. Equ.S Oper. Theory **64**, 1–20 (2009)
3. Delfour, M.C., Mitter, S.K.: Controllability and observability for infinite dimensional systems. SIAM J. Control. **10**, 329–333 (1972)
4. Dolecki, S., Russell, D.L.: A general theory of observation and control. SIAM J. Control. Optim. **15**(2), 185–220 (1977)
5. Fortin, M., Glowinski, R.: Augmented Lagrangian methods: applications to the numerical solution of boundary-value problems. North-Holland, Amsterdam (1983)
6. A. EL JAI, M.C SIMON AND E. ZERRIK, *Regional observability and sensors structures*, Sensors and Actuators Journal, Vol. **39**, 95–102, 1993
7. Kalman, R.E.: On the general theory of control systems, Proceedings of the 1st World Congress of the International Federation of Automatic Control, vol. 481–493., Moscow (1960)
8. J.L. LIONS AND E. MAGENES, *Problèmes aux limites non homogènes et applications*, Vol **1** & **2**, Dunod, Paris, 1968
9. Matei, A.: Weak solvability via Lagrange multipliers for two frictional contact models. Ann. Univ. Bucharest. **4**, 179–191 (2013)
10. Pazy, A.: Semigroups of linear operators and applications to partial differential equations. Springer, New York (1990)
11. Rockafellar, R.T.: Lagrange multipliers and optimality. SIAM Rev. **35**(2), 183–238 (1993). https://doi.org/10.1137/1035044
12. Russell, D.L., Weiss, G.: A general necessary condition for exact observability. SIAM J. Control. Optim **32**, 1–23 (1994)
13. Zeidler, E.: Nonlinear functional analysis and its applications II/A linear applied functional analysis. Springer, Berlin (1990)
14. Zerrik, E., Badraoui, L., El Jai, A.: Sensors and regional boundary state reconstruction of parabolic systems. Sens.S Actuators J. **75**, 102–117 (1999)
15. Zerrik, E., Bourray, H., Boutoulout, A.: Regional boundary observability: A numerical approach. Int. J. Appl. Math. Compat. Sci **12**, 143–151 (2002)
16. E. ZERRIK AND F. GHAFRANI, *Optimal control of wastewater biological treatment*, International Journal of Engineering and Innovative Technology, Vol. **3**, Issue **1**, 2013
17. H. ZOUITEN, F.Z. EL ALAOUI AND A. BOUTOULOUT, *Regional boundary observability with constraints: A numerical approach*, International Review of Automatic Control (I.RE.A.CO.), Vol. **8**, 357–361, 2015

Printed in the United States
By Bookmasters